CLASSICAL AND NONCLASSICAL LOGICS

CLASSICAL AND NONCLASSICAL LOGICS
An Introduction to the Mathematics of Propositions

Eric Schechter

Princeton University Press
Princeton and Oxford

Copyright © 2005 by Princeton University Press
Published by Princeton University Press, 41 William Street, Princeton, New Jersey 08540

In the United Kingdom: Princeton University Press, 3 Market Place, Woodstock, Oxfordshire OX20 1SY

All Rights Reserved

Library of Congress Cataloging-in-Publication Data

Schechter, Eric, 1950–
    Classical and nonclassical logics : an introduction to the mathematics of propositions / Eric Schechter.
       p. cm.
    Includes bibliographical references and index.
    ISBN-13: 978-0-691-12279-3 (acid-free paper)
    ISBN-10: 0-691-12279-2 (acid-free paper)
    1. Mathematics—Philosophy. 2. Proposition (Logic). I. Title.

QA9.3.S39 2005
160—dc22   2004066030

British Library Cataloging-in-Publication Data is available

The publisher would like to acknowledge the author of this volume for providing the camera-ready copy from which this book was printed.

Printed on acid-free paper. ∞

pup.princeton.edu

Printed in the United States of America

10 9 8 7 6 5 4 3 2 1

Contents

A Preliminaries   1

1 Introduction for teachers   3
• Purpose and intended audience, 3 • Topics in the book, 6 • Why pluralism?, 13 • Feedback, 18 • Acknowledgments, 19

2 Introduction for students   20
• Who should study logic?, 20 • Formalism and certification, 25 • Language and levels, 34 • Semantics and syntactics, 39 • Historical perspective, 49 • Pluralism, 57 • Jarden's example (optional), 63

3 Informal set theory   65
• Sets and their members, 68 • Russell's paradox, 77 • Subsets, 79 • Functions, 84 • The Axiom of Choice (optional), 92 • Operations on sets, 94 • Venn diagrams, 102 • Syllogisms (optional), 111 • Infinite sets (postponable), 116

4 Topologies and interiors (postponable)   126
• Topologies, 127 • Interiors, 133 • Generated topologies and finite topologies (optional), 139

5 English and informal classical logic   146
• Language and bias, 146 • Parts of speech, 150 • Semantic values, 151 • Disjunction (or), 152 • Conjunction (and), 155 • Negation (not), 156 • Material implication, 161 • Cotenability, fusion, and constants

(postponable), 170 • Methods of proof, 174 • Working backwards, 177 • Quantifiers, 183 • Induction, 195 • Induction examples (optional), 199

6 Definition of a formal language   206
• The alphabet, 206 • The grammar, 210 • Removing parentheses, 215 • Defined symbols, 219 • Prefix notation (optional), 220 • Variable sharing, 221 • Formula schemes, 222 • Order preserving or reversing subformulas (postponable), 228

B Semantics   233

7 Definitions for semantics   235
• Interpretations, 235 • Functional interpretations, 237 • Tautology and truth preservation, 240

8 Numerically valued interpretations   245
• The two-valued interpretation, 245 • Fuzzy interpretations, 251 • Two integer-valued interpretations, 258 • More about comparative logic, 262 • More about Sugihara's interpretation, 263

9 Set-valued interpretations   269
• Powerset interpretations, 269 • Hexagon interpretation (optional), 272 • The crystal interpretation, 273 • Church's diamond (optional), 277

10 Topological semantics (postponable)   281
• Topological interpretations, 281 • Examples, 282 • Common tautologies, 285 • Nonredundancy of symbols, 286 • Variable sharing, 289 • Adequacy of finite topologies (optional), 290 • Disjunction property (optional), 293

Contents

11 More advanced topics in semantics — 295
• Common tautologies, 295 • Images of interpretations, 301 • Dugundji formulas, 307

C Basic syntactics — 311

12 Inference systems — 313

13 Basic implication — 318
• Assumptions of basic implication, 319 • A few easy derivations, 320 • Lemmaless expansions, 326 • Detachmental corollaries, 330 • Iterated implication (postponable), 332

14 Basic logic — 336
• Further assumptions, 336 • Basic positive logic, 339 • Basic negation, 341 • Substitution principles, 343

D One-formula extensions — 349

15 Contraction — 351
• Weak contraction, 351 • Contraction, 355

16 Expansion and positive paradox — 357
• Expansion and mingle, 357 • Positive paradox (strong expansion), 359 • Further consequences of positive paradox, 362

17 Explosion — 365

18 Fusion — 369

19 Not-elimination — 372
• Not-elimination and contrapositives, 372 • Interchangeability results, 373 • Miscellaneous consequences of not-elimination, 375

20 Relativity   **377**

E   Soundness and major logics   381

21 Soundness   **383**

22 Constructive axioms: avoiding not-elimination   **385**
• Constructive implication, 386 • Herbrand-Tarski Deduction Principle, 387 • Basic logic revisited, 393 • Soundness, 397 • Nonconstructive axioms and classical logic, 399 • Glivenko's Principle, 402

23 Relevant axioms: avoiding expansion   **405**
• Some syntactic results, 405 • Relevant deduction principle (optional), 407 • Soundness, 408 • Mingle: slightly irrelevant, 411 • Positive paradox and classical logic, 415

24 Fuzzy axioms: avoiding contraction   **417**
• Axioms, 417 • Meredith's chain proof, 419 • Additional notations, 421 • Wajsberg logic, 422 • Deduction principle for Wajsberg logic, 426

25 Classical logic   **430**
• Axioms, 430 • Soundness results, 431 • Independence of axioms, 431

26 Abelian logic   **437**

F   Advanced results   441

27 Harrop's principle for constructive logic   **443**
• Meyer's valuation, 443 • Harrop's principle, 448 • The disjunction property, 451 • Admissibility, 451 • Results in other logics, 452

28 Multiple worlds for implications 454
• Multiple worlds, 454 • Implication models, 458 • Soundness, 460 • Canonical models, 461 • Completeness, 464

29 Completeness via maximality 466
• Maximal unproving sets, 466 • Classical logic, 470 • Wajsberg logic, 477 • Constructive logic, 479 • Non-finitely-axiomatizable logics, 485

References 487

Symbol list 493

Index 495

Part A

Preliminaries

Chapter 1

Introduction for teachers

Readers with no previous knowledge of formal logic will find it more useful to begin with Chapter 2.

PURPOSE AND INTENDED AUDIENCE

1.1. *CNL* (*Classical and Nonclassical Logics*) is intended as an introduction to mathematical logic. However, we wish to immediately caution the reader that the topics in this book are

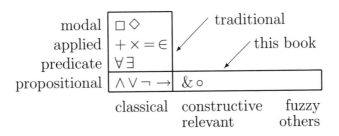

not the same as those in a conventional introduction to logic. *CNL* should only be adopted by teachers who are aware of the differences and are persuaded of this book's advantages. Chiefly, *CNL* trades some depth for breadth:

- A traditional introduction to logic covers *classical logic only*, though possibly at several levels — propositional, predicate, modal, etc.

- *CNL* is *pluralistic*, in that it covers classical and several nonclassical logics — constructive, quantitative, relevant, etc. — though almost solely at the propositional level.

Of course, a logician needs both depth and breadth, but both cannot be acquired in the first semester. The depth-first approach is prevalent in textbooks, perhaps merely because classical logic developed a few years before other logics. I am convinced that a breadth-first approach would be better for the students, for reasons discussed starting in 1.9.

1.2. *Intended audience.* This is an introductory textbook. No previous experience with mathematical logic is required. Some experience with algebraic computation and abstract thinking is expected, perhaps at the precalculus level or slightly higher. The exercises in this book are mostly computational and require little originality; thus *CNL* may be too elementary for a graduate course. Of course, the book may be used at different levels by different instructors.

CNL was written for classroom use; I have been teaching undergraduate classes from earlier versions for several years. However, its first few chapters include sufficient review of prerequisite material to support the book's use also for self-guided study. Those chapters have some overlap with a "transition to higher mathematics" course; *CNL* might serve as a resource in such a course.

I would expect *CNL* to be used mainly in mathematics departments, but it might be adopted in some philosophy departments as well. Indeed, some philosophers are very mathematically inclined; many of this book's mathematical theorems originated on the chalkboards of philosophy departments. Colleagues have also informed me that this book will be of some interest to students of computer science, but I am unfamiliar with such connections and have not pursued them in this book.

1.3. *In what sense is this new?* This book is a work of exposition and pedagogy, not of research. All the main theorems of this book have already appeared, albeit in different form, in research

Purpose and intended audience 5

journals or advanced monographs. But those articles and books were written to be read by experts. I believe that the present work is a substantially new *selection and reformulation of results*, and that it will be more accessible to beginners.

1.4. *Avoidance of algebra.* Aside from its pluralistic approach (discussed at much greater length later in this chapter), probably *CNL*'s most unusual feature is its attempt to avoid higher algebra.

In recent decades, mathematical logic has been freed from its philosophical and psychological origins; the current research literature views different logics simply as different kinds of algebraic structures. That viewpoint may be good for research, but it is not a good prescription for motivating undergraduate students, who know little of higher algebra.

CNL attempts to use as little algebra as possible. For instance, we shall use topologies instead of Heyting algebras; they are more concrete and easier to define. (See the remark in 4.6.i.)

1.5. *Rethinking of terminology.* I have followed conventional terminology for the most part, but I have adopted new terminology whenever a satisfactory word or phrase was not available in the literature. Of course, what is "satisfactory" is a matter of opinion.

It is my opinion that there are far too many objects in mathematics called "regular," "normal," etc. Those words are not descriptive — they indicate only that some standard is being adhered to, without giving the beginner any assistance whatsoever in identifying and assimilating that standard. Whenever possible, I have attempted to replace such terms with phrases that are more descriptive, such as "truth-preserving" and "tautology-preserving."

A more substantive, and perhaps more controversial, example of rejected terminology is "intuitionistic logic." That term has been widely used for one particular logic since it was introduced in the early 20th century by Brouwer, Heyting, and Kolmogorov. To call it anything else is to fight a strong tradition. But the word "intuitionistic" has connotations of subjectivity and mysticism that may drive away some scientifically inclined students. There is nothing subjective, mystical, or unscientific about this interesting logic, which we develop in Chapters 10, 22, 27, 28, and part of 29.

Moreover, not all mathematicians share the same intuition. Indeed, aside from logicians, most mathematicians today are schooled only in classical logic and find all other logics to be nonintuitive. It is only a historical accident that Brouwer, Heyting and Kolmogorov appropriated the word "intuitionistic" for their system. The term "BHK logic," used in some of the literature, is less biased, but it too is descriptive only to someone who already knows the subject.

A more useful name is "constructive logic," because BHK logic is to a large extent the reasoning system of constructive mathematics (discussed in 2.42–2.46). Mathematicians may not be entirely in agreement about the importance of constructivism, but at least there is consensus on what the term "constructive" means. Its meaning in mathematics is quite close to its meaning outside mathematics, and thus should be more easily grasped by beginning students.

1.6. *What is not covered.* This book is intended as an introductory textbook, not an encyclopedia — it includes enough different logics to illustrate some basic ideas of the subject, but it does not include all major logics. Derivations in *CNL* follow only the Hilbert style, because in my opinion that is easiest for beginners to understand. The treatment of quantifiers consists of only a few pages (sections 5.40–5.51), and that treatment is informal, not axiomatic. Omitted entirely (or mentioned in just a sentence or two) are $\Box$, $\Diamond$, formal predicate logic, Gentzen sequents, natural deduction, modal logics, Gödel's Incompleteness Principles, recursive functions, Turing machines, linear logic, quantum logic, substructures logics, nonmonotonic logics, and many other topics.

TOPICS IN THE BOOK

1.7. *Order of topics.* I have tried to arrange the book methodically, so that topics within it are not hard to find; but I have also provided frequent cross-referencing, to facilitate reading the book in other orders than mine.

Chapter 2 gives an overview of, and informal introduction to, the subject of logic. The chapter ends with a detailed discussion (2.42–2.46) of constructivism and Jarden's Proof, surely the simplest example of the fact that a different philosophy can require a different logic.

Chapters 3 and 4 give a brief introduction to naive set theory and general topology. Chapter 5 gives a more detailed introduction to informal classical logic, along with comments about how it compares with nonclassical logics and with ordinary nonmathematical English. Particular attention is given to the ambiguities of English.

Chapters 2–5 may be considered "prerequisite" material, in the sense that their content is not part of logic but will be used to develop logic. Different students will need different parts of this prerequisite material; by including it I hope to make the book accessible to a wide variety of students. Admittedly, these introductory chapters take up an unusually large portion of the book, but they are written mostly in English; the remainder of the book is written in the more concise language of mathematics.

Finally, in Chapter 6 we begin formal logic. This chapter presents and investigates a formal language that will be used throughout the remainder of the book. Among the terms defined in this chapter are "formula," "rank of a formula," "variable sharing," "generalization," "specialization," and "order preserving" and "order reversing."

There are several feasible strategies for ordering the topics after formal language. The most obvious would be to present various logics one by one — e.g., classical logic, then constructive logic, then relevant logic, etc. This strategy would juxtapose related results — e.g., constructive semantics with constructive syntactics — and perhaps it is the most desirable approach for a reference book. But I have instead elected to cover all of semantics before beginning any syntactics. This approach is better for the beginning student because semantics is more elementary and concrete than syntactics, and because this approach juxtaposes related *techniques* — e.g., constructive semantics and relevant semantics.

Semantics is introduced in Chapter 7, which defines "valuation," "interpretation," and "tautology." Then come some examples of interpretations — numerically valued in Chapter 8, set-valued in Chapter 9, and topological in Chapter 10. In the presentation of these examples, one recurring theme is the investigation of relevance: If A and B are formulas that are unrelated in the sense that they share no propositional variable symbols, and $\underline{A} \to B$ is a tautology in some interpretation, does it follow that $\overline{A}$ or B are tautologies? Our conclusions are summarized in one column of the table in 2.37.

The aforementioned chapters deal with examples of semantic systems, one at a time. Chapter 11, though not lacking in examples, presents more abstract results. Sections 11.2–11.7 give shortcuts that are often applicable in verifying that a formula is tautologous. Sections 11.8–11.12 give sufficient conditions for one interpretation to be an extension of another. Sections 11.13–11.17 show that, under mild assumptions, the Dugundji formula in n symbols is tautological for interpretations with fewer than n semantic values, but not for interpretations with n or more semantic values; as a corollary we see that (again under mild assumptions) an infinite semantics cannot be replaced by a finite semantics.

Syntactics is introduced in Chapter 12, which defines "axiom," "assumed inference rule," "derivation," "theorem," etc. The chapters after that will deal with various syntactic logics, but in what order should those be presented? My strategy is as follows.

The logics of greatest philosophical interest in this book are classical, constructive (intuitionist), relevant, and fuzzy (Zadeh and Łukasiewicz), shown in the upper half of the diagram below. These logics have a substantial overlap, which I call *basic logic*;[1] it appears at the bottom of the diagram. To reduce repetition, our syntactic development will begin with basic logic and then

---

[1] That's my own terminology; be cautioned that different mathematicians use the word "basic" in different ways.

Topics in the book

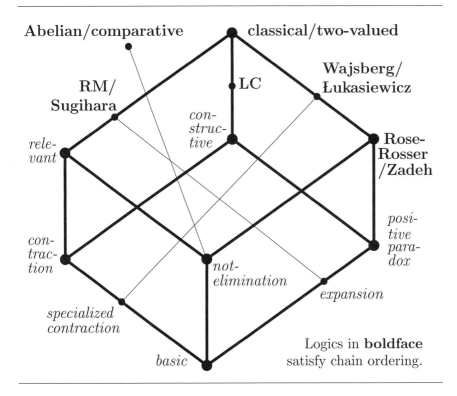

gradually add more ingredients.

Chapter 13 introduces the assumptions of basic implication,

$$\{A, A \to B\} \vdash B, \qquad \vdash (G \to H) \to [(I \to G) \to (I \to H)],$$
$$\vdash C \to C, \qquad \vdash [D \to (E \to F)] \to [E \to (D \to F)],$$

and investigates their consequences. One elementary but important consequence is the availability of detachmental corollaries; that is, $\vdash A \to B \Rightarrow A \vdash B$. Chapter 14 adds the remaining assumptions of basic logic,

$$\{A, B\} \vdash A \wedge B, \qquad \vdash A \to (A \vee B), \qquad \vdash (A \wedge B) \to A,$$
$$\vdash (A \to \overline{B}) \to (B \to \overline{A}), \qquad \vdash B \to (A \vee B), \qquad \vdash (A \wedge B) \to B,$$
$$\vdash [(A \to B) \wedge (A \to C)] \to [A \to (B \wedge C)],$$
$$\vdash [(B \to A) \wedge (C \to A)] \to [(B \vee C) \to A],$$
$$\vdash [A \wedge (B \vee C)] \to [(A \wedge B) \vee C],$$

and then investigates their consequences. One consequence is a substitution principle: if s is an order preserving or order reversing function, then ⊢ $A \to B$ implies, respectively, ⊢ $s(A) \to s(B)$ or ⊢ $s(B) \to s(A)$.

Next come several short chapters, each investigating a different one-axiom extension of basic logic:

Chapter		axiom added to basic logic
15	Contraction	⊢ $(A \to (A \to B)) \to (A \to B)$,
16	Expansion and positive paradox	⊢ $(A \to B) \to (A \to (A \to B))$, ⊢ $A \to (B \to A)$
17	Explosion	⊢ $(A \land \overline{A}) \to B$ (conjunctive) or ⊢ $A \to (\overline{A} \to B)$ (implicative)
18	Fusion	⊢ $[(A \& B) \to X] \leftrightarrow [A \to (B \to X)]$,
19	Not-elimination	⊢ $\overline{\overline{A}} \to A$,
20	Relativity	⊢ $((A \to B) \to B) \to A$.

Those extensions are considered independently of one another (i.e., results of one of those chapters may not be assumed in another of those chapters), with this exception:

$$\text{relativity} \Rightarrow \text{not-elimination} \Rightarrow \text{fusion}.$$

Anticipating the discussion below, we mention a few more one-axiom extensions :

24.5	Implicative disjunction	⊢ $((A \to B) \to B) \to (A \lor B)$
15.3.a	Specialized contraction	⊢ $(A \to (A \to \overline{A})) \to (A \to \overline{A})$
8.37.f	Centering	⊢ $(A \to A) \leftrightarrow \overline{A \to A}$

The preceding chapters have shown that various expressions are derivable. Chapter 21 introduces soundness, a new tool that will finally enable us to show that certain expressions are *not* derivable, a fact that we have only hinted at in earlier chapters. A pairing of an interpretation (semantic) with an axiomatization (syntactic) is *sound* if

$$\left\{ \begin{array}{c} \text{theorems of the} \\ \text{axiomatization} \end{array} \right\} \subseteq \left\{ \begin{array}{c} \text{tautologies of the} \\ \text{interpretation} \end{array} \right\}.$$

If equality holds, we have *completeness*, but that's much harder to establish and doesn't come until the end of the book. Soundness is introduced at this point because we can put it to good use in the next few chapters.

The next few chapters investigate our "major" logics:

Chapter	assumptions: basic logic plus ...
22 Constructive	positive paradox, contraction, explosion;
23 Relevant	not-elimination and contraction;
24 Fuzzy	positive paradox, implicative disjunction, specialized contraction, not-elimination;
25 Classical	all of the above;
26 Abelian	relativity and centering.

These chapters include, among other things, several deduction principles, converses to the detachmental corollary procedure; see 2.37.

Chapter 27 proves the propositional version of Harrop's admissibility rule for constructive logic. The proof is via Meyer's "metavaluation," a computational device that is a sort of mixture of semantics and syntactics. Two corollaries are the Disjunction Property and Mints's Admissibility Rule. The latter is the most elementary example of an admissibility rule that is not also a derivable inference rule. (That admissibility and derivability are the same in classical logic is proved in 29.15.)

Finally, we prove some completeness pairings. Proofs for constructive implication and relevant implication, in Chapter 28, use Kripke-style "multiple worlds" interpretations. Proofs for classical, fuzzy, and constructive logic are presented in Chapter 29, all using what I call "maximal Z-unproving sets" — i.e., sets S that are maximal for the property that $S \nvdash Z$.

1.8. *What to cover; what to skip.* That's up to the individual instructor's own taste, but here are a few suggestions and hints based on my own teaching experience.

Different students are motivated in different ways, so the first few chapters cover several different kinds of introductions — historical, linguistic, etc. In my own lectures, I skip large parts of the first few chapters, leaving those parts as recommended reading for students with weak backgrounds. But my lectures generally *do* include Jarden's example (2.45), sets of sets (3.3–3.6), Russell's Paradox (3.11–3.12), Venn diagrams (3.47–3.55.a), topologies and interiors (4.1–4.18), and quantifiers (5.40–5.51).

I do *not* lecture on every section in the book. I merely hit the highlights and some crucial parts; I expect the students to read the rest on their own. I permit my students to use their textbooks and notes on all tests and quizzes after midsemester, because I do not expect students to memorize vast collections of formulas. I encourage students to write their own summaries, or "crib sheets," for use during those tests and quizzes; preparing such summaries is an excellent way to study.

I tend to spend most of my semester on those parts of the book that have exercises. The exercises may seem repetitious and mechanical to advanced readers, but they are crucial for bringing the beginner into frequent and close contact with basic ideas. I assign many of the exercises as homework, and give tests made of similar problems. Note that the chapters vary greatly in length,[2] and the exercises do not appear at a constant rate, so a syllabus cannot be planned by a rule like "a chapter every week."

Though there are occasional exceptions, the general trend in the book is from elementary topics to topics of greater mathematical sophistication. Graduate students might skip the first few chapters of the book; at the other extreme, my undergraduate classes generally do not reach the last few chapters. Still, I do state some of the results of those chapters. Even without their proofs, the results on admissibility and completeness play an essential role in tying the book together conceptually, so they are mentioned frequently throughout the book.

One simple way to abridge the book is to leave out all the

---

[2]It has been my experience that students sometimes become confused about which formula schemes or inference rules are permitted as justifications in any particular homework problem. I have arranged the book so that I can answer, "anything from Chapters such-and-such, plus anything from the beginning of the current chapter to our current location in that chapter." This consideration outweighed the unpleasantness of having chapters with vastly different lengths.

Why pluralism? 13

topological material, and all the material on constructive logic that is unavoidably topological — but retain the nontopological parts of constructive logic. This means skipping Chapters 4 and 10, as well as some parts of Chapters 22 and 29. Chapter 27 does not depend on topology, so the instructor might decide near the end of the semester whether to cover it or jettison it, depending on how much time is left. Its proofs are slightly tedious, but they do make good exercises, and the conclusion of that chapter is one of the juiciest bits of the book.

WHY PLURALISM?

1.9. It has been my experience that many mathematicians, even including some accomplished logicians, are unfamiliar with the pluralist approach. I believe that many of them would like it if they gave it a try.

This subchapter is, quite frankly, a sales pitch. I envision an instructor of logic standing in a bookstore or library or at a conference exhibit table, leafing through textbooks and trying to choose one for his/her course. In the next few pages I will attempt to persuade that instructor that pluralism is pedagogically superior to the traditional, classical-only approach — i.e., that it will bring beginning students to a better understanding of logic.

1.10. The traditional and pluralist approaches to introductory logic share the goal of conveying to students not just one or a few separate logics, but also certain deeper ideas that are common to all logics:

> interpretations, derivations, soundness, completeness, independence of axioms, redundancy of connectives, sharing of propositional variable symbols, the finite model property, etc.

Though the two approaches share deeper ideas, they differ greatly on the surface level — i.e., in their choice of examples. Indeed, consider again the diagram in 1.1; the small box in the lower left corner represents the overlap between traditional depth and pluralist breadth. Because the overlap is small, it would not be meaningful to statistically compare the two approaches by giving the same exam to students from the two courses. Instead I will advocate pluralism using arguments based on everyday experience and commonly held principles of teaching.

1.11. *The classical-only approach is unnatural and artificial,* whereas the pluralist approach is motivated by the students' own nonmathematical experiences. Classical logic presented by itself doesn't really make much sense; it embraces non sequiturs such as

> if today is Tuesday then the earth is round

— true for a classical logician, but nonsense for anyone else. The student is left wondering *why* implication is defined the way it is defined; it seems rather arbitrary.

Classical logic is computationally the simplest of all the main logics, and most mathematicians are comfortable with it as a method for presenting mathematical proofs. But it does not closely resemble the way that we actually *think* most of the time. Human thought — even that of mathematicians — is a mixture of many logics, not just classical. Human thought may be too complicated to be fully understood, but some of its ingredients are simple enough to analyze. The few logics in *CNL* were selected from the much wider variety in the literature, in part for their computational simplicity, but also for their philosophical and/or psychological significance.

Admittedly, the student's everyday experience is not mathematical, and it is not precise enough to actually be used as a justification in any mathematical *proofs*. But that experience provides intuition and motivation, which are invaluable to beginners.

Some teachers, familiar only with classical logic, may fear that pluralism will open the floodgate of cultural relativism: If all logics are permitted, then no one of them is of any particular interest or value. But just the opposite is true. Accepting an arbitrary-seeming definition as the only correct one deprives us

Why pluralism?

of any meaningful choice, whereas pluralism explores the different advantages enjoyed by different logics.

1.12. *Classical propositional semantics is too easy.* Later in the semester, after students in the traditional course have become accustomed to classical logic and resigned to its artificiality, they learn to use true/false tables. For instance, they can determine that $(P \to \neg Q) \to (Q \to \neg P)$ is true by plugging in the four combinations TT, TF, FT, FF for P and Q.

But then students may not see any need for syntactic derivations. Why bother to prove something, when we already know it is true? (An analogous pedagogical problem arises in the traditional, Euclidean-only course on geometry; see 2.29. Students may use pictures to learn isolated facts about lines and circles without understanding the proofs that connect those facts.)

And even if we persuade them that proofs are somehow worthwhile, students may still have difficulty understanding the criteria that determine whether a proof is *correct*. An omitted step will hardly be noticed, when we already know the conclusion is true. The choice of steps may seem dictated by arbitrary ritual rather than logical necessity.

1.13. *Reasoning requires doubt, which requires plausible contrasting alternatives.* The traditional approach to logic presents classical results — e.g., excluded middle or the Herbrand-Tarski deduction principle — as absolute truths, with no plausible alternatives.

To teach students to reason mathematically, we must encourage them to doubt everything until it is proven. The teacher should be a guide who helps find the right questions, not an authority who dictates the answers. The student should learn to always ask,

what if this is not true?

That question is hypothetical and abstract, and requires more sophistication than is possessed by many beginning students. They need plausible concrete alternatives — i.e., they can ask the easier question

> what happens in one of the other examples that we've been studying?

And those examples are useful for more advanced students too, and even for researchers. But to ask about other examples, first we must *have* other examples.

1.14. Even students who have already learned to doubt will gain insights from the additional examples. One of the most widely held principles of teaching is that

> *whenever possible, an abstract idea should be accompanied by several different examples.*

One example — in this case, classical logic — hardly suffices to explain abstract notions such as the law of the excluded middle, the deduction principle, variable sharing, or explosion. See the table in 2.37 for a summary of how those four abstract notions fare in five different logics studied in this book.

Many traditional textbooks follow a route that builds up toward Gödel's Incompleteness Principles. But I think that many of the students studying from such a textbook are lost by the end of the semester, largely because the incompleteness principles are preceded by too few examples of completeness. *CNL* proves completeness for several logics and describes it for several others; see the table in 2.26.

1.15. *The traditional emphasis on predicate logic sometimes obscures basic ideas.* The most fundamental ideas of logic — derivation, interpretation, etc. — are already present, in simpler but still nontrivial form, in propositional logic. Predicate logic complicates the presentation but in many cases does not enrich the ideas.

A good example of this is the classical Herbrand-Tarski deduction principle.[3] The predicate logic version (found in most logic textbooks) is complicated and hard to understand:

---

[3]Known as the "Deduction Theorem" in most of the literature; see 2.18.

Why pluralism?

> Let F and G be formulas, and let $\mathcal{H}$ be a set of formulas. Assume $\mathcal{H} \cup \{F\} \vdash G$. Then $\mathcal{H} \vdash F \to G$ ...
>
> ... *provided that*, in the given derivation $\mathcal{H} \cup \{F\} \vdash G$, none of the steps involves a substitution in which a free individual variable being replaced is one that appears in F.

In effect, the hypothesis "$\mathcal{H} \cup \{F\} \vdash G$" is much more complicated than it at first appears. We need to know not only that G is derivable from $\mathcal{H} \cup \{F\}$, but also *how* it is derivable, all the way down to nitty-gritty details about scopes of quantifiers.

Now contrast that with the propositional logic version. In propositional logic, there are no free or bound individual variables, so we can omit the entire fine print clause. The resulting principle is shorter and simpler, but still retains much of the power of the predicate logic version.

Admittedly, predicate logic is too important to omit altogether. A first course in logic — for most students, the *only* course in logic — should include at least a brief, informal introduction to quantifiers, as *CNL* does in 5.40–5.51. But that topic does not need to skew the entire semester; we don't necessarily have to carry $\forall$ and $\exists$ through our formal development.

1.16. In all fairness, I must also mention one seeming disadvantage of pluralism. Classical logic is computationally the simplest of all the major logics. To compare it with anything else, we must accept some complications.

For instance, teachers who are familiar with classical logic's short formulation (three axioms, two definitions, and one inference rule, given in 25.1.d) may be dismayed to see a dozen axioms in nonclassical logics. Those longer lists are unavoidable. In classical logic we could define two of $\lor, \land, \to, \neg$ in terms of the other two, but for some nonclassical logics we need all four connectives as primitives. (See 10.9.) More primitives require more axioms to govern them.

Still, the dozen axioms need not all be swallowed at one time. They can be digested in several courses, each quite tasty by itself. That is the plan we have described in 1.7.

1.17. *Pluralism is more modern; the prepluralistic view is rather antiquated.* Traditional, classical-only textbooks are still largely built around Gödel's important (but not elementary) discoveries of the 1930s.

I do not advocate teaching to our undergraduates whatever is the latest discovery. Thousands of new mathematical discoveries are published in research journals every year, but most of those discoveries are too advanced and specialized to deserve the attention of beginning students.

Still, when a fundamental (important and/or elementary) development does comes along, we should not overlook it. Such a development has occurred gradually during the late 20th century, in the work of Kripke, Anderson, Belnap, Zadeh, and others. Pluralism (many logics considered simultaneously) has become predominant in the research literature — not as a focus of attention, but as a fact in the background, an accepted and assimilated part of the paradigm. Moreover, enough elementary examples have accumulated in recent decades to make pluralism feasible in the beginners' classroom. I believe that beginning logic students would benefit from pluralism, for reasons indicated in the preceding pages.

1.18. *Pluralism has greater applicability to computer science.* Or at least, so I have been told. I know too little of that subject, so it will not be pursued in the present edition of this book.

FEEDBACK

1.19. Despite the best efforts of myself and the editors at Princeton University Press, I'm sure a few errors remain in this edition. I hope they're small ones, but at any rate I'll list them on a web page when they're reported. I may also post some shorter proofs, if I learn of any. The main page for this book is at

http://www.math.vanderbilt.edu/~schectex/logics/

and links from that page will lead to the errata and addenda pages. Also I invite suggestions for alterations for a second edition, which might or might not eventually happen. Email me at eric.schechter@vanderbilt.edu.

Acknowledgments 19

(Yes, the URL and email are spelled differently.) A second edition, if it happens, probably will have shorter introductory chapters, less repetition, more exercises, and perhaps (if I can find sufficiently elementary ones) proofs of some of the results that are stated without proof in this edition.

Acknowledgments

1.20. This book benefited from discussions with Greg Bush, Elvira Casal, Dave Easley, Mark Ellingham, Jonathan Farley, Isidore Fleischer, Klaus Glashoff, Peter Jipsen, Bjarni Jónsson, Christian Khoury, Ayan Mahalanobis, Ralph McKenzie, Charles Megibben, Peter Nyikos Dave Renfro, Fred Richman, Peter Suber, Constantine Tsinakis, and others. Special thanks go to Norm Megill, who first introduced me to matrix interpretations, and to Robert Meyer, who answered many questions.

This book would not have been written without the help of free Latex software from Donald Knuth, Leslie Lamport, Christian Schenk, Sven Wiegand, and others. Several of the examples in this book were located using MaGIC (MAtrix Generator for Implication Connectives), a computer program made available by John Slaney of the Automated Reasoning Group at Australian National University; my thanks to Norm Megill for porting the program to Windows. I am also grateful to Jonathan Balsam, Nikolaos Galatos, Norm Megill, and the people at Princeton University Press, whose careful reading of previous versions caught many errors. Of course, any errors that remain are my own.

Most of all, I am indebted to the students who permitted me to experiment on them. I hope their pains will be justified by benefits to later students.

Chapter 2

Introduction for students

2.1. This is a textbook about propositional logics. A more detailed overview of the contents can be found in 1.7.

This chapter, and to a lesser extent Chapter 5, are intended as an overview of logic. *Caution:*

- These introductory chapters are not intended to be mathematically precise. Arguments sketched in these chapters should not be viewed as proofs, and will not be used as ingredients in the rigorous proofs developed later in the book.

- These introductory chapters are not typical of the style of the whole book. Formal logic, which begins in Chapter 6, is much more computational. The student who is undecided about registering for or dropping the course should look at some of the later chapters.

WHO SHOULD STUDY LOGIC?

2.2. The subjects of logic, algebra, and computer science are enough alike that a previous course in any one of these three subjects may aid the student in pursuing the other two — partly because of specific results, but more because of a familiarity with the general style of thinking.

Regardless of previous background, however, any liberal arts undergraduate student might wish to take at least one course in logic, for the subject has great philosophical significance: It tells

Who should study logic? 21

us something about our place in the universe. It tells us how we think — or more precisely, how we think we think. Mathematics contains eternal truths about number and shape; mathematical logic contains truths about the nature of truth itself.

The grandeur of logic's history (sketched later in this chapter) and the power and beauty of its ideas can be appreciated by any student in a first course on logic, but only a few students will greatly enjoy its computations. It is those few who may choose to study beyond the first course (i.e., beyond this book) and become logicians. I would expect that most students taking a first course in mathematical logic are simply liberal arts students fulfilling a mathematics requirement and seeking a sampling of our culture. Logic is an important part of that culture, and I will try to present it in a way that does not require all the background of a mathematician.

2.3. Logic and set theory are sometimes called the *foundations* of mathematics, because they are used as a basis for other branches of mathematics; those other branches are then called *ordinary mathematics*.

- Logic is a foundation because the logician studies the kinds of *reasoning* used in the individual steps of a proof. An ordinary mathematical proof may be long and complicated, involving many steps, but those steps involve only a few different kinds of reasoning. The logician studies the nature of those few kinds of reasoning.
- Set theory (also studied briefly in this book) is a foundation for the *objects* that mathematicians reason about. For instance, 3 and $\sqrt{2}$ are members of the *set* of real numbers. Likewise, $\pi_1 \to \pi_2$ and $\pi_2 \wedge \neg \pi_1$ are members of the *set* of all formulas in the formal language studied in this book.

Logic and set theory are fundamental but not central in mathematics. Knowing a little about the foundations may make you a better mathematician, but knowing a lot about the foundations will only make you a better logician or set theorist.

That disclaimer surprises some people. It might be explained by this analogy: If you want to write The Great American Novel, you'll need to know English well; one or two grammar books might help. But studying dozens of grammar books will make you a grammarian, not a novelist. Knowing the definitions of "verb" and "noun" is rather different from being able to use verbs and nouns effectively. Reading many novels might be better training for a novelist, and writing many novels — or rewriting one novel many times — may be the best training. Likewise, to become proficient in the kinds of informal proofs that make up "ordinary" mathematics, you probably should read and/or write more of "ordinary" mathematics.

For that purpose, I would particularly recommend a course in general topology, also known as "point set topology." We will use topology as a tool in our development of formal logic, so a very brief introduction to topology is given in Chapter 4 of this book, but mathematics majors should take one or more courses on the subject.

General topology has the advantage that its results are usually formulated in sentences rather than equations or other mathematical symbols. Also, it has few prerequisites — e.g., it does not require algebra, geometry, or calculus.

General topology is one of the subjects best suited for the *Moore method*. That is a teaching method in which the students are supplied with just definitions and the statements of theorems, and the students must discover and write up the proofs. In my opinion, that is the best way to learn to do proofs.

2.4. A few students — particularly young adults — may begin to study logic in the naive hope of becoming better organized in their personal lives, or of becoming more "logical" people, less prone to making rash decisions based on transient emotions. I must inform those students that, unfortunately, a knowledge of mathematical logic is *not* likely to help with such matters. Those students might derive greater understanding of their lives by visiting a psychotherapist, or by reading biochemists' accounts of the hormones that storm through humans during adolescence.

Logic was misrepresented by the popular television series *Star Trek*. The unemotional characters Spock (half human, half Vulcan) and Data (an android) claimed to be motivated solely by logic. But that claim was belied by their behavior. Logic can only be used as an aid in acting on one's values, not in choosing those values. Spock's and Data's very positive values, such as respect for other sentient beings and loyalty to their crew mates, must have been determined by extralogical means. Indeed, a few episodes of the program pitted Spock or Data against some "logical" villains — evil Vulcans, evil androids — who exhibited just as much rationality in carrying out quite different values. (See the related remarks about logic and ethics, in 2.19.d.)

Logicians are as likely as anyone to behave illogically in their personal lives. Kurt Gödel (1906–1978) was arguably the greatest logician in history, but he was burdened with severe psychiatric problems for which he was sometimes hospitalized. In the end, he believed he was being poisoned; refusing to eat, he starved to death. In the decades since Gödel's time, medical science has made great progress, and many afflictions of the mind can now be treated successfully with medication or by other means. Any reader who feels emotionally troubled is urged to seek help from a physician, not from this logic book.

2.5. *Why should we formalize logic?* What are the advantages of writing $\wedge$ instead of "and," or writing $\exists$ instead of "there exists"? These symbols are seldom used in other branches of mathematics — e.g., you won't often find them in a research paper on differential equations.

One reason is if we're interested in logic itself. Pursuit of that subject will lead us to logical expressions such as

$$\left\{ (\forall x) \left[W(x) \to R(x) \right] \right\} \vee \left\{ (\forall x) \left[R(x) \to W(x) \right] \right\},$$

much more complicated than any arising from merely abbreviating the reasoning of, say, differential equations. Clearly, without symbols an expression like this would be difficult to analyze.

But even students who do not intend to become logicians would benefit from using these symbols for at least a while. Basic properties of the symbols, such as the difference between $\forall x \, \exists y$ and $\exists y \, \forall x$ are easier to understand if presented symbolically.

Moreover, even if the symbols do not show up explicitly elsewhere, the underlying concepts are implicit in some other definitions (particularly in mathematics). For instance, the expression

"$\limsup_{t\to\infty} f(t)$" might appear somewhere in a paper on differential equations. One way to define limsup is as follows: It is a number with the property that, for numbers r,

$$\limsup_{t\to\infty} f(t) \leq r \quad \Longleftrightarrow \quad (\forall \varepsilon > 0)(\exists u \in \mathbb{R})(\forall t > u)\ f(t) < r+\varepsilon.$$

(Other definitions of limsup say essentially the same thing, but may differ in appearance.) Researchers in differential equations generally think of limsup in terms of its familiar properties rather than in terms of this definition, but they first must learn it in terms of its definition — either in words or symbolically. In effect, there are $\forall$'s and $\exists$'s in a paper on differential equations, but they are hidden inside the limsup's and other tools of that subject.

2.6. *Teaching students how to think* — how to analyze, how to question, etc. — is sometimes cited as one of the goals of a mathematics course, especially a mathematical logic course. But that justification for logic may be erroneous. Though "how to think" and mathematical logic do have some overlap, they are two different subjects. Techniques of thinking — e.g., look for a similar but more familiar problem, think about conditions that would hold if the problem were already solved, draw a diagram — can be found, not in logic books, but in a book such as Polya's *How to Solve It*.

Though logic does formalize some thinking techniques, the use of those techniques and the study of the formalizations occur on very different levels.

- The lower level consists of equations, formulas, and computational techniques. For instance, *CNL* shows that the "proof by contradiction" formula $(A \to B) \to ((A \to \overline{B}) \to \overline{A})$ is a theorem in classical, relevant, and constructive logics, and shows how that formula is used in the derivations of other formulas. Practice should make the student adept at computations of this sort.

- The higher level consists of paragraphs of reasoning *about*

Formalism and certification

those symbolic formulas. Section 5.35 discusses proof by contradiction in informal terms, and the technique is applied in paragraph-style arguments in 3.70, 7.7, 7.14, 22.16, 27.6, 29.6, 29.14, and a few other parts of the book. All of the higher-level reasoning in this book is in an informal version of classical logic (see discussion of metalogic in 2.17). However, this book is not written for use with the Moore method — the exercises in this book are mostly computational and do not require much creativity.

2.7. Granted that you have decided to study logic, is *this* book the right one for you? This book is unconventional, and does not follow the standard syllabus for an introductory course on logic. It may be unsuitable if you need a *conventional* logic course as a prerequisite for some other, more advanced course. The differences between this book and a conventional treatment are discussed in 1.1, 1.10, and at the end of 2.36.

FORMALISM AND CERTIFICATION

2.8. To describe what logic is, we might begin by contrasting formal and informal proofs. A *formal proof* is a purely computational derivation using abstract symbols such as $\vee, \wedge, \neg, \rightarrow, \vdash, \forall, \exists, \Rightarrow, =, \in, \subseteq$. All steps must be justified; nothing is taken for granted. Completely formal proofs are seldom used by anyone except logicians. *Informal* reasoning may lack some of those qualifications.

2.9. Here is an example. *Informal set theory* is presented without proofs, as a collection of "observed facts" about sets. We shall review informal set theory in Chapter 3, before we get started on formal logic; we shall use it as a tool in our development of formal logic.

In contrast, *formal set theory* (or *axiomatic set theory*) can only be developed *after* logic. It is a rigorous mathematical theory in which all assertions are proved. In formal set theory, the symbols $\in$ ("is a member of") and $\subseteq$ ("is a subset of") are stripped of their traditional intuitive meanings, and given only the meanings determined by consciously chosen axioms. Indeed, $\subseteq$ is not taken as a primitive symbol at all; it is defined in terms of $\in$. Here is its definition: $x \subseteq y$ is an abbreviation for

$$(\forall z) \ [\, (z \in x) \to (z \in y) \,].$$

That is, in words: For each z, if z is a member of x, then z is a member of y.

In informal set theory, a "set" is understood to mean "a collection of objects." Thus it is *obvious* that

(a) there exists a set with no elements, which we may call the *empty set*; and

(b) any collection of sets has a union.

But both of those "obvious" assertions presuppose some notion of what a "set" is, and assertion (b) also presupposes an understanding of "union." That understanding is based on the usual meanings of words in English, which may be biased or imprecise in ways that we have not yet noticed.

Axiomatic set theory, in contrast, rejects English as a reliable source of understanding. After all, English was not designed by mathematicians, and may conceal inherent erroneous assumptions. (An excellent example of that is Russell's Paradox, discussed in 3.11.) Axiomatic set theory takes the attitude that we do *not* already know what a "set" is, or what "$\in$" represents. In fact, in axiomatic set theory we don't even use the word "set," and the meaning of the symbol "$\in$" is specified only by the axioms that we choose to adopt. For the existence of empty sets and unions, we adopt axioms something like these:

(a) $(\exists x)\,(\forall y) \ \neg\,(y \in x)$.

(b) $(\forall x)\,(\exists y)\,(\forall z) \ [\, (z \in y) \leftrightarrow (\exists w\,(z {\in} w) \wedge (w {\in} x)) \,]$.

That is, in words,

(a) there exists a set x with the property that for every set y, it is not true that y is a member of x; and

(b) for each set x there exists a set y with the property that the members of y are the same as the members of members of x.

The issue is not whether empty sets and unions "really exist," but rather, what consequences can be proved about some abstract objects from an abstract system of axioms, consisting of (a) and (b) and a few more such axioms *and nothing else* — i.e., with no "understanding" based on English or any other kind of common knowledge.

Because our natural language is English, we cannot avoid thinking in terms of phrases such as "really exist," but such phrases may be misleading. The "real world" may include three apples or three airplanes, but the abstract concept of the number 3 itself does not exist as a physical entity anywhere in our real world. It exists only in our minds.

If some sets "really" do exist in some sense, perhaps they are not described accurately by our axioms. We can't be sure about that. But at least we can investigate the properties of any objects that *do* satisfy our axioms. We find it convenient to call those objects "sets" because we believe that those objects correspond closely to our intuitive notion of "sets" — but that belief is not really a part of our mathematical system.

Our successes with commerce and technology show that we are in agreement about many abstract notions — e.g., the concept of "3" in my mind is essentially the same as the concept of "3" in your mind. That kind of agreement is also present for many higher concepts of mathematics, but not all of them. Our mental universes may differ on more complicated and more abstract objects. See 3.10 and 3.33 for examples of this.

2.10. *Formalism* is the style or philosophy or program that formalizes each part of mathematics, to make it more clear and less prone to error. *Logicism* is a more extreme philosophy that advocates reducing all of mathematics to consequences of logic.

The most extensive formalization ever carried out was *Principia Mathematica*, written by logicists Whitehead and Russell near the beginning of the 20th century. That work, three volumes totaling nearly 2000 pages, reduced all the fundamentals of mathematics to logical symbols. Comments in English appeared occasionally in the book but were understood to be outside the formal work. For instance, a comment on page 362 points out that the main idea of $1 + 1 = 2$ has just been proved!

In principle, all of known mathematics can be formulated in terms of the symbols and axioms. But in everyday practice, most ordinary mathematicians do not completely formalize their work; to do so would be highly impractical. Even partial formalization of a two-page paper on differential equations would turn it into a 50-page paper. For analogy, imagine a cake recipe written by a nuclear physicist, describing the locations and quantities of the electrons, protons, etc., that are included in the butter, the sugar, etc.

Mathematicians generally do not formalize their work completely, and so they refer to their presentation as "informal." However, this word should not be construed as "careless" or "sloppy" or "vague." Even when they are informal, mathematicians do check that their work is *formalizable* — i.e., that they have stated their definitions and theorems with enough precision and clarity that any competent mathematician reading the work *could* expand it to a complete formalization if so desired. Formalizability is a requirement for mathematical publications in refereed research journals; formalizability gives mathematics its unique ironclad certainty.

2.11. Complete formalization *is* routinely carried out by computer programmers. Unlike humans, a computer cannot read between the lines; every nuance of intended meaning must be spelled out explicitly. Any computer language, such as Pascal or C++, has a very small vocabulary, much like the language of formal logic studied in this book. But even a small vocabulary can express complicated ideas if the expression is long enough; that is the case both in logic and in computer programs.

In recent years some researchers in artificial intelligence have begun carrying out complete formalizations of *mathematics* —

Formalism and certification

they have begun "teaching" mathematics to computers, in excruciating detail. A chief reason for this work is to learn more about intelligence — i.e., to see how sentient beings think, or how sentient beings *could* think. However, these experiments also have some interesting consequences for mathematics. The computer programs are able to check the correctness of proofs, and sometimes the computers even produce new proofs. In a few cases, computer programs have discovered new proofs that were shorter or simpler than any previously known proofs of the same theorems.

Aside from complete formalizations, computers have greatly extended the range of feasible proofs. Some proofs involve long computations, beyond the endurance of mathematicians armed only with pencils, but not beyond the reach of electronic computers. The most famous example of this is the *Four Color Theorem*, which had been a conjecture for over a century: Four colors suffice to color any planar map so that no two adjacent regions have the same color. This was finally proved in 1976 by Appel and Haken. They wrote a computer program that verified 1936 different cases into which the problem could be classified.

In 1994 another computer-based proof of the Four Color Theorem, using only 633 cases, was published by Robertson, Sanders, Seymour, and Thomas. The search for shorter and simpler proofs continues. There are a few mathematicians who will not be satisfied until/unless a proof is produced that can actually be read and verified by human beings, unaided by computers. Whether the theorem is actually *true* is a separate question from what kind of proof is acceptable, and what "true" or "acceptable" actually mean. Those are questions of philosophy, not of mathematics.

2.12. One of the chief attractions of mathematics is its iron-clad *certainty*, unique among the fields of human knowledge. (See Kline [1980] for a fascinating treatment of the history of certainty.) Mathematics can be certain only because it is an artificial and finite system, like a game of chess: All the ingredients in a mathematical problem have been put there by the mathematicians who formulated the problem. In contrast, a problem in physics or chemistry may involve ingredients that have not yet been detected by our imperfect instruments; we can never be sure that our observations have taken everything into account.

The absolute certainty in the mental world of pure mathematics is intellectually gratifying, but it disappears as soon as we try to apply our results in the physical world. As Albert Einstein said,

> As far as the laws of mathematics refer to reality, they are not certain; and as far as they are certain, they do not refer to reality.

Despite its uncertainty, however, applied mathematics has been extraordinarily successful. Automobiles, television, the lunar landing, etc., are not mere figments of imagination; they are real accomplishments that used mathematics.

This book starts from the uncertain worlds of psychology and philosophy, but only for motivation and discovery, not to justify any of our reasoning. The rigorous proofs developed in later chapters will not depend on that uncertain motivation.

Ultimately this is a book of *pure* mathematics. Conceivably our results could be turned to some real-world application, but that would be far beyond the expertise of the author, so no such attempt will be made in this volume.

2.13. Some beginning students confuse these two aspects of mathematics:

(*discovery*) How did you find that proof?
(*certification*) How do you know that proof is correct?

Those are very different questions. This book is chiefly concerned with the latter question, but we must understand the distinction between the two questions.

2.14. Before a mathematician can write up a proof, he or she must first discover the ideas that will be used as ingredients in that proof, and any ideas that might be related. During the discovery phase, all conclusions are uncertain, but all methods are permitted. Certain heuristic slogans and mantras may be followed — e.g., "First look for a similar, simpler problem that we *do* know how to solve."

Correct reasoning is not necessary or even highly important. For instance, one common discovery process is to *work backwards* from what the mathematician is trying to prove, even though this is known to be a faulty method of reasoning. (It is discussed further starting in 5.36.)

Another common method is *trial-and-error*: Try something, and if it doesn't work, try something else. This method is more suitable for some problems than others. For instance, on a problem that only has eight conceivable answers, it might not take long to try all eight. The method may be less suitable for a problem that has infinitely many plausible answers. But even there we may gain something from trial-and-error: When a trial fails, we may look at *how* it fails, and so our next guess may be closer to the target. One might call this "enlightened trial-and-error"; it does increase the discoverer's understanding after a while.

The discovery process may involve experimenting with numerous examples, searching through books and articles, talking to oneself and/or to one's colleagues, and scribbling rough ideas on hamburger sacks or tavern napkins. There are many false starts, wrong turns, and dead ends. Some mathematicians will tell you that they get their best inspirations during long evening walks or during a morning shower. They may also tell you that the inspiration seems to come suddenly and effortlessly, but only after weeks of fruitless pacing and muttering. Though it is not meant literally, mathematicians often speak of "banging one's head against a wall for a few weeks or months."

The discovery phase is an essential part of the mathematician's work; indeed, the researcher devotes far more time to discovery than to writing up the results. But discovery may be more a matter of psychology than of mathematics. It is personal, idiosyncratic, and hard to explain. Consequently, few mathematicians make any attempt to explain it.

Likewise, this book will make little attempt to explain the discovery process; this book will be chiefly concerned with the certification process of mathematical logic. To solve this book's exercises, we recommend to the student the methods of working backward and enlightened trial-and-error. Students with a knowledge of computers are urged to use those as well; see 1.20.

The need for discovery may not be evident to students in pre-college or beginning college math courses, since those courses mainly demand mechanical computations similar to examples given in the textbook. Higher-level math courses require more discovery, more original and creative thought.

Actually, for some kinds of problems a mechanical, computational procedure *is* known to be available, though explaining that procedure may be difficult. For other problems it is known that there *cannot* be a mechanical procedure. For still other problems it is not yet known whether a mechanical procedure is possible. The subject of such procedures is *computability theory*. But that theory is far too advanced for this introductory course.

2.15. After discovery, the mathematician must start over from scratch, and rewrite the ideas into a legitimate, formalizable proof following rigid rules, to *certify* the conclusions. Each step in the proof must be justified. "Findings" from the discovery phase can be used as suggestions, but not as justifications.

The certification process is more careful than the discovery process, and so it may reveal gaps that were formerly overlooked; if so, then one must go back to the discovery phase. Indeed, I am describing discovery and certification as two separate processes, to emphasize the differences between them; but in actual practice the two processes interact with each other. The researcher hops back and forth between discovery and certification; efforts on one yield new insights on the other.

In the presentation of a proof, any comment about intuition or about the earlier discovery of the ideas is a mere aside; it is not considered to be part of the actual proof. Such comments are optional, and they are so different in style from the proof itself that they generally seem out of place. Skilled writers may sometimes find a way to work such comments into the proof, in hopes of assisting the reader if the proof is hard to understand. But it is more commonplace to omit them altogether.

Formalism and certification 33

Indeed, it is quite common to rewrite the proof so that it is more orderly, even though this may further remove any traces of the discovery process. (An example of this is sketched in 5.39.) Consequently, some students get the idea that there is little or no discovery process — i.e., that the polished proof simply sprang forth from the mathematician's mind, fully formed. These same students, when unable to get an answer quickly, may get the idea that they are incapable of doing mathematics. Some of these students need nothing more than to be told that they need to continue "banging their heads against the the walls" for a bit longer — that it is perfectly normal for the ideas to take a while to come.

2.16. Overall, what procedures are used in reasoning? For an analogy, here are two ways behavior may be prescribed in religion or law:

(a) A few *permitted* activities ("you may do this, you may do that") are listed. Whatever is not expressly allowed by that list is forbidden.

(b) Or, a few *prohibited* activities ("thou shalt nots") are listed. Whatever is not explicitly forbidden by that list is permitted.

Real life is a mixture of these two options — e.g., you *must* pay taxes; you must *not* kill other people. But we may separate the two options when trying to understand how behavior is prescribed.

Theologians and politicians have argued over these options for centuries — at least as far back as Tertullian, an early Christian philosopher (c. 150–222 A.D.). The arguments persist. For instance, pianos are not explicitly mentioned in the bible; should they be permitted or prohibited in churches? Different church denominations have chosen different answers to this question. (See Shelley [1987].)

In the *teaching* of mathematics, we discuss both

- *permitted* techniques, that can and should be used, and

- common errors, based on unjustified techniques — i.e., *prohibited* techniques.

Consequently, it may take some students a while to understand that mathematical *certification* is based on *permissions only* — or more precisely, a modified version of permissions:

(a′) We are given an explicit list of axioms (permitted statements), and an explicit list of rules for deriving further results from those axioms (permitted methods). Any result that is not among those axioms, and that cannot be derived from those axioms via those rules, is prohibited.

LANGUAGE AND LEVELS

2.17. *Levels of formality.* To study logic mathematically, we must reason about reasoning. But isn't that circular? Well, yes and no.

Any work of logic involves at least two levels of language and reasoning; the beginner should take care not confuse these:

a. The *inner system*, or *object system*, or *formal system*, or *lower system*, is the formalized theory that is being studied and discussed. Its language is the *object language*. It uses symbols such as

$$(, \;), \quad \wedge, \quad \neg, \quad \forall, \quad \rightarrow, \quad \pi_1, \quad \pi_2, \quad \pi_3, \quad \ldots$$

that are specified explicitly; no other symbols may be used. Also specified explicitly are the grammatical rules and the axioms. For instance, in a typical grammar,

$$\pi_1 \rightarrow (\pi_1 \vee \pi_2) \text{ is a formula;} \qquad \pi_1)(\rightarrow \pi_1 \pi_2 \vee \text{ is not.}$$

(Precise rules for forming formulas will be given in 6.7.) Each inner system is fixed and unchanging (though we may study different inner systems on different days). The inner system is entirely artificial, like a language of computers or of robots.

Language and levels 35

The individual components of that language are simpler and less diverse than those of a natural language such as English. However, by combining a very large number of these simple components, we can build very complicated ideas.

b. The *outer system*, or *metasystem*, or *higher system*, is the system in which the discussion is carried out. The reasoning methods of that system form the *metalogic* or *metamathematics*. The only absolute criteria for correctness of a metamathematical proof are that the argument must be clear and convincing. The *metalanguage* may be an informal language, such as the slightly modified version of English that mathematicians commonly use when speaking among themselves. ("Informal" should not be construed as "sloppy" or "imprecise"; see 2.10.) For instance,

> assume $\mathcal{L}$ is a language with
> only countably many symbols

is a sentence *in* English *about* a formal language $\mathcal{L}$; here English is the informal metalanguage.

c. (Strictly speaking, we also have an intermediate level between the formal and informal languages, which we might call the "level of systematic abbreviations." It includes symbols such as $\vdash$, $\vDash$, $\Rightarrow$, and $A, B, C, \ldots$, which will be discussed starting in 2.23.)

Throughout this book, we shall use classical logic for our metalogic. That is, whenever we assert that one statement about some logic *implies* another statement about some logic, the "implies" between those two statements is the "implies" of classical logic. For instance, Mints's Admissibility Rule for Constructive Logic can be stated as

$$\vdash \overline{S} \to (Q \vee R) \quad \Rightarrow \quad \vdash (\overline{S} \to Q) \vee (\overline{S} \to R);$$

this rule is proved in 27.13. Here the symbols "$\to$" are implications in the object system, which in this case happens to be constructive logic. The symbols "$\vdash$" mean "is a theorem" — in this case, "is a theorem of constructive logic." The symbol

$\Rightarrow$ is in the metasystem, and represents implication in *classical* logic. The whole line can be read as

if $\overline{S} \to (Q \vee R)$ is a theorem of constructive logic,
then $(\overline{S} \to Q) \vee (\overline{S} \to R)$ is too.

The words "if" and "then," shown in italic for emphasis here, are the classical implication ($\Rightarrow$) of our metalogic.[1]

The metalogic is classical not only in this book, but in most of the literature of logic. However, that is merely a convenient convention, not a necessity. A few mathematicians prefer to base their outer system on more restrictive rules, such as those of the relevantist or the constructivist (see 2.41 and 2.42). Such an approach requires more time and greater sophistication; it will not be attempted here.

Returning to the question we asked at the beginning of this section: Is our reasoning circular? Yes, to some extent it is, and unavoidably so. Our outer system is classical logic, and our inner system is one of several logics — perhaps classical. How can we justify using classical logic in our study of classical logic?

Well, actually it's two different kinds of classical logic.

- Our inner system is some formal logic — perhaps classical — which we study without any restriction on the complexity of the formulas.

- Our outer system is an informal and simple version of classical logic; we use just fundamental inference rules that we feel fairly confident about.

It must be admitted that this "confidence" stems from common sense and some experience with mathematics. We assume a background and viewpoint that might not be shared by all readers.

---

[1] Actually, even that last explanation of Mints's rule is an abbreviation. In later chapters we shall see that $\overline{S} \to (Q \vee R)$ is not a formula, but a formula scheme, and it is *not* a theorem scheme of constructive logic as stated. A full statement of Mints's rule is the following:

Suppose that $\mathcal{X}$, $\mathcal{Y}$, and $\mathcal{Z}$ are some formula schemes. Suppose that some particular substitution of formulas for the metavariables appearing in $\mathcal{X}$, $\mathcal{Y}$, and $\mathcal{Z}$ makes $\overline{\mathcal{Z}} \to (\mathcal{X} \vee \mathcal{Y})$ into a theorem. Then the same substitution also makes $(\overline{\mathcal{Z}} \to X) \vee (\overline{\mathcal{Z}} \to Y)$ into a theorem.

Language and levels 37

There is no real bottom point of complete ignorance from which we can begin.

An analogous situation is that of the student in high school who takes a course in English.[2] The student already must know *some* English in order to understand the teacher; the course is intended to extend and refine that knowledge. Likewise, it is presumed that the audience of this book already has some experience with reasoning in everyday nonmathematical situations; this book is intended to extend and refine that knowledge of reasoning.

2.18. To reduce confusion, this book will use the terms *theorem* and *tautology* for truths in the inner system, and *principle* and *rule* for truths on a higher level. However, we caution the reader that this distinction is not followed in most other books and papers on logic. In most of the literature, the words "principle," "theorem," and "tautology" are used almost interchangeably (and outside of logic, the word "rule" too); the word "theorem" is used most often for all these purposes. In particular, what I have called the Deduction Principle, Glivenko's Principle, and Gödel's Incompleteness Principles are known as "theorems" in most of the literature. One must recognize from the context whether inner or outer is being discussed.

I will use the term *corollary* for "an easy consequence," in all settings — inner and outer, semantic and syntactic. See also 13.16.

2.19. *Adding more symbols to the language.* One major way of classifying logics is by what kinds of symbols they involve.

    a. This book is chiefly concerned with *propositional logic*, also known as *sentential logic*. That logic generally deals with $\vee$ (or), $\wedge$ (and), $\neg$ (not), $\rightarrow$ (implies), and propositional variable symbols such as $\pi_1, \pi_2, \pi_3, \ldots$. A typical formula in propositional logic is $(\pi_1 \vee \pi_2) \rightarrow \pi_3$. (The symbols $\leftrightarrow$, $\circ$, $\rightsquigarrow$, $\&$, $\bot$ are also on this level but are used less often in this book.)

    b. *Predicate logic* adds some symbols for "individual variables" $x, y, \ldots$ to the language, as well as the *quantifier* symbols $\forall$ and $\exists$. Propositions $\pi_1, \pi_2, \ldots$ may be functions of those variables. For instance, $\forall x\, \exists y\, \pi_2(x, y)$ says that "for each x there exists at least one y (which may depend on x) such that

---

[2] Replace "English" with whatever is your language, of course.

$\pi_2(x,y)$ holds." This book includes a brief study of quantifiers in informal classical logic, starting in 5.41.

c. An *applied logic*, or *first-order theory*, adds some *nonlogical symbols*. For instance,

$$= \text{ (equals)}, \qquad + \text{ (plus)}, \qquad \times \text{ (times)}$$

are symbols, not of logic, but of arithmetic, another subject. To obtain the theory of arithmetic we add to predicate logic those nonlogical symbols, plus a few axioms governing the use of those symbols. A typical theorem about the set $\mathbb{Z}$ of integers (but not about the positive integers $\mathbb{N}$) is

$$\forall x \; \forall y \; \exists z \; x + z = y,$$

which says that for each choice of x and y there exists at least one z such that $x + z = y$. (In other words, we are able to subtract.) Another important applied logic is axiomatic set theory, described in 2.9. Applied logics will not be studied in this book.

d. *Modal logic* (also not studied in this book) adds some modal operators to one of the logics described above. Here are some examples of modal operators:

In this logic	the symbol "□" means	the symbol "◇" means
alethic	it is necessary that	it is possible that
deontic	it is obligatory that	it is permitted that
epistemic	it is known that	it is not known to be false that

(See Copeland [1996] and references therein.) These modal operators occur in dual pairs, satisfying $\neg(\Diamond x) = \Box(\neg x)$. For instance,

> it is not permitted that we kill people

says the same thing as

> it is obligatory that we do not kill people.

Each modal logic has many variants. For instance, different ethical systems can be represented by the deontic logics stemming from different axiom systems. We cannot prove that one ethical system is "more correct" or "more logical" or "better" than another, but in principle we should be able to calculate the different consequences of those different ethical systems.

SEMANTICS AND SYNTACTICS

2.20. The logician studies the form and the meanings of language separately from each other. For a nonmathematical example of this separation, consider Lewis Carroll's nonsense poem *Jabberwocky*, which begins

> 'Twas brillig, and the slithy toves
> Did gyre and gimble in the wabe;
> All mimsy were the borogoves,
> And the mome raths outgrabe.

We don't know exactly what "slithy" or "tove" means, but we're fairly certain that "slithy" is an adjective and "tove" is a noun — i.e., that "slithy" specifies some particular *kind* of "tove" — because the verse seems to follow the grammatical rules of ordinary English.

In much the same fashion, in mathematical logic we can study the grammar of formulas separately from the meanings of those formulas. For instance, the expression "$(A \vee B) \to C$" conforms to grammatical rules commonly used in logic, while the expression "$A)(\vee \to BC$" does not. All the logics studied in this book use the same grammatical rules, which we will study in Chapter 6.

2.21. *Two faces of logic.* Logic can be classified as either semantic or syntactic.[3] We will study these separately at first, and then together.

---

[3] Van Dalen refers to these as the "profane" and "sacred" sides of logic. That description may be intended humorously but has some truth to it.

Semantics is the concrete or applied approach, which initially may be more easily understood. We study examples, and investigate which statements are true as computational facts. A typical computation is

$$[\![\pi_1 \vee \neg \pi_1]\!] = [\![\pi_1]\!] \mathbin{\underline{\vee}} \left(\underline{\neg}[\![\pi_1]\!]\right) = \max\left\{[\![\pi_1]\!], 1 - [\![\pi_1]\!]\right\} \in \Sigma_+,$$

which would be read as follows: The expression $\pi_1 \vee \neg \pi_1$ is an uninterpreted string of symbols. We are interested in finding its value, $[\![\pi_1 \vee \neg \pi_1]\!]$, in some valuation, in some particular interpretation. We first expand that expression to $[\![\pi_1]\!] \mathbin{\underline{\vee}} \underline{\neg}[\![\pi_1]\!]$ by replacing each uninterpreted connective symbol, $\vee$ or $\neg$, with an interpretation of that symbol. Different logics give different meanings to those symbols; in the fuzzy interpretation which we are taking as an example, disjunction is the maximum of two numbers, and negation is subtraction from 1. Finally, we evaluate those numbers, and determine whether the resulting value is a member of Σ_+, the set of true values. The formula $\pi_1 \vee \neg \pi_1$ always evaluates to "true" in the two-valued interpretation and in some other interpretations; we write that fact as $\vDash \pi_1 \vee \neg \pi_1$. It evaluates sometimes to true and sometimes to false, in other interpretations, such as the fuzzy interpretation; we write that fact as $\nvDash \pi_1 \vee \neg \pi_1$. Semantic logic can also prove deeper results; for instance, it is shown in 9.12 that if some two formulas A and B satisfy $\vDash A \to B$ in the crystal interpretation, then the formulas A and B must have at least one π_j in common — they cannot be unrelated formulas like $\pi_1 \to \pi_6$ and $\pi_2 \vee \neg \pi_2$.

Syntactics is the abstract or theoretical approach. In syntactic logics, we do not have a set Σ_+ of true values or a set Σ_- of false values; we do not evaluate a formula to semantic values at all. Instead we start from some axioms (assumed formulas) and assumed inference rules, and investigate what other formulas and inference rules can be proved from the given ones. As a typical example, we shall now present a proof of the *Third Contrapositive Law*, $\vdash (\overline{A} \to B) \to (\overline{B} \to A)$, though the notation used here is less concise than that used later in the book. Of course, the proof makes use of results that are established prior to the Third Contrapositive Law.

Semantics and syntactics

#	formula	justification
(1)	$(A\to B)\to (\overline{\overline{B}\to \overline{A}})$	This is a specialization of the Second Contrapositive Law, which is proved earlier.
(2)	$\overline{\overline{A}}\to A$	This is an axiom (an assumed formula).
(3)	$(\overline{\overline{B}\to \overline{A}})\to (\overline{B}\to A)$	This follows by applying the detachmental corollary of $\to$-Prefixing to step (2).
(4)	$(A\to B)\to (\overline{B}\to A)$	This follows by applying Transitivity, proved earlier, to steps (1) and (3).

2.22. *Example (optional): the real numbers.* Though the theory of $\mathbb{R}$ will not be studied in this book, we sketch the definition of $\mathbb{R}$ briefly here, because it makes a good example of the relation between semantics and syntactics.

In lower-level mathematics courses, the real numbers are presented as infinite decimal expansions or as points on a line. Those presentations are concrete and appeal to the intuition, and they will suffice for the needs of this book as well, but they do not really suffice to support proofs of deeper theorems about the real number system. A freshman college calculus book proves some of its theorems but leaves many unproved, stating that their proofs can be found in more advanced courses; those proofs actually require a more rigorous definition. For instance:

> *Intermediate Value Theorem.* If f is continuous on $[a,b]$ and $f(a) < m < f(b)$, then there is at least one number c in (a,b) such that $f(c) = m$.

> *Maximum Value Theorem.* If f is continuous on $[a,b]$, then there is at least one number c in $[a,b]$ with $f(c) = \max\{f(t) : a \leq t \leq b\}$.

These results are evident from pictures (using the "points on a line" explanation of the real numbers), but the pictures do not really constitute a rigorous proof, and the proof is omitted from such books. The rigorous proof *must* be omitted, because it requires notions that cannot be seen in pictures and that are too advanced to be discussed in any detail in freshman calculus books. The proof requires the completeness of the reals (discussed in 2.27), or equivalently the *least upper bound property*: Every subset of $\mathbb{R}$ that has an upper bound in $\mathbb{R}$, also has a *least* upper bound in $\mathbb{R}$. Some calculus books mention that property, but they don't follow through with the proofs.

A more advanced course, usually called "Introduction to Real Analysis," is taken by some mathematics majors in their senior year of college. It investigates the real number system more carefully. It lists the axioms for a *complete ordered field*; these include the least upper bound property and rules such as $x + y = y + x$. The axioms give us a *syntactic* system.

Some introductory textbooks on real analysis simply *assume* that the real numbers satisfy those axioms, but that really is evading part of the question. What is this thing that satisfies those axioms? How do we know that there *is* a thing satisfying those axioms? What are the "real numbers," really? More mature mathematicians take a slightly longer route to the real numbers:

One can prove that the infinite decimal expansions satisfy these axioms and thus form a complete ordered field. However, it is much easier to prove that those axioms are satisfied by *Dedekind cuts of rationals* or by Cantor's *equivalence classes of Cauchy sequences of rationals*; these are two "constructions" of the real numbers from the rational numbers. Decimal expansions, Dedekind cuts, and equivalence classes of Cauchy sequences are three different *semantic* systems that satisfy the axioms for a complete ordered field.

In a still more advanced course on algebra, one can prove that any two complete ordered fields are *isomorphic*.[4] For instance, each infinite decimal expansion corresponds to one Dedekind cut, or to one equivalence class of Cauchy sequences. Thus it doesn't really matter *which* complete ordered field we use. All such fields have the same algebraic properties, and thus they are just different manifestations or representations or relabelings of the same algebraic system.

Finally, one can *define* the real number system to *be* that algebraic system. Once we've done all this, we can put aside the semantic examples (decimal expansions, Dedekind cuts, equivalence classes), and just concentrate on the syntactic axioms. This may seem excessively abstract, but ultimately it focuses on the useful part. For developing abstract theorems such as the Intermediate Value Theorem and the Maximum Value Theorem, we are more concerned with what the real numbers do than with what they look like. In the case of real numbers, the axioms (syntactics) give us a more useful description or explanation than do the examples (semantics).

Actually, the usual axioms for the real numbers do not make a "first-order theory." Most of the axioms, such as $x+y = y+x$, are about numbers; but the least upper bound property is an axiom about numbers and about

---

[4]The meaning of the term "isomorphic" varies from one part of mathematics to another. In general, we say that two mathematical objects are isomorphic if one is just a relabeling of the other — i.e., if there exists a mapping between them that preserves all structures currently of interest. The precise definition varies because different mathematicians may find different structures to be of interest. For instance, the rational numbers and the integers are isomorphic if we're just interested in cardinality, but not if we're interested in multiplication and division.

Semantics and syntactics 43

sets of numbers. It is not possible to describe the real numbers as a first-order theory. One might expect that a simpler system, such as the integers, can be fully explained by a first-order theory, but Gödel showed that even that is not possible. See 2.34.

2.23. *Formulas* are expressions like $\pi_1 \to (\pi_2 \wedge \pi_1)$ or $\pi_1 \vee \neg \pi_1$. The Greek letters $\pi_1, \pi_2, \ldots$ are propositional variable symbols. A formula such as (for instance) $\pi_1 \vee \neg \pi_1$ may be "valid" in a couple of different ways:

- Semantics is concerned with values and meanings of formulas. For instance, the expression $\vDash \pi_1 \vee \neg \pi_1$ means "the formula $\pi_1 \vee \neg \pi_1$ is a *tautology* — i.e., it is an 'always-true' formula; it is true in all the valuations (examples) in the interpretation that we're currently considering."

- In a syntactic logic, $\vdash \pi_1 \vee \neg \pi_1$ means "the formula $\pi_1 \vee \neg \pi_1$ is a *theorem* — i.e., it is a *provable* formula; it is derivable from the assumptions of the inference system that we're currently considering." (Derivations are explained in greater detail starting in 12.5.) If the formula *is* one of those assumptions, we would also say it is an *axiom*. (Thus, each axiom is a theorem; it has a one-line derivation.)

Formula *schemes* are expressions like $A \to (B \vee A)$; the letters are metavariables. Each formula scheme represents infinitely many formulas, because each metavariable can be replaced by any formula. For instance, the formula scheme $A \vee B$ includes as instances the formulas

$$\pi_1 \vee \pi_2, \qquad (\pi_2 \to \pi_3) \vee (\pi_1 \wedge \neg \pi_5), \qquad \pi_1 \vee (\pi_2 \vee \pi_3)$$

and infinitely many others. Those individual formulas are part of the object system, but the formula scheme $A \vee B$ is at a slightly higher level of conceptualization. Our reasoning will lead us to draw conclusions about $A \vee B$, but this might be best understood as a way of summarizing conclusions drawn about the individual formulas. See 6.27 for further discussion.

A *tautology scheme* or *theorem scheme* is a formula scheme whose instances are all tautologies or theorems, respectively.

2.24. At a higher level, an *inference rule* is an expression involving two or more formulas, related by a turnstile symbol ($\vdash$ or $\vDash$). Again, inference rules come in two flavors:

- In a semantic logic, $\{A \to B,\ B \to C\} \vDash A \to C$ means that (regardless of whether the formulas involved are tautologies) in each valuation where the hypotheses $A \to B$ and $B \to C$ are true, the conclusion $A \to C$ is also true. See 7.10 for further discussion of such inference rules.

- In syntactic logic, $\{A \to B,\ B \to C\} \vdash A \to C$ means that (regardless of whether the formulas involved are theorems) from the hypotheses $A \to B$ and $B \to C$, we can derive the conclusion $A \to C$. See 12.5 for further discussion of derivations.

In this book, any inference rule must have at least one hypothesis — i.e., there must be at least one formula in the set on the left of the turnstile. That is part of this book's definition of "inference rule." (Permitting inference rules with no hypotheses — i.e., classifying tautologies and theorems as special cases of "inference rules" — would make sense, but would complicate some parts of our exposition.)

2.25. In addition to tautologies/theorems and inference rules, we will also study a few *higher principles*, such as admissibility rules. Here are three different levels of consequences:

Tautology $\vDash A \to B$	The formula $A \to B$ is always true.
Inference $A \vDash B$	Whenever A is true, then B is true.
Admissibility $\vDash A \Rightarrow \vDash B$	If some substitution of formulas for metavariables makes A an always-true formula, then the same substitution makes B always-true.

Beginners grasp the semantic notions of "always true" and "whenever true" with little difficulty, but are more challenged by the syntactic analogues, which are more abstract:

Theorem $\vdash A \to B$	The formula $A \to B$ can be derived from nothing (i.e., from just the logic's assumptions).

Semantics and syntactics 45

> Inference $A \vdash B$ From formula A we can derive formula B.
>
> Admissibility   If some substitution of formulas for metavariables makes A into a formula that can be derived from nothing, then the same substitution also makes B derivable from nothing.
> $\vdash A \Rightarrow \vdash B$

These three notions can easily be confused with one another by students, and even by experienced logicians who are new to pluralism. The distinction is not strongly supported by our language; it requires phrasing that is sometimes awkward. That may stem from the fact that in classical logic, the three notions turn out to be equivalent, and so they are used interchangeably in much of the literature. But in some other logics they are not equivalent. Moreover, even in classical logic the proof of equivalence (given in 22.5 and 29.15) is far from trivial; we must not treat the notions as equivalent until after we have proceeded through that proof.

The best way to alleviate the beginner's confusion is to look at several examples in different logics. We will do that in later chapters; following is a brief preview.

- In most logics of interest, $\vdash A \rightarrow B$ implies $A \vdash B$. Thus we can go from an implication theorem to an inference rule; we can move A leftward past the turnstile. This is essentially just *detachment* (13.2.a), an assumed inference rule in all the logics considered in this book.

But what about moving A to the *right* past the turnstile? That is the subject of three deduction principles[5] listed below:

- *Constructive version.* $A \vdash B \;\Rightarrow\; \vdash A \rightarrow B$. For this to hold, constructive implication is both necessary and sufficient; we prove that in 22.5 and 22.6.
- *Lukasiewicz version.* $A \vdash B \;\Rightarrow\; \vdash A \rightarrow (A \rightarrow B)$ in the three-valued Lukasiewicz logic; see 24.26.
- *Relevant version.* $A \vdash B \;\Rightarrow\; \vdash \overline{A} \vee B$ in the comparative

---

[5]Called "Theorems" in most of the literature. See 2.18.

and Sugihara interpretations and in relevant logic; see 8.30.c and 23.6.

All three of these principles are valid in classical logic. Indeed, they coalesce into one principle, since the three formulas $A \to B$, $A \to (A \to B)$, and $\overline{A} \lor B$ are equivalent in classical logic.

2.26. *Relating the two faces.* After we study some syntactic systems and semantic systems separately, we will study the interaction between them. In particular, by *completeness* we will mean a pairing of some syntactic system with some semantic system such that $\vdash A \Leftrightarrow \vDash A$. That is, these two sets of formulas are equal:

$$\left\{\begin{array}{c}\text{theorems of the}\\\text{syntactic system}\end{array}\right\} = \left\{\begin{array}{c}\text{tautologies of the}\\\text{semantic system}\end{array}\right\}.$$

In effect, the abstract axioms "explain" the concrete interpretation, and we obtain a complete analysis of the set of truths:

any statement can be proved (P) by an abstract derivation or disproved (D) by a concrete counterexample.

That dichotomy is illustrated by the table below, which considers five syntactic logics (classical, Wajsberg, constructive, RM, or Abelian) and the corresponding five semantic logics (two-valued, Łukasiewicz, topological, Sugihara, comparative). For

	syntactics	class. 25.1	Waj. 24.14	constr. 22.1	RM 23.13	Abel. 26.1
	semantics	2-val. 8.2	Łuk. 8.17	topol. 10.1	Sugi. 8.38	comp. 8.28
	complete?	29.12	29.20	29.29	(23.11)	(26.8)
$A \lor \neg A$		P	D	D	P	P
$(\neg\neg A) \to A$		P	P	D	P	P
$(A \to \neg A) \to \neg A$		P	D	P	P	D
$A \to (B \to A)$		P	P	P	D	D
$(A \land \neg A) \to B$		P	D	P	D	D
$((A \to B) \to B) \to A$		D	D	D	D	P

Semantics and syntactics 47

the first three of these five pairings, completeness is proved in this book. The last two pairings are also complete, but their completeness proofs are too advanced for this book, and are merely mentioned in the sections indicated by the italicized and parenthesized numbers.

A much longer list of completeness pairings is given in the next table below; again, italicized reference numbers indicate discussion in lieu of proof.

syntactics		*semantics*		*complete?*
classical	25.1	two-valued	8.2	29.12
classical	25.1	powerset	9.3	11.11
classical	25.1	six-valued	9.4	11.12
constructive	22.1	all topologies	10.1	29.29
constructive	22.1	finite topologies	10.1	29.29
constructive	22.1	no finite functional		22.16
constructive	22.1	$\mathbb{R}$ topology	10.1	(*29.29*)
Dummett	22.18	upper sets	4.6.h	(*22.18*)
relevant	23.1	Church chain	9.13	sound
relevant	23.1	Ch. diamond	9.14	sound
relevant	23.1	no finite functional		23.11.a
Brady	23.11.b	crystal	9.7	(*23.11.b*)
RM	23.13	Sugihara	8.38	(*23.11.c*)
RM	23.13	no finite functional		23.11.a
implications		worlds	28.1	28.13
Wajsberg	24.14	Łukasiewicz	8.16	29.20
Rose-Rosser	24.1	Zadeh	8.16	(*24.2*)
Abelian	26.1	comparative	8.28	(*26.8*)
not finite		Dziobiak		(*29.31*)
not findable		arithmetic		(*2.34*)

We caution the student that the pairs in this introductory book were selected in part for their simplicity, and so are atypical; "most" logics are more complicated. In particular, relevant logic (23.1) is one of the most important and interesting logics studied in this book, but its characterizing semantics are algebraic structures too complicated for us even to describe in this book. (Those semantics can be found in Anderson and Belnap [1975] and Dunn [1986].) The logic RM (relevant plus mingle) is of interest because it has some of the flavor of relevant logic and yet has a very simple semantics, but the proof of that completeness pairing is too advanced for this book to do more than state it; see 23.11.c.

Advanced books and papers on logic often use the terms "theorem" and "tautology" interchangeably, because they assume that the reader is already familiar with some completeness pairing. We may follow that practice in these introductory/preview chapters; but we will cease that practice when we begin formal logic in Chapter 6, and we will not resume it until after we prove completeness near the end of the book.

Remarks. More precisely, the pairing $\vdash A \Leftrightarrow \vDash A$ is called *weak completeness* in some of the literature. *Strong completeness* says that the syntactic and semantic systems have the same inference rules; that is, $\mathcal{S} \vdash A \Leftrightarrow \mathcal{S} \vDash A$ for any formula A and set of formulas $\mathcal{S}$. We will consider both kinds of completeness; see particularly 21.1.

2.27. The word "completeness" has many different meanings in math. Generally, it means "nothing is missing — all the holes have been filled in"; but different parts of mathematics deal with different kinds of "holes."

For instance, the rational number system has a hole where $\sqrt{2}$ "ought to be," because the rational numbers

$$1, \quad 1.4, \quad 1.41, \quad 1.414, \quad 1.4142, \quad 1.41421, \quad 1.414213, \quad \ldots$$

get closer together but do not converge to a rational number. If we fill in all such holes, we get the real number system, which is the completion of the rationals.

Even within logic, the word "completeness" has a few different meanings. Throughout this book, it will usually have the meaning sketched in 2.26, but some other meanings are indicated in 5.15, 8.14, and 29.14.

2.28. The two halves of completeness have their own names:

Adequacy means {tautologies} ⊆ {theorems}. That is, our axiomatic method of reasoning is *adequate* for proving all the true statements; no truths are missing.

Soundness means {theorems} ⊆ {tautologies}. That is, our method of reasoning is *sound*; it proves *only* true statements.

Historical perspective 49

In most pairings of interest, soundness is much easier to establish. Consequently, soundness is sometimes glossed over, and some mathematicians use the terms "completeness" and "adequacy" interchangeably.

HISTORICAL PERSPECTIVE

2.29. Prior to the 19th century, mathematics was mostly empirical. It was viewed as a collection of precise observations about the physical universe. Most mathematical problems arose from physics; in fact, there was no separation between math and physics. Every question had a unique correct answer, though not all the answers had yet been found. Proof was a helpful method for organizing facts and reducing the likelihood of errors, but each physical fact remained true by itself regardless of any proof. This pre-19th-century viewpoint still persists in many textbooks, because textbooks do not change rapidly, and because a more sophisticated viewpoint may require higher learning.

2.30. Prior to the 19th century, Euclidean geometry was seen as the best known description of physical space. Some non-Euclidean axioms for geometry were also studied, but not taken seriously; they were viewed as works of fiction. Indeed, most early investigations of non-Euclidean axioms were carried out with the intention of proving those axioms *wrong*: Mathematicians hoped to prove that Euclid's parallel postulate was a consequence of Euclid's other axioms, by showing that a denial of the parallel postulate would lead to a contradiction. All such attempts were unsuccessful — the denial of the parallel postulate merely led to bizarre conclusions, not to outright contradictions — though sometimes errors temporarily led mathematicians to believe that they had succeeded in producing a contradiction.

However, in 1868 Eugenio Beltrami published a paper showing that some of these geometries are not just fictions, but skewed views of "reality" — they are *satisfied* by suitably peculiar *interpretations* of Euclidean geometry. For instance, in "double elliptic geometry," any two straight lines in the plane must meet.

This axiom is satisfied if we interpret "plane" to mean the surface of a sphere and interpret "straight line" to mean a great circle (i.e., a circle whose diameter equals the diameter of the sphere).

By such peculiar interpretations, mathematicians were able to prove that certain non-Euclidean geometries were *consistent* — i.e., free of internal contradiction — and thus were legitimate, respectable mathematical systems, not mere hallucinations. These strange interpretations had an important consequence for conventional (nonstrange) Euclidean geometry as well: We conclude that the Euclidean parallel postulate can *not* be proved from the other Euclidean axioms.[6]

After a while, mathematicians and physicists realized that we don't actually know whether the geometry of our physical universe is Euclidean, or is better described by one of the non-Euclidean geometries. This may be best understood by analogy with the two-dimensional case. Our planet's surface appears flat, but it is revealed to be spherical if we use delicate measuring instruments and sophisticated calculations. Analogously, is our three-dimensional space "flat" or slightly "curved"? And is it curved positively (like a sphere) or negatively (like a horse saddle)? Astronomers today, using radio-telescopes to study "dark matter" and the residual traces of the original big bang, may be close to settling those questions.

2.31. *The formalist revolution.* Around the beginning of the 20th century, because of Beltrami's paper and similar works, mathematicians began to change their ideas about what is mathematics and what is truth. They came to see that their symbols can have different interpretations and different meanings, and consequently there can be multiple truths. Though some branches of math (e.g., differential equations) continued their close connections with physical reality, mathematics as a whole has been freed from that restraint, and elevated to the realm

---

[6]These consistency results are actually *relative*, not *absolute*. They show that *if* Euclidean geometry is free of contradictions, *then* certain non-Euclidean geometries are also free of contradictions, and the parallel postulate cannot be proved from the other Euclidean axioms.

Historical perspective

of pure thought.[7] Ironically, most mathematicians — even those who work regularly with multiple truths — generally retain a Platonist attitude: They see their work as a human investigation of some sort of objective "reality" which, though not in our physical universe, nevertheless somehow "exists" independently of that human investigation.

In principle, any set of statements could be used as the axioms for a mathematical system, though in practice some axiom systems might be preferable to others. Here are some of the criteria that we might wish an axiom system to meet, though generally we do not insist on all of these criteria. The system should be:

Meaningful. The system should have some uses, or represent something, or make some sort of "sense." This criterion is admittedly subjective. Some mathematicians feel that if an idea is sufficiently fundamental — i.e., if it reveals basic notions of mathematics itself — then they may pursue it without regard to applications, because some applications will probably become evident later — even as much as a century or two later. This justification-after-the-theory has indeed occurred in a few cases.

Adequate (or "complete"). The axioms should be sufficient in number so that we can prove all the truths there are about whatever mathematical object(s) we're studying. For instance, if we're studying the natural number system, can we list as axioms enough properties of those numbers so that all the other properties become provable?

Sound. On the other hand, we shouldn't have too many axioms. The axioms should not be capable of proving any *false* statements about the mathematical object(s) we're studying.

Independent. In another sense, we shouldn't have too many axioms: They should not be redundant; we shouldn't include any axioms that could be proved using other axioms. This criterion affects our efficiency and our aesthetics, but it does not really affect the "correctness" of the system; repetitions of axioms are tolerable. Throughout most of the axiom systems studied in this book, we will *not* concern ourselves about independence. This

---

[7]Or reduced to a mere game of marks on paper, in the view of less optimistic mathematicians.

book is primarily concerned with comparing different logics, but an axiom that works adequately in studying several logics is not necessarily optimal for the study of any of them.

Consistent. This is still another sense in which we shouldn't have too many axioms: Some mathematicians require that the axioms do not lead to a contradiction. In some kinds of logic, a contradiction can be used to prove *anything*, and so any one contradiction would make the notion of "truth" trivial and worthless. In paraconsistent logics, however, a contradiction does not necessarily destroy the entire system; see discussion in 5.16.

Recursive. This means, roughly, that we have some algorithm that, after finitely many steps, will tell us whether or not a given formula is among our axioms. (Nearly every logic considered in this book is defined by finitely many axiom schemes, and so it is recursive.)

2.32. Over the last few centuries, mathematics has grown, and the confidence in mathematical certainty has also grown. During the 16th–19th centuries, that growth of certainty was part of a wider philosophical movement known as the Enlightenment or the Age of Reason. Superstitions about physical phenomena were replaced by rational and scientific explanations; people gained confidence in the power of human intellect; traditions were questioned; divine right monarchies were replaced by democracies. Isaac Newton (1643–1727) used calculus to explain the motions of the celestial bodies. Gottfried Wilhelm von Leibniz (1646–1716) wrote of his hopes for a universal mathematical language that could be used to settle all disputes, replacing warfare with computation.

The confidence in mathematics was shown at its peak in David Hilbert's famous speech[8] in 1900. Here is an excerpt from the beginning of that speech:

---

[8]Hilbert (1862–1943) was the leading mathematician of his time. In his address to the International Congress of Mathematicians in 1900, he described 23 well-chosen unsolved problems. Attempts to solve these famous "Hilbert problems" shaped a significant part of mathematics during the 20th century. A few of the problems still remain unsolved.

Historical perspective

> This conviction of the solvability of every mathematical problem is a powerful incentive to the worker. We hear within us the perpetual call: There is the problem. Seek its solution. You can find it by pure reason, for in mathematics there is no "unknowable."

Hilbert was the leader of the formalists. He advocated working to put all of mathematics on a firm axiomatic foundation; this plan is now known as *Hilbert's program*. He and other mathematicians were encouraged by some successes in the 1920s, and particularly by Gödel's proof of completeness of first-order logic in 1930.

2.33. In 1931, however, Gödel proved his Incompleteness Principles (discussed below), thereby demolishing Hilbert's program: Gödel proved that, in fact, mathematics does have some "unknowable" propositions. At nearly the same time, Heisenberg formulated his Uncertainty Principle of particle physics. Evidently, there are inherent limitations on what we can know, in any field of investigation. However, these discoveries mark only a boundary, not a refutation, of the Enlightenment. Despite its limitations, reason remains a powerful tool.

2.34. *Gödel's First Incompleteness Principle* is a result of applied (first-order) logic, not propositional logic, so its detailed explanation is beyond the scope of this book; but we will describe it briefly:

> It is not possible to specify a set of axioms that specify precisely the truths of $\mathbb{N}$.

By "the truths of $\mathbb{N}$" we mean facts about the positive integers, including advanced facts about prime numbers.

A subtlety must be pointed out here: "not possible to specify" is not the same thing as "there does not exist." For instance, for our set of axioms, we could simply use the set of *all* true statements about $\mathbb{N}$. That set exists, but we can't actually *find* it. A set of axioms that we can "specify" means a recursive set — a set that is finite, or an infinite set of axioms, for which membership or nonmembership can be determined by some explicit algorithm.

The system ℕ is particularly important because most parts of mathematics can be expressed in countable languages — i.e., only a sequence of symbols is needed. Consequently, most statements of mathematics can be encoded as statements about ℕ. (Indeed, statements could be encoded in terms of sequences of 0's and 1's; that is the language used inside modern electronic computers.) If we could axiomatize ℕ, then we could axiomatize most of mathematics.

After Gödel's work, other examples of incompleteness were found — i.e., mathematical theories which, if viewed as semantic systems, could not be suitably axiomatized. In the late 20th century, Gregory Chaitin showed that *most* of mathematics cannot be suitably axiomatized — that mathematics is riddled with holes. As Chaitin himself put it,

> most of mathematics is true for no particular reason.

Thus, a reasonable goal for mathematicians is not to explain everything (Hilbert's program, discussed in 2.32), but just to find some parts of mathematics that *can* be explained and that have some usefulness or other interest.

2.35. *Gödel's Second Incompleteness Principle* was concerned with *consistency*, mentioned in 2.31 as a minimal requirement for an axiom system. Hilbert hoped to prove that each of the main axiom systems of mathematics was consistent. This was in fact accomplished for some elementary axiom systems.

Any proofs about a logical "inner" system depend on the use of other mathematics in the "outer" system (see terminology in 2.17), but Hilbert hoped that such proofs could be arranged so that the outer mathematics used in the proof was more elementary than the inner system whose consistency was being proved. Thus we could "bootstrap" our way up, from systems so simple that they were obviously consistent, to more complicated systems. Gödel himself is credited with one of the most successful of these bootstrapping operations: In 1931 he proved the consistency of first-order logic, assuming only much more elementary and more obvious systems.

Historical perspective 55

But Gödel's Second Incompleteness Principle showed limitations in this bootstrapping process: It cannot be applied to some axiomatic systems a bit higher than first-order logic. For some of those systems, consistency can only be proved by assuming the consistency of other systems that are at least as complicated and as subject to doubt. Thus an absolute proof of consistency is not possible.

One of the axiomatic systems to which Gödel's incompleteness result is applicable is set theory, the "foundation" of the objects of modern mathematics. Set theory is *empirically* consistent: It is now over a century old; if it had any contradictions, probably we would have found one by now. But we're still not certain of that, and Gödel has made it clear that we never will be. Even a mathematician must accept some things on faith or learn to live with uncertainty.

2.36. Though Gödel demonstrated the impossibility of Hilbert's *program*, Hilbert's *style* of formalism and formalizability continued to spread throughout mathematics, and it remains dominant to this day. One of the last branches of mathematics to embrace the formalist revolution was logic itself. Just as "geometry" meant "Euclidean geometry" until near the end of the 19th century, so too logic was dominated by a classical-only viewpoint until late in the 20th century.

Early in the 20th century, a few pioneering mathematicians began to investigate nonclassical logics. The resulting proofs have the same ironclad certainty as any other mathematical results, but the investigators themselves were motivated by philosophical beliefs. Many of them were not pluralists (espousing many logics), but were enamored of one particular nonclassical logic, and advocated it as "the one true logic," the one correct approach to thinking. Fighting against the majority view made them controversial figures.

One particularly interesting example was constructivism, introduced briefly in 2.42 and in greater detail in later chapters. Errett Bishop's book, *Foundations of Constructive Analysis* [1967], was not actually a book on logic — in fact, it barely mentioned formal logic — but it was nevertheless a radical breakthrough affecting logic. The book presented a substantial portion of analysis (a branch of mathematics) at the level of an advanced undergraduate or beginning graduate course, covering fairly standard material but in a constructivist style. This had never been done before, nor had previous mathematicians even believed it could be done.

Most of Bishop's book was just mathematics — objective and indisputable. However, Bishop explained his unconventional viewpoint in a preface titled "A Constructivist Manifesto." This told not only *how* his methods of reasoning worked (an objective, mathematical matter) but also *why* he felt these methods were preferable (a more subjective matter). Here is one of the more colorful passages:

> Mathematics belongs to man, not to God. We are not interested in properties of the positive integers that have no descriptive meaning for finite man. When a man proves a positive integer to exist, he should show how to find it. If God has mathematics of his own that needs to be done, let him do it himself.

(See the footnote in 29.2 for an example of God's mathematics.)

During the last decades of the 20th century, logic finally accepted the formalist revolution. Mathematical logic became a respectable subject in its own right, no longer dependent on philosophy or even particularly concerned about it. Researchers saw different logics merely as different types of abstract algebraic structures. (For instance, the rules in 7.5.c(ii) show that a functional valuation $[\![\ ]\!]$ is nothing other than a homomorphism from the algebra of formulas to the algebra of semantic values.) With this philosophy-free view, it was inevitable that research logicians would become pluralists; the many-logic viewpoint is now taken for granted in the research literature of logic. But that replacement of philosophy with algebra also made the research literature less accessible to beginners. Textbooks lagged behind, continuing to emphasize classical logic.

So far, only a few textbooks have been written with the pluralist approach; this book is one of them. Though pluralistic in content, this book is old-fashioned in style: It is intended for beginners, so we use philosophy for motivation and we attempt to keep the algebra to a minimum.

PLURALISM

2.37. How can there be different kinds of logics? Isn't logic just common sense? No, it's not that simple. Different situations may call for different kinds of reasoning. As we already indicated in 2.26, different logics have different truths (i.e., theorems or tautologies). They also have different inference rules and higher-level principles, as indicated in the table below. We may understand these principles better if we study the contrasts between the different logics; that is one of the chief strategies of this book. Some of our main logics are briefly introduced in the next few pages.

property $\rightarrow$ logic $\downarrow$	Excluded middle or a variant of it	Deduction principle: If $A \vdash B \ldots$	If $\vdash A \rightarrow B$ with A, B unrelated	Explosion of some sort
relevant	$\vdash A \vee \overline{A}$	$\vdash \overline{A} \vee B$	can't be	none
integer-valued	$\vdash A \vee \overline{A}$	$\vdash \overline{A} \vee B$	$\vdash \overline{A}$ and $\vdash B$	$\vdash (A \wedge \overline{A}) \rightarrow (B \vee \overline{B})$
Wajsberg	$\vdash A \vee \overline{A} \vee \widetilde{A}$	$\vdash A \rightarrow (A \rightarrow B)$	no consequences	$A \wedge \overline{A} \vdash B$
constructive	$(\vdash A \vee B) \Rightarrow$ $(\vdash A$ or $\vdash B)$	$\vdash A \rightarrow B$	$\vdash \overline{A}$ or $\vdash B$	$\vdash (A \wedge \overline{A}) \rightarrow B$
classical	$\vdash A \vee \overline{A}$	all of the above	$\vdash \overline{A}$ or $\vdash B$	$\vdash (A \wedge \overline{A}) \rightarrow B$

2.38. So-called *classical logic* is the logic developed by Frege, Russell, Gödel, and others. Among commonly used logics, it is computationally the simplest, and it is adequate for the needs of most mathematicians. In fact, it is the only logic with which most mathematicians (other than logicians) are familiar. The main ideas of propositional classical logic can be summarized by the simple true/false table given here; a similar table can

inputs		outputs			
P	Q	¬P	$P \wedge Q$	$P \vee Q$	$P \to Q$
0	0	1	0	0	1
0	1	1	0	1	1
1	0	0	0	1	0
1	1	0	1	1	1

be found as early as the works of Philo of Megara, one of the Greek Stoics (approximately 200 B.C.). For compatibility with other parts of this book, we shall abbreviate $0 =$ "false" and $1 =$ "true."

We emphasize that $P \to Q$ is considered to be true (1) in all cases except the case where P is true (1) and Q is false (0). Some noteworthy corollaries are

(a) a false statement implies anything;

(b) anything implies a true statement; and

(c) $P \to Q$ is true precisely when $P \leq Q$.

Exercise. Extend the table, giving output columns for the three formulas $Q \to P$, $(P \to Q) \vee (Q \to P)$, and $(P \to Q) \to (Q \to P)$.

Remarks. The "noteworthy corollary" (c) will generalize to most of the nonclassical logics in this book, but (a) and (b) may fail in logics that have several kinds of "true" or several kinds of "false." An example is examined in detail in 2.40.

2.39. Aristotle objected to the Megarian approach. He pointed out that if every statement — even a statement about a future event — is already either "true" or "false" (perhaps without our knowing which), then the future is predetermined. For instance, one of these statements is true, and the other is false:

- There will be a sea-battle tomorrow.
- There will not be a sea-battle tomorrow.

Pluralism 59

Łukasiewicz (1878–1956; pronounced woo-kah-sheay-vitch) extended this objection; if the future is predetermined then we have no free will. So in the 1920s he began studying *multivalued logics*. His earliest work involved only three semantic values — true (1), false (0), and indeterminate (1/2) — but later he considered *quantitative logics* with a whole range of values, e.g., all the numbers in the interval [0, 1]. A statement such as

most old men are *mostly* bald

is *mostly* true. We can make this precise by assigning numbers between 0 and 1 to "old," "bald," and the three *most*'s.

Closely related to Łukasiewicz's logics are the fuzzy logics studied by L. A. Zadeh in the 1960s. Zadeh observed that mechanical sensory devices such as thermostats cannot supply information with perfect precision, and even high precision is prohibitively expensive. In the real world, data are unavoidably imprecise. Thus, the circuits using that information must be designed to make the best precise use of imprecise data. Fuzzy logics are now used in the design of control circuitry in dishwashers, clothes dryers, automobile cruise controls, and other devices.

The term "fuzzy" may be unfortunate, for it has picked up a rather negative connotation in our society in recent years; a person who uses "fuzzy thinking" is a person who does not think clearly. But the two notions are nearly opposites: Fuzzy thinking is unnecessarily imprecise reasoning, while fuzzy logic is precise reasoning about imprecise information.

Fuzzy logic is investigated further in 8.16–8.26, Chapter 24, and 29.16–29.20.

2.40. Aristotle also mentioned a type of *comparative logic* that is incompatible with classical logic. He wrote[9]

(a) if there are two things both more desirable than something, the one which is more desirable to a greater

---

[9]This can be found in *Topics* III, 3, 118b2–3 of Aristotle [1984]. The analysis given here is modified from Casari [1989]; our development of this logic later in this book is based largely on Meyer and Slaney [1989, 2002].

degree is more desirable than the one more desirable to a less degree.

That's admittedly a bit hard to parse, but an example may be easier. Suppose that I'm serving coffee, tea, and punch. Then

(a′) if "the coffee is hotter than the punch" is more true than "the tea is hotter than the punch," then the coffee is hotter than the tea.

How can we restate this mathematically?

Assume that I've picked some particular temperature, and I call anything above that temperature "hot." This "hot" cutoff is a sort of absolute mark on our temperature scale, but it is a sliding absolute: We only use it temporarily, and our reasoning ultimately will not depend on where we put this cutoff point. Our reasoning will depend only on the differences in temperatures — i.e., how the coffee, tea, and punch stand *relative* to each other. (This explanation may justify the use of the word "relativity" in 20.2.)

Now, let us abbreviate the three propositions:

$$\begin{aligned} C &= \text{"the coffee is hot,"} \\ T &= \text{"the tea is hot,"} \\ P &= \text{"the punch is hot."} \end{aligned}$$

As in 2.38(c), we understand that a less true statement implies a more true statement. For instance, saying that

(b) the coffee is hotter than the punch

is the same as saying that, no matter what we have chosen for the cutoff temperature in our definition of "hot,"

(b′) if the punch has a high enough temperature to be called "hot," then the coffee also has a high enough temperature to be called "hot,"

or, in other words,

(b″) $P \to C$.

Pluralism

Similarly, "the tea is hotter than the punch" can be restated as $P \to T$. Putting these statements together, since a less true statement implies a more true statement, "$P \to C$ is more true than $P \to T$" can be restated as $(P \to T) \to (P \to C)$. Finally, condition (a′) can be restated as

(a″) $\quad [(P \to T) \to (P \to C)] \to (T \to C)$.

This "prefix cancellation" formula is always true in *comparative logic*, a mathematical logic developed further in 8.28–8.37 and Chapters 20 and 26.

But the prefix cancellation formula is not always true in classical logic — i.e., classical logic cannot be used to compare beverages in the fashion indicated above. For instance, suppose that the tea is hot, and the punch and coffee are not. Then $T = 1$ and $P = C = 0$. In two-valued classical logic, this yields

$$\underbrace{\left[\underbrace{(P \to T)}_{1} \to \underbrace{(P \to C)}_{1}\right]}_{\underbrace{1\ (*)}_{0}} \to \underbrace{(T \to C)}_{0}$$

so prefix cancellation is false. Evidently, classical logic is not a good method for analyzing statements such as (a′).

How does classical logic go astray? The error first appears at the step marked by the asterisk (∗) in the evaluation above. The statements $P \to T$ and $P \to C$ are both true, so classical logic calls $(P \to T) \to (P \to C)$ true. But $P \to T$ is *more true* than $P \to C$, so comparative logic calls $(P \to T) \to (P \to C)$ false.

2.41. In classical logic, $P \to Q$ is true whenever at least one of $\overline{P}$ or Q is true — regardless of how P and Q are related. Thus,

> if the earth is square then today is Tuesday

is true to a classical logician (since the earth is *not* square); but it is nonsense to anyone else. *Relevant logic* is designed to avoid implications between unrelated clauses.

Most mathematicians believe that they are using classical logic in their everyday work, but that is only because they are unfamiliar with relevant logic. Here is an example:

A new Pythagorean theorem. If $\lim_{\theta \to 0} \theta \csc \theta = 1$, then the sides of a right triangle satisfy $a^2 + b^2 = c^2$.

A theorem of this type would not be accepted for publication in any research journal. The editor or referee might respond that "one of the hypotheses of the theorem has not been used," or might use much stronger language in rejecting the paper. A superfluous hypothesis does not make the theorem false (classically), but it does make the theorem unacceptably weak; to even *think* of publishing such a theorem would be in very poor taste. Most mathematicians, not being familiar with relevant logic, do not realize that it describes much of their "good taste."

Relevant logic is investigated further in 5.29, 8.31, 8.43, 8.44, 9.12, and Chapters 23 and 28.

2.42. In the classical viewpoint, mathematics is a collection of statements; but to constructivists,[10] mathematics is a collection of procedures or constructions. The rules for combining procedures are slightly different from the rules for combining statements, so they require a different logic.

For instance, in classical logic, $A \lor \neg A$ is always true (as evidenced by the truth table in 2.38); either a thing is or it is not. That is known as the *Law of the Excluded Middle*. Thus, most mathematicians would agree that

> Goldbach's conjecture is true or the negation of Goldbach's conjecture is true.

But constructivists feel that it is meaningless to talk about the truthfulness of Goldbach's conjecture separately from the proof of Goldbach's conjecture. They point out that it is *not* presently true that

---

[10]The student is cautioned not to confuse these similar-sounding words: converse (5.25), contrapositive (5.27), contradiction (5.35), contraction (15.4.c), and constructive (2.42 and 22.1).

we know how to prove Goldbach's conjecture or we know how to disprove Goldbach's conjecture.

Thus the formula $A \vee \neg A$ does not always represent a construction, and so it is not a theorem of *constructive* logic.

Constructivism (or the lack of it) is illustrated further by Jarden's example (2.44–2.46), by the notion of inhabited sets (3.10), and by the Axiom of Choice (3.33). Constructive logic is investigated further in Chapters 10, 22, and 27–29.

2.43. *Goldbach's Conjecture*, mentioned in the preceding example, is a question of number theory, not logic; but as an unsolved problem it is useful for illustrating certain ideas of logic. It involves the *prime numbers* (2, 3, 5, 7, 11, 13, ...) — that is, the integers greater than 1 that are evenly divisible by no positive integers other than themselves and one. In 1742 Goldbach observed that

$$4 = 2+2, \quad 6 = 3+3, \quad 8 = 3+5, \quad 10 = 3+7, \quad 12 = 5+7, \quad \ldots \quad .$$

He conjectured that this sequence continues — i.e., that every even number greater than 2 can be written as the sum of two primes in at least one way. In a quarter of a millennium, Goldbach's Conjecture has not yet been proved or disproved, despite the fact that it is fairly simple to state.

JARDEN'S EXAMPLE (OPTIONAL)

2.44. *Prerequisites for Jarden's proof*

a. The number $\sqrt{2}$ is irrational. *Proof.* Assume (for contradiction) that $\sqrt{2}$ is rational; say $\sqrt{2} = p/q$ where p and q are positive integers. Choose p as small as possible — i.e., let p be the smallest positive integer whose square is equal to twice a square. Then $p^2 = 2q^2$, hence $p > q$. Also then, p^2 is even, hence p is even, hence $p = 2r$ for some positive integer r. But then $q^2 = 2r^2$. Thus q's square is twice a square — i.e., q is smaller than p but satisfies the condition for which p was supposed to be smallest.

b. The number $\log_2 9$ is irrational. *Proof.* Assume that $\log_2 9 = p/q$ where p and q are positive integers. Then algebra yields $9^q = 2^p$. But the left side of that equation is a product of 3's and is not divisible by 2; the right side is a product of 2's and is not divisible by 3.

2.45. *Jarden's theorem.* There exist positive, irrational numbers p and q such that q^p is rational.

Jarden's nonconstructive proof. Consider two cases:

(a) If $\sqrt{2}^{\sqrt{2}}$ is rational, use $p = q = \sqrt{2}$.
(b) If $\sqrt{2}^{\sqrt{2}}$ is irrational, use $p = \sqrt{2}$ and $q = \sqrt{2}^{\sqrt{2}}$. Then a bit of algebra shows that q^p is a rational number. (*Exercise.* Find that number.)

In either case we have demonstrated the existence of p and q.

2.46. *Discussion of Jarden's proof.* The proof in the preceding paragraph is peculiar: After you read the proof, you may feel convinced that the desired numbers p and q exist, but you still don't know what they are. (More precisely, you don't know what q is.) We have not *constructed* the pair (p, q).

This peculiarity stems from the fact that we know $\sqrt{2}^{\sqrt{2}}$ is rational or irrational, but we don't know which. Take A to be the statement that $\sqrt{2}^{\sqrt{2}}$ is rational; then Jarden's proof relies on the fact that (classically) we know $A \vee \overline{A}$, the Law of the Excluded Middle, even though we do not know A and we do not know $\overline{A}$.

It is only Jarden's proof, not his theorem, that is nonconstructive. Some assertions (such as 3.33) are inherently nonconstructive, but for other theorems such as Jarden's a nonconstructive proof can be replaced by a (often longer) constructive proof. For instance, it can actually be shown that $\sqrt{2}^{\sqrt{2}}$ is irrational, using the Gel'fond-Schneider Theorem (a very advanced theorem whose proof will not be given here); hence we can use case (b) of Jarden's proof. Or, for a more elementary proof, use $p = \log_2 9$ and $q = \sqrt{2}$ it is easy to show that those numbers are irrational (see 2.44) and a bit of freshman algebra yields the value of q^p, which turns out to be a rational number. (*Exercise.* Find that number.)

Chapter 3

Informal set theory

Instructors are encouraged to skip whatever parts of this chapter are review for their classes; students with weaker backgrounds may read those parts on their own.

3.1. *Notations.* This chapter will introduce both sets and sequences; they should not be confused with one another. We will use

braces	$\{a,b\}$	for sets,
parentheses	(a,b)	for sequences,
angle brackets	$\langle a,b \rangle$	for multisets,
brackets / parentheses	$[a,b)$	for intervals in $\mathbb{R}$,
all of the above	$[(\neg A) \to B]$	for grouping,
doubled brackets	$[\![(\neg A) \to B]\!]$	for valuations.

(We caution the student that not all mathematical books and papers follow the delimiter conventions listed above.)

  a. A *sequence* is a list of objects; it keeps track of both repetitions and order. For instance,

  (chair, stool, stool, couch, stool, chair, couch, chair)

  might be the historical sequence of locations I occupied last night when I had insomnia. This list has length eight; there were eight events in my nighttime journey. Changing the order or repetitions in this presentation of events would change the history being conveyed.

An *ordered pair* is a sequence of length 2, such as $(5, 17)$. An *ordered triple* is a sequence of length 3, such as (shoe, shirt, tie). A sequence may also be infinite in length; for instance,
$$\bigl(1,\ 0,\ 4,\ 0,\ 9,\ 0,\ 16,\ 0,\ 25,\ \ldots \bigr)$$
is the sequence obtained by alternating zeros with the squares of integers.[1]

Unfortunately, parentheses are also used to indicate intervals, so parentheses are sometimes ambiguous. See 3.7.f.

b. A *set* is an unordered collection of distinct objects. For instance, refer to the previous example; on which pieces of furniture did I rest last night? We could answer without thinking — we could simply change the parentheses to braces. The answer is

{chair, stool, stool, couch, stool, chair, couch, chair}.

But this is a set of only three objects; it is *equal to* the set

{chair, stool, couch}.

Repetitions do not affect the value of a set. But this example is admittedly contrived; it does not indicate why anyone would *want* to repeat any members in a representation of a set. That will be better explained by an example in 3.24.a.

Sets also disregard order; the set above is also equal to {chair, couch, stool}.

c. (Postponable.) A *multiset* is a collection that keeps track of repetitions but not ordering; thus it is a compromise between sets and sequences.

For instance, suppose I want to buy a pack of chewing gum, or something else that is small and just requires a few coins. I reach into my pocket to see what coins I have, and I take out two quarters, a dime, and a penny. Those coins could be listed as

⟨quarter, quarter, dime, penny⟩.

---

[1] This example is given by an explicit rule or pattern. Some students may get the idea that every sequence must be described by a pattern or rule. But that is not the case, if we adhere to the conventions of mainstream mathematics. In fact, *most* sequences cannot be described explicitly — but each of our *examples* of sequences must be describable. See 3.69–3.71.

Informal set theory

Admittedly, if I thought about it for a while I could find a way to describe how the coins are arranged in my hand; thus they are in *some* order. But that order is not relevant to the application I have in mind for the coins, and mentioning their order would just be a distraction. Thus, multisets disregard order:

$$\langle \text{quarter, quarter, dime, penny} \rangle = \langle \text{quarter, dime, quarter, penny} \rangle.$$

If we change the angle brackets to parentheses, we get two different sequences:

$$(\text{quarter, quarter, dime, penny}) \neq (\text{quarter, dime, quarter, penny}).$$

We shall say that those are two different *sequential representations* of the same multiset. That terminology will be briefly useful in 13.23 and 28.12, when we consider repetitions of mathematical formulas.

Of course, two mathematical formulas may be identical, but no two quarters are actually identical. They may have different years or different states printed on them, or perhaps one is shinier than the other. Or even if the coins appear identical to the unaided eye, microscopes and other tools would eventually reveal some difference: One coin has a tiny scratch where the other does not; one coin has a very slightly higher concentration of (for instance) aluminum; etc. Thus the two quarters are not really identical, and I could list my four distinct but unordered coins as a set with four elements:

$$\{\text{dime, penny, quarter}_1, \text{quarter}_2\}.$$

However, the differences between the two quarters are of no importance to the applications I have in mind for the coins. Thus, I probably will disregard the differences, omit the subscripts, and think in terms of a multiset.

d. All delimiters — parentheses, brackets, braces, and other symbols — are used for grouping, to indicate order of operations. A familiar example from arithmetic is in the distributivity of multiplication over addition:

$$6 \times (3 + 5) = (6 \times 3) + (6 \times 5).$$

Similarly, we use grouping symbols in complicated logical expressions, such as the "self-distribution" formula for implication:

$$\vdash [A \to (B \to C)] \to [(A \to B) \to (A \to C)].$$

We will discuss the use of grouping symbols in more detail starting in 6.7 and 6.15. Fortunately, there is little chance of confusing grouping with sets, multisets, sequences, or intervals, because the delimiters are used in different ways in those contexts.

e. Doubled brackets will be used for valuations, starting in 7.4.

SETS AND THEIR MEMBERS

3.2. Definition. A *set* is an unordered collection of objects; those objects are called its *elements* or *members* (or sometimes *points*, regardless of whether any geometry is involved). Generally we indicate a set with braces { }.

To indicate that x is an element of the set S, we may write $x \in S$ or $S \ni x$ (though the latter notation is less common). We write $x \notin S$ to indicate that x is *not* an element of S.

Two sets are equal if they have the same members. In other words, $S = T$ means that

$$x \in S \quad \text{if and only if} \quad x \in T.$$

Some consequences of this definition of equality are that

$$\{2,3,5,8\} \;=\; \{5,2,3,8\} \;=\; \{5,2,3,8,8\};$$

those are three different representations of the same set, since they all have the same four members: $2, 3, 5, 8$. Thus, we are permitted to change the order of the members, or repeat members, in our representation of a set. Why anyone would *want* to represent a set with numbers out of order, or with repetitions, will be explained by examples in 3.24.

In this book we shall denote by $|S|$ the number of members of a set S; it is called the *cardinality* of S. The meaning of this number should be clear at least when S is a finite set. (Cardinality of infinite sets turns out to be more complicated; we will discuss it starting in 3.62.)

We may say that set S is *strictly larger* than set T if $|S| > |T|$. However, the terms "larger" and "strictly larger" have another, different meaning for sets; see 3.13.

Sets and their members

3.3. *Examples*

a. $S = \{5, 3, 8, 2\}$ is the set whose four members are the numbers 5, 3, 8, and 2. Here $|S| = 4$. We may write $5 \in S$ and $3 \in S$ and $6 \notin S$.

b. $T = \{\{1, 4\}, [6, 7], 3\}$ is a set with just three members; thus $|T| = 3$. Its three members are

$$\text{the set } \{1,4\}, \quad \text{the interval } [6,7], \quad \text{and the number } 3,$$

which are, respectively,

- a set with two members,
- a set with infinitely many members, and
- not a set.[2]

We emphasize that

$$4 \in \{1, 4\} \in T, \quad 6 \in [6, 7] \in T, \quad \text{but} \quad 4, 6 \notin T.$$

In other words, 4 and 6 are *members of members* of T, but 4 and 6 are not *members* of T.

3.4. *Technical note (optional).* The set theory used throughout this book is "naive (or informal), classical set theory with atoms."

- The term "classical" refers to the fact that this is the set theory of mainstream mathematics — in contrast with (for instance) constructivist set theory. See 3.10 and 3.33.
- The terms "naive" or "informal" refer to the fact that we base our set theory on commonly held intuition and stated "facts," rather than axioms and proofs about sets. See 2.9.
- An *atom* is an object that is not a set. Set theorists usually prefer a nonatomic set theory, where *everything* is made of sets, as in the footnote in 3.3.b. But that approach is less intuitive for beginners, and will not be followed in this book.

---

[2]In this book, 3 is not a set. However, in some books on set theory, the positive integers are built as sets: 0 is an abbreviation for the empty set, and $n + 1$ is an abbreviation for the set $\{0, 1, 2, \ldots, n\}$. Thus 3 is the set $\{\varnothing, \{\varnothing\}, \{\varnothing, \{\varnothing\}\}\}$ in such books.

3.5. The distinction between a set and its members confuses some beginning students, perhaps because in their previous non-mathematical experience they have mainly had uses for the individual members of sets, and not for the sets themselves. Beginners may be aided by the following metaphors (though experienced mathematicians may feel these are silly).

When the semester begins, my class contains about 30 students, all strangers to one another. By the end of the semester, some of them have formed friendships; they make up a society. A society is more than just the individuals that it contains. Think of { and } as representing friendships. There, now, aren't you getting a warm feeling about mathematics?

Here is a second, more complicated example: At the grocery store, you purchase a transparent plastic bag containing three slightly different apples. But the cashier rings up only one price for the bag and its contents, not three prices for the three apples. The bag together with its contents counts as one item in the cashier's mind. Think of { } as a plastic bag. The numbers $1, 2, 3$ are three separate objects, but the set $\{1, 2, 3\}$ is one object — it is one *set* that contains three numbers.

Admittedly, we may have to stretch the metaphor a bit, to deal with more complicated sets. For instance, the set $\{1, \{2, 3\}\}$ is like a large bag, containing both a peach and a smaller bag, which in turn contains two apples.

But then how do we explain that the expressions "$\{1, 2\}$" and "$\{1, 1, 2\}$" represent the same set? Well, perhaps they're identical in the cashier's mind, if the grocery store is having a "two for the price of one" sale. But I'll admit that this is stretching the metaphor beyond its safe limits; we can't really explain all of set theory with plastic bags. Use the metaphor to get you started, but after that just rely on the math, which does *not* correspond perfectly to plastic bags or other real-world objects.

3.6. *More examples*

  a. I am a member of the Mathematical Association of America (MAA). I am also a member of the American Mathematical Society (AMS). Both of those organizations are, in turn,

Sets and their members 71

members of the American Association of Professional Organizations (AAPO), which provides services such as printing of calendars and buttons, reserving of hotels for conferences, etc. Thus, in two different ways, I am a member of a member of the AAPO. However, I am not a member of the AAPO. One of their requirements for membership is that a member must contain at least 1000 people. Personally, I do not contain 1000 people, so I am not eligible to join the AAPO.

b. $U = \{2, \{6\}, \{5, \{7\}\}\}$ is a set with three members; thus $|U| = 3$. Indeed, the three members of U are

- the number 2,
- the set $\{6\}$, and
- the set $\{5, \{7\}\}$ — that is, the set whose two members are
  - the number 5 and
  - the set $\{7\}$, whose only member is 7.

We have $2 \in U$ and $\{6\} \in U$ but $6 \notin U$. This kind of nesting of sets-within-sets is used all the time in advanced set theory. We'll need just a little of it later in this book.

c. $V = \{\{1\}, \{1, 2\}, \{1, 2, 3\}, \{1, 2, 3, 4\}, \ldots\}$ is an infinite set whose members are finite sets.

d. A set with exactly one member is a *singleton*; a set with exactly two members is a *doubleton*. For instance:

- $\{3\}$ is a singleton; its only member is the number 3.
- $\{3, 5\}$ is a doubleton.
- At first glance, $\{3, 3\}$ and $\{3, 12/4\}$ look like doubletons, but they are actually two more representations of the singleton $\{3\}$.
- $\{\{3, 5\}\}$ is a singleton, whose only member is the doubleton $\{3, 5\}$.

Note that x and $\{x\}$ are different mathematical objects. They should not be written interchangeably, but unfortunately in some contexts mathematicians occasionally use "x"

as an abbreviation for "$\{x\}$." (See 2.24, for instance.) Experienced mathematicians usually can figure out what is going on from the context, but such abbreviations may confuse beginners.

Exercise. When is $\{x, y\}$ a singleton? Discuss.

e. $\varnothing$ and $\{\ \}$ are two different, interchangeable notations for the *empty set*, occasionally called the *null set*; it is the set with no members. Note that any two empty sets are equal to each other, since they have the same members. See 2.9, 3.9, and 3.10 for some subtleties about the empty set.

It must be emphasized that $\varnothing \neq \{\varnothing\}$. The set $\varnothing$ has no members; the set $\{\varnothing\}$ has one member. (Using the metaphor of 3.5, $\varnothing$ is an empty bag, while $\{\varnothing\}$ is a bag containing another bag.)

The empty set is sometimes an exception to rules. Consequently, whenever you see a statement that begins with "for every set S," you should check whether the statement really ought to be saying "for every nonempty set S."

3.7. *Notations.* Sets may be represented in several ways:

a. As an actual list of the elements — for instance, $\{1, 2, 3\}$.
b. As an implied list of the elements — for instance, the finite set $\{1, 2, 3, \ldots, 99, 100\}$ or the infinite set $\{1, 2, 3, \ldots\}$.
c. As a description of the elements — for instance, $\{$even integers between 11 and 49$\}$.
d. By some special name. Following are some standard abbreviations, which we shall use freely hereafter:

$$\begin{aligned}
\mathbb{N} &= \{\text{natural numbers}\} = \{1, 2, 3, \ldots\}, \\
\mathbb{Z} &= \{\text{integers}\} = \{\ldots, -3, -2, -1, 0, 1, 2, 3, \ldots\}, \\
\mathbb{Q} &= \{\text{rational numbers}\} = \{\text{quotients of two integers}\}, \\
\mathbb{R} &= \{\text{real numbers}\} = \{\text{points on the number line}\}.
\end{aligned}$$

We define $\mathbb{N} = \{1, 2, 3, \ldots\}$ in this book, as do many other books, but remark that some books define $\mathbb{N} = \{0, 1, 2, 3, \ldots\}$ instead. Note $\mathbb{N} \subseteq \mathbb{Z} \subseteq \mathbb{Q} \subseteq \mathbb{R}$, as illustrated by the Venn diagram below.

Sets and their members

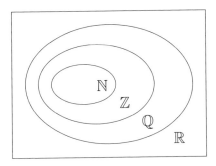

Technical notes: Discussions of subsets ($\subseteq$) and Venn diagrams begin in 3.13 and 3.47, respectively. A more advanced definition of real numbers is given in 2.22. The reader should not confuse $\mathbb{R}$ (real numbers) with R (relevant logic — see 23.1).

e. *Set-builder notation.* When S is a given set, then

$$T = \{x \in S : x \text{ has property } P\}$$

denotes the collection of those members of S that satisfy a given property P. (Some mathematicians use a vertical bar instead of a colon.) Here are some examples:

- $\{x \in \mathbb{Z} : 2 < x < 8\}$ can also be written in list format as $\{3, 4, 5, 6, 7\}$.
- $\{x \in \mathbb{N} : x \text{ is a prime number and } x < 15\}$ can also be written in list format as $\{2, 3, 5, 7, 11, 13\}$.
- $\{x \in \mathbb{R} : 3 < x \leq 8\}$ is the set of all the real numbers that are greater than 3 and less than or equal to 8. That's $(3, 8]$, in the interval notation introduced below.

If the choice of S is understood, then it can be omitted from our notation. For instance, the second example above could be rewritten as $\{x : x \text{ is a prime number and } x < 15\}$. But that sort of abbreviation should only be used with caution. It is at the heart of *Russell's Paradox*; see 3.11.

f. *Intervals.* The phrase "the interval from a to b" is ambiguous; it may refer to any one of the first four sets graphed below. For comparison we have added a fifth picture which is not an interval, because some beginners confuse the doubleton $\{a, b\}$ with one of the intervals from a to b. In all of these graphs, it is customary to use a solid dot to indicate that an endpoint of the interval is included in the set being graphed, and a hollow dot to indicate that the endpoint is not in the

$$(a, b) = \{x \in \mathbb{R} : a < x < b\},$$
$$(a, b] = \{x \in \mathbb{R} : a < x \leq b\},$$
$$[a, b) = \{x \in \mathbb{R} : a \leq x < b\},$$
$$[a, b] = \{x \in \mathbb{R} : a \leq x \leq b\},$$
$$\{a, b\} = \{x \in \mathbb{R} : x = a \text{ or } x = b\}.$$

set.[3] We shall use this graphing system in 10.3, for instance.

We permit a or b to be infinite. For instance, the interval $[0, \infty)$ is the set $\{x \in \mathbb{R} : x \geq 0\}$, and the interval $[-\infty, +\infty]$ is the *extended real line* — that is, $\mathbb{R}$ plus two more objects, $\pm \infty$.

Some mathematicians consider a singleton to be an interval; for instance, $\{6\} = [6, 6]$. Also, some mathematicians consider the empty set to be an interval — albeit a degenerate one. Any interval containing at least two points is actually an infinite set — for instance, $(1, 2)$ contains 1.1 and 1.69 and $\sqrt{2}$ and all the other numbers between 1 and 2.

Unfortunately, an expression such as $(1, 2)$ is ambiguous: it may mean an ordered pair or an interval. Generally the reader can determine which meaning is intended, from the context. A few additional words can remove ambiguity if necessary — for instance, we might talk about "the ordered pair $(1, 2)$" or "the interval $(1, 2)$." Some mathematicians, especially in Europe, avoid this ambiguity by using a slightly different notation; they write our four intervals as

$]a, b[$, $]a, b]$, $[a, b[$,   and $[a, b]$,   respectively.

g. If X, Y, Z are any sets, then we define their *products*
$$X \times Y = \{(x, y) : \quad x \in X, \ y \in Y\},$$
$$X \times Y \times Z = \{(x, y, z) : \quad x \in X, \ y \in Y, \ z \in Z\},$$
etc. — the set of ordered pairs, the set of ordered triples, etc. We abbreviate $X \times X = X^2$ and $X \times X \times X = X^3$, and so on.

---

[3] It is better not to call these "black" and "white" dots, since the color scheme is reversed on a chalkboard; the terms "solid" and "hollow" work properly both on chalkboard and on paper.

Sets and their members

For instance, $\mathbb{R}^2$ denotes the set of all ordered pairs of real numbers; this is a commonly used notation for the Euclidean plane.

If X and Y are finite sets, then clearly $|X \times Y| = |X||Y|$. Thus, cardinality preserves multiplication.

h. *Still more notations.* Sets can be built from other sets using various set operations, described starting in 3.34. Also, sets can be specified in terms of functions, using the "image" and "inverse image" notations of 3.24.

3.8. *Exercises.* How many members does each of these sets have? (Your answer should be a nonnegative integer or "infinity.")

- **a.** $(3, 8]$.
- **b.** $\{5, 6, 7, \ldots, 99\}$.
- **c.** $\{1, \{1, 2\}\}$.
- **d.** $\{1, \{1\}, 2\}$.
- **e.** $\{\varnothing, \{\varnothing\}, \{\{\varnothing\}\}\}$.
- **f.** $\{\varnothing, \{\varnothing, \{\varnothing\}\}\}$.
- **g.** $\{1, 2, 3, \ldots, 100\}$.
- **h.** $\{1, 4, 9, 16, 25, \ldots, 100\}$.
- **i.** {even integers between 11 and 49}.
- **j.** $\{1, \{1\}, \{1, \{1\}\}, \{\{1\}, \{1\}\}, 2\}$.
- **k.** $\{x \in \mathbb{Z} : 2 < x < 8\}$.
- **l.** $\{x \in \mathbb{R} : 2 < x < 8\}$.

3.9. *Joke.* Nothing is better than true love. However, a ham sandwich is better than nothing. Therefore, a ham sandwich is better than true love.

Explanation. The joke's conclusion is wrong. It is arrived at by using the word "nothing" in two different ways and then treating those two words as if they were the same.

The second sentence is fairly straightforward: I would rather have a ham sandwich than have nothing. In symbols

$$\text{ham sandwich} \quad > \quad \varnothing,$$

where ">" is temporarily being used as a symbol for "is preferable to."

However, "nothing is better than true love" requires more careful interpretation. It should *not* be interpreted as "$\varnothing >$ true love." Rather, it should be read as: "There aren't any things that are better than true love." That is,

$$\varnothing = \{x : x > \text{true love}\}.$$

Thus, we can *not* conclude that a ham sandwich is better than true love. But if we're just a little more careful, our two premises can be put together to yield this correct conclusion:

$$\text{ham sandwich} > \{x : x > \text{true love}\}.$$

That is,

> I would rather have a ham sandwich than hold in my hands all the things in the world that are better than true love (because then my hands would be empty).

3.10. *Nonempty sets versus inhabited sets.* Throughout this book, we use classical set theory, so either a set S is empty or it's not; there are only two possibilities. But constructivists make a finer distinction, yielding *three* cases:

- S is nonempty in this strong sense: We are actually able to *find* at least one particular example of a member of S. (Some constructivists then say that S is *inhabited*.) Classical set theorists usually are not concerned about this additional information.
- S is nonempty in this weak sense: We have enough information to convince ourselves that S does contain something, but we do not necessarily have within our grasp any particular examples of members of S. This is the "nonempty" generally used by classical set theorists. — For instance, Jarden's nonconstructive proof (2.45, without 2.46) is enough to convince us that the set of ordered pairs

$$\{(p,q) \in \mathbb{R}^2 : p,q > 0, \quad p,q \text{ irrational}, \quad q^p \text{ rational}\}$$

  is nonempty in this weaker sense.
- S is empty.

The weaker, classical notion of "nonempty" is the one used throughout this book. All of our metamathematical ("outer system") work in this book uses classical reasoning. When this book says that a set is nonempty, we will not necessarily know how to produce a sample member of that set.

Russell's paradox

3.11. Can you find the fallacy in the following reasoning?

Some sets are members of themselves. For example:

- An *infinite set* is a set with infinitely many members. There are many examples of infinite sets — for instance, the set of all integers, the set of all real numbers, the set of all sequences of 0's and 1's. In fact, there are infinitely many different infinite sets. Hence the set of all infinite sets is itself an infinite set. Hence *the set of all infinite sets* is a member of itself.

- We shall say that a set S is *simple* if it can be described in English in 100 words or less. For instance, the empty set is simple, and the set of all integers is simple. Our definition of "simple set" took only a few words, so the set of all simple sets also takes only a few words. Thus the set of all simple sets is a simple set — i.e., *the set of all simple sets* is a member of itself.

- *The set of all sets* certainly is a set, hence a member of itself.

On the other hand, some sets are not members of themselves. For instance:

- Let A be the set of all the pages in this book. Then A is not a page, so A is not a member of itself. (Using the metaphor of 3.5, a bag full of pages is not a page.)

- Let B be the set of all the students in my class. Then B is not a student, so B is not a member of itself. (Using the metaphor of 3.5, a bag full of students is not a student.)

Let us call a set *self-inclusive* if it is a member of itself, or *non-self-inclusive* otherwise. Let $\mathcal{N}$ be the collection of all non-self-inclusive sets. Thus $\mathcal{N}$ is the set defined by this condition:

(∗)   for every set S, $S \in \mathcal{N} \Leftrightarrow S \notin S$.

Finally, here is the paradox: Is $\mathcal{N}$ a member of itself? Think about that. It is if it isn't, and it isn't if it is. We get a contradiction either way.

Some readers may understand the contradiction more easily in symbolic form: $\mathcal{N}$ is a particular set, but $(*)$ holds for *every* set S. In particular, it holds when we substitute $S = \mathcal{N}$. But that yields $\mathcal{N} \in \mathcal{N} \Leftrightarrow \mathcal{N} \notin \mathcal{N}$.

What has gone wrong here? Ponder that for a few minutes, before you read the next section.

3.12. *The paradox explained.* Russell's Paradox arises from using our language too freely. Some things that we can describe nevertheless do not exist.[4]

Axiomatic set theory (described briefly in 2.9) was designed to avoid such paradoxes. However, we can also avoid such paradoxes by simple and informal methods. Our immediate needs will be satisfied by this rule:

we shall prohibit self-referencing

— i.e., we shall prohibit definitions and statements that refer to themselves. In particular, we must slightly amend the definition of "set" given in 3.2. In that section, we said that a set was "a collection of objects"; instead we shall say that

a set is a collection of *already formed* objects.

Each of the objects that we collect must each already be formed before we assemble the objects into the collection; hence the collection being formed cannot be one of those objects. With this newly revised definition of "set," it is easy to see that

a set cannot be a member of itself

— that is, there are *no* "self-inclusive" sets. In particular,

the collection of all sets is not a set.

Self-referencing also can arise without set theory. For instance, write these two statements on the two sides of a sheet

---

[4]Whereas some indescribable objects do exist; see 3.71.

of paper:

| The statement on the other side of this paper is false. | The statement on the other side of this paper is true. |

Then turn the paper over and over until you figure out what's what. The paradox can be simplified further, though perhaps it then becomes too obvious to be amusing. Eubulides of Miletus (4th century B.C.) reduced the paradox to just one short statement: "This statement is false."

Self-referencing is the explanation for another paradox given in an exercise elsewhere in this book. Can you find it?

Actually, a ban on self-referencing is merely the easiest solution to Russell's Paradox, not the only solution. Systems *with* self-referencing are investigated in Barwise and Etchemendy [1987] and Barwise and Moss [1996]. Such systems may even have practical applications. For instance, they can be used to analyze a computer program that is editing a file that contains a copy of the very computer program that is running; in fact, the program may edit itself. Some people have suggested that self-editing is a trait of intelligence, found in humans and sought in artificial intelligence. However, the analysis of self-referencing systems is complicated and not presently part of the mainstream of mathematical logic.

There are other ways to avoid paradox as well. Indeed, in Quine's "New Foundations" system, the set of all sets *is* a set. But for simplicity and for compatibility with conventional set theory, we shall simply prohibit self-referencing in this book.

SUBSETS

3.13. Definition. Let S and T be sets. We say that S is a *subset* of T, written $S \subseteq T$, if every member of S is also a member of T. Equivalently, T is a *superset* of S; that can also be written $T \supseteq S$. (See also the formal definition in 2.9.) Here are a few examples:

$$\{3,4\} \subseteq \mathbb{R}, \qquad \{3,4\} \subseteq \{1,2,3,4\}, \qquad \{3,4\} \subseteq \{3,4\}.$$

That last example might look wrong to some beginners, but it is correct — it does satisfy our definition. In fact,

every set is a subset of itself.

If we wish to not only state that $S \subseteq$, but emphasize that the two sets are different, we say that S is a *proper subset* of T, or T is a *proper superset* of S, written respectively as $S \subsetneq T$ or $T \supsetneq S$.

$$\{3,4\} \subsetneq \mathbb{R}, \qquad \{3,4\} \subsetneq \{1,2,3,4\}.$$

Remarks. Some students get confused because the mathematical convention for making comparisons differs from the convention of nonmathematical English.

- In nonmathematical English, most comparisons are like $<$. When we say that one apple is larger than another, we understand that the two apples are not the same size.

- But in the language of mathematicians, many comparisons are more like $\leq$. For instance, when we say that one topology is stronger than another, we do not exclude the possibility that the two topologies have the same strength. In mathematics, "stronger than" is often understood to mean "at least as strong as." This grammatical structure applies to many comparisons in mathematics. To indicate that the two things being compared are not equal, we may have to add the word *proper* or *strict*.

3.14. When applied to sets, the comparative term *"larger"* may have two possible meanings.

- In some contexts, "S is larger than T" means that the cardinality of S is greater than the cardinality of T. That is, $|S| > |T|$, or perhaps (depending on the context) $|S| \geq |T|$. For instance, $\{3,4,6,7\}$ is larger than $\{1,2\}$ in this sense. We may say "strictly larger" to convey $|S| > |T|$ without ambiguity.

- In other contexts, "S is larger than T" means that S is a superset of T. That is, $S \supseteq T$, or perhaps (depending on the context) $S \supsetneq T$. For instance, $\{3,4,6,7\}$ is larger than $\{3,4\}$ in this sense. When this meaning is in use, the *largest* member of a collection of sets would be a member that contains all the other members; thus it would be the union of all the members of the collection. See 4.10 for an example.

Subsets

The two notions of "larger" are related: Whenever $S \supseteq T$, then $|S| \geq |T|$. However, the converse of that implication does not always hold. For instance, we will see later in this chapter that the intervals $[0,1]$ and $[0,2]$ have the same cardinality (i.e., it's the same kind of infinity); thus $|[0,1]| = |[0,2]|$. Hence $|[0,1]| \geq |[0,2]|$, but it is not true that $[0,1] \supseteq [0,2]$.

Without clarification, the word "larger" may be ambiguous — it may refer to supersets or to cardinality. But this is nothing new; we already have this phenomenon in English. For instance, who is larger: A man who is 6 feet tall and weighs 175 pounds, or a man who is 5 feet 10 inches tall and weighs 200 pounds? It depends on what we're measuring.

The word "smaller" is used analogously (with all the relations reversed, of course). Obvious meanings also apply to "largest" and "smallest."

3.15. *More notation.* Unfortunately, the symbol $\subset$ is ambiguous:

- some mathematicians use it as we are using $\subseteq$, and
- some mathematicians use it as we are using $\subsetneq$.

Because of this ambiguity, I will not use the symbol $\subset$ at all, and I urge you to avoid it too.

The words "contains" and "includes" are also ambiguous. Each of the statements

$$T \text{ contains } S \qquad\qquad T \text{ includes } S$$

can mean either "$T \supseteq S$" or "$T \ni S$." The ambiguity here is not so severe, since the distinction between $T \supseteq S$ and $T \ni S$ is often evident from the context. For instance, if we know that S and T are subsets of $\{1, 2, 3, 4, 5\}$, then one meaning is clear; if we know that S is a number and T is an interval, then the other meaning is clear.

3.16. The empty set is a subset of every set. That is:

$$\text{For every set } S, \text{ we have } \varnothing \subseteq S.$$

How can this be true? This says that for each x,

$$\text{if } x \in \varnothing, \quad \text{then} \quad x \in S.$$

But there aren't any x's in $\varnothing$, so the hypothesis "x is in the empty set" is never true. Thus there aren't any cases for us to verify, so the implication (the "if ... then" mentioned above) is true in "all the cases." We say that the implication $x \in \varnothing \Rightarrow x \in S$ is *vacuously* true; that term is discussed in 5.30.

3.17. *Examples.* These statements are true:

$$\{7\} \in \{3, \{7\}\}, \quad \{\{7\}\} \subseteq \{3, \{7\}\}, \quad \{3, \{7\}\} \subseteq \{3, \{7\}\},$$

but these statements are false:

$$\{7\} \subseteq \{3, \{7\}\}, \quad \{\{7\}\} \in \{3, \{7\}\}, \quad \varnothing \in \{\{7\}\}.$$

Exercise. Is each of the following statements true or false? (Count nonsense statements as "false.") To reduce the paper-grader's work, everyone in the class should write their answers as T or F in a 4×4 array, following the same ordering and arrangement as the 16 statements below.

$3 \in \{3\}$ $3 \in \{3, \{7\}\}$ $3 \subseteq \{3\}$ $3 \subseteq \{3, \{7\}\}$

$\{3\} \in \{3\}$ $\{3\} \in \{3, \{7\}\}$ $\{3\} \subseteq \{3\}$ $\{3\} \subseteq \{3, \{7\}\}$

$\varnothing \in \{3\}$ $\varnothing \in \{3, \{7\}\}$ $\varnothing \subseteq \{3\}$ $\varnothing \subseteq \{3, \{7\}\}$

$\{7\} \in \{\{7\}\}$ $\{\{7\}\} \in \{\{7\}\}$ $\{7\} \subseteq \{\{7\}\}$ $\{\{7\}\} \subseteq \{\{7\}\}$

3.18. *Exercises*

a. Define $A = \{1, 2\}$, $B = \{1, 2, 3, 4\}$, $C = \{\{1, 2\}, 3, 4\}$, and $D = \varnothing$. List all true statements that can be made of the form $\underline{\quad} \subsetneq \underline{\quad}$ or of the form $\underline{\quad} \in \underline{\quad}$, by replacing the blanks with any of A, B, C, D.

b. Define the sets $D = \varnothing$, $E = \{\varnothing\}$, $F = \{\varnothing, \{\varnothing\}\}$, and $G = \{\varnothing, \{\varnothing\}, \{\varnothing, \{\varnothing\}\}\}$. List all true statements of the form $\underline{\quad} \subsetneq \underline{\quad}$ or of the form $\underline{\quad} \in \underline{\quad}$ that can be made by replacing the blanks with any of D, E, F, G.

Subsets

3.19. Definition. It sometimes happens that all the sets currently being studied are subsets of one set Ω. Then that set Ω is called the *universe*, or *universe of discourse*, or *universal set*. Certain conveniences of notation then become available; see especially the uses of $\complement$ starting in 3.43.

For instance, during a week when we're just studying the real numbers and nothing else, we might find that the only sets we're considering are the subsets of $\mathbb{R}$. In that situation, $\mathbb{R}$ would be the universal set.

It should be emphasized that a universal set can only be used in very specific and temporary contexts. There does not exist one set Ω that is a superset of *all* sets. That fact is evident from our explanation of Russell's Paradox, in 3.12.

Any letter can be used for a universal set, but in this book we'll most often use Ω for that purpose.

3.20. *Optional: Extra for algebraists.* A *partially ordered set*, or *poset*, is a set X equipped with a relation $\preccurlyeq$ that satisfies

$$\begin{array}{ll}\text{(reflexive property)} & a \preccurlyeq a \\ \text{(antisymmetric property)} & \text{If } a \preccurlyeq b \text{ and } b \preccurlyeq a \text{ then } a = b \\ \text{(transitive property)} & \text{If } a \preccurlyeq b \text{ and } b \preccurlyeq c \text{ then } a \preccurlyeq c\end{array}$$

Some examples:

a. Let X be the set of real numbers (or any subset of the real numbers), and let $\preccurlyeq$ be just ordinary $\leq$.

b. Let X be the set of positive integers (or any subset of the positive integers), and let $a \preccurlyeq b$ mean that "a is a factor of b" — that is, $au = b$ for some positive integer u.

c. Let X be a collection of sets, and let $a \preccurlyeq b$ mean $a \subseteq b$.

Actually, *every* poset can be represented in the form of that last example. Indeed, it can be shown that

If $(X, \preccurlyeq)$ is any poset, then there exists a collection $\mathcal{C}$ of sets such that $(X, \preccurlyeq)$ is order isomorphic to $(\mathcal{C}, \subseteq)$

— i.e., the two structures are identical, except for relabeling. (*Sketch of proof.* Let $\varphi(a) = \{x \in X : x \preccurlyeq a\}$, and let $\mathcal{C}$ be the range of the set-valued function φ.)

3.21. *Optional: Still more algebra.* A *chain ordering* on a set X is a partial ordering $\preccurlyeq$ (defined as in 3.20) with the additional property that

for each a and b in X, at least one of $a \preccurlyeq b$ or $b \preccurlyeq a$ holds.

A *chain* is a set equipped with such an ordering. *Examples:*

- Any subset of the real number system is a chain when equipped with the usual ordering ($\leq$). As a consequence we shall see in Chapter 8 that the formula scheme $(A \to B) \vee (B \to A)$ is tautological in any numerically valued semantic interpretation.

- A collection of sets is not necessarily a chain when ordered by inclusion ($\subseteq$). As a consequence, we shall see in Chapters 9 and 10 that the formula scheme $(A \to B) \vee (B \to A)$ is not tautological in some set-valued semantic interpretations.

FUNCTIONS

3.22. *Definition.* Let S and T be sets. A *function* from S to T is a rule f that assigns, to each $s \in S$, some particular $t \in T$. We then write $t = f(s)$ to indicate that this t was assigned to that s; we say t is the *value* taken by the function f at the *argument* s.

We are generally concerned with the results of the function, not the description of the function. For instance, $s^2 + 2s + 1$ and $(s+1)^2$ are different strings of symbols, but they represent the same function from $\mathbb{R}$ to $\mathbb{R}$.

Other words for "function" are *map* or *mapping*. The statement "f is a function from S to T" is abbreviated as $f : S \to T$. We say that S is the *domain* and T is the *codomain* of the function.

The two expressions

$$f : S \to T \quad \text{(``}f \text{ is a function from } S \text{ to } T\text{'')}$$
and $$S \to T \quad \text{(``}S \text{ implies } T\text{'')}$$

are entirely unrelated uses of the arrow symbol. They should be easily distinguished from one another by the contexts in which they are used.

3.23. The *range* of a function $f : S \to T$ is the set of all the values that it takes; that is,

$$\text{Range}(f) \;=\; \{t \in T \,:\, t = f(s) \text{ for some } s \in S\}.$$

Some students confuse codomain with range, but it's not the

Functions

same thing. The codomain can be any set that contains the range. Thus, the codomain is less precisely specified, and perhaps that causes some of the confusion; but that is also one of the chief advantages of the codomain.

For instance, consider the function $f(x) = x^4 - x^3$, with domain $\mathbb{R}$. We can see at a glance that f is a *real-valued function*. For many applications, we don't need to know more precisely what the range is, so we may as well save ourselves some trouble. If we're trying to just give a quick, easy description of the function $f(x) = x^4 - x^3$, we might make any of the following statements (all of which mean exactly the same thing):

f is a real-valued function defined on all of $\mathbb{R}$,

f is a function with domain $\mathbb{R}$ and codomain $\mathbb{R}$,

f is a function from $\mathbb{R}$ to $\mathbb{R}$,

$f : \mathbb{R} \to \mathbb{R}$.

But we usually would not say

f is a function with domain $\mathbb{R}$ and range $\left[\frac{-27}{256}, \infty\right)$.

That last statement is true, but in most cases it's more effort and more information than we need or want.

The codomain can also be used to intentionally restrict the range. For instance, I might say "let any function $f : S \to [0, 1]$ be given." This is a convenient way of saying that I'm considering functions f that have domain equal to S, and that have range equal to some subset of $[0, 1]$. The range might be all of $[0, 1]$, or it might be some proper subset of $[0, 1]$. Sometimes we want to work with a collection of functions that is most conveniently described in that fashion.

3.24. Here are more notations for sets, to be added to the list of notations that we gave in 3.7.

- As we noted above, the *range* of a function $f : S \to T$ is the set

$$\{t \in T \; : \; t = f(s) \text{ for some } s \in S\}.$$

That set can also be written more briefly as
$$f(S) = \{f(s) : s \in S\}.$$
In this context, the domain S is also called an *index set*.

- More generally, if A is any subset of the domain, then the *image* of A under the function is the set
$$f(A) = \{f(s) : s \in A\}.$$

- Also, if B is any subset of the codomain T, then
$$f^{-1}(B) = \{s \in S : f(s) \in B\}$$
is called the *inverse image* of B.

Here are a couple of examples of ranges:

a. We considered the numerical function $x^4 - x^3$ in 3.23. Its range is $\{x^4 - x^3 : x \in \mathbb{R}\}$; here $\mathbb{R}$ is the index set. Note that this representation involves some *repetitions* — for instance, 0 shows up as the value of $x^4 - x^3$ at more than one choice of x. A repetition-free representation is possible but would require much more work. For most uses of this set, we can skip that additional work; repetitions are ignored in the representations of sets, as we noted in 3.1.b and 3.2.

b. Consider these two representations of one set:
$$A = \{-3x + 7 : x \in [3, 5]\} = [-8, -2].$$
In the first representation, the index set is $[3, 5]$. Just what is or is not a member of the set A is not affected by the order in which x runs through the index set, but $[3, 5]$ happens to be a convenient way to represent the index set. We might imagine x increasing from 3 to 5; then $-3x + 7$ is actually *decreasing* from -2 to -8. That might be the most convenient way to view A, if we're studying the function $-3x + 7$. On the other hand, the set A is also equal to the interval $[-8, -2]$; that presentation suggests that the members of A increase from -8 to -2. In effect, the two representations list the members of A in two different orders. It's still the same set, in either case. Which representation is more convenient depends on what purpose we're going to use A for.

Functions

3.25. Exercise. Suppose $f : X \to Y$ is a function, $B \subseteq Y$, and $A \subseteq f^{-1}(B)$. Then $f(A) \subseteq B$.

The *proof* should consist of several complete sentences. It should begin with "Suppose w is a member of $f(A)$ ..." and end with "... and so $w \in B$."

3.26. Exercise. Rewrite each of the following sets as a list (as in 3.7.a) or as an interval (as in 3.7.f).

a. $\{2x + 5 : x \in \mathbb{Z}, \ x^2 < 9\}$.
b. $\{2x + 5 : x \in \mathbb{N}, \ x^2 < 9\}$.
c. $\{2x + 5 : x \in \mathbb{R}, \ x^2 < 9\}$.
d. $\{\sqrt{x} : x = 3, 4, 5, 6\}$.
e. $\{\sqrt{x} : x \in [3, 6]\}$.
f. $\{x \in \mathbb{Z} : -10 < x < 10 \text{ and } x \text{ is odd}\}$.

3.27. (Technical, postponable.) Suppose some universal set Ω has been selected or is understood (as in 3.19). Let $S \subseteq \Omega$. The *characteristic function* of S is then the function $1_S : \Omega \to \{0, 1\}$ defined by

$$1_S(x) = \begin{cases} 1 & \text{if } x \in S, \\ 0 & \text{if } x \in \Omega \setminus S. \end{cases} \quad \text{for } x \in \Omega.$$

This is also called the *indicator function*. Some other common notations for this function are χ_S, φ_S, and I_S. Note that this function does depend on Ω, but that dependence is not displayed in the notation "1_S"; the choice of Ω must be understood. We will be using characteristic functions in 4.26 and 29.9.

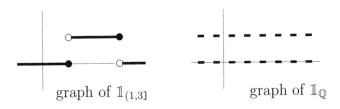

graph of $1_{(1,3]}$ graph of $1_\mathbb{Q}$

Examples. Let $\Omega = \mathbb{R}$. The characteristic function of the interval $(1, 3]$ is simply a step-function; see first illustration. The characteristic function of the set of rational numbers is harder to graph, since every interval of positive length contains infinitely many rational numbers and infinitely many irrational numbers. The graph of $1_\mathbb{Q}$ usually is represented as a couple of dashed lines to help the reader's imagination (see second illustration).

3.28. Let f be a real-valued function, defined on $\mathbb{R}$ or on some subset of $\mathbb{R}$. In this book we shall say that

- f is *order preserving* (or *isotone*) if $x \leq y \Rightarrow f(x) \leq f(y)$. Some examples of such functions $f(x)$ are x itself, e^x, $\ln x$, x^2, $\sqrt{x}$, and the *greatest integer function*.

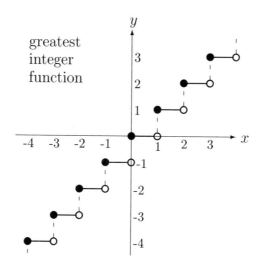

- f is *order reversing* (or *antitone*) if $x \leq y \Rightarrow f(x) \geq f(y)$. Some examples are $-x$ on $\mathbb{R}$, $1/x$ on $(0, +\infty)$, and $\cos x$ on $[0, \pi]$.

A function that is either isotone or antitone is called *monotone*.[5]

Any constant function (with graph just a horizontal line) is *both* order preserving and order reversing. But most[6] functions are neither order preserving nor order reversing. For instance,

---

[5]Order preserving functions are also known in some of the literature as *weakly increasing* functions. A *strongly increasing* or *strictly increasing* function is one that satisfies the condition $x < y \Rightarrow f(x) < f(y)$. We emphasize that this stronger condition is not part of our definition of "order preserving" — for instance, this condition is not satisfied by the greatest integer function. The term *increasing* usually means weakly increasing, but some mathematicians use it to mean strongly increasing.

[6]How can I talk about "most" of an infinite set? Well, I won't try to make that precise, but here's an example: If you pick seven numbers at random

Functions

$\sin(x)$ is order preserving on some intervals and order reversing on other intervals; overall it is not monotone.

What about the composition of two or more monotone functions? Let m and n be nonnegative integers. If we compose

$$m \text{ order preserving and } n \text{ order reversing functions}$$

in any order, the result is

- order preserving if n is even,
- order reversing if n is odd,

regardless of the value of m, because each *pair of reversals* cancels out. For instance, $\sqrt[3]{-e^{-x}}$ is order preserving, as it is the composition of the two order reversing functions $y = -x$ and $y = -x$ and some number of order preserving functions (in this example, the two functions $y = \sqrt[3]{x}$ and $y = e^x$).

We can also describe *functions of several variables* as monotone in one variable when the other variables are held fixed. For instance,

- $f(x, y) = x - e^y$ is order preserving function in x (i.e., when y is held fixed),
- $f(x, y) = x - e^y$ is order reversing function in y (i.e., when x is held fixed),
- $x + y$ is order preserving in x and y.

In 3.45 we define analogous notions for operations on sets. In 6.35 we define analogous notions for logic, to be used later in this book — particularly in 8.44 and 14.16.

3.29. The *minimum* of some numbers is the lowest of those numbers; the *maximum* is the highest. For instance,

$$\min\{2, 3, 5, 7\} = 2; \qquad \max\{2, 3, 5, 7\} = 7.$$

When S is a finite nonempty subset of $\mathbb{R}$, then both $\min(S)$ and $\max(S)$ always exist. For an infinite set, the min and max might

---

and use them for the coefficients of a sixth degree polynomial, usually the polynomial will not be a monotone function.

or might not exist. For instance, max$[0, 1] = 1$, but max$(0, 1)$ does not exist. Indeed, by taking numbers such as 0.9 and 0.9999, we can get closer and closer to where the highest member of $(0, 1)$ "would be" — but for any number x in $(0, 1)$ there is a higher number y in $(0, 1)$. *Exercise*: Prove that $y = (1 + x)/2$ is such a higher number.

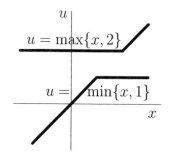

Both of the functions min$\{x, y\}$ and max$\{x, y\}$ are order preserving in both variables. To illustrate this, pictured here are the graphs of $u = \max\{x, 2\}$ and $u = \min\{x, 1\}$. Observe that both are order preserving.

The functions min$\{x, y\}$ and max$\{x, y\}$ will be used extensively starting in Chapter 8, so we now briefly review some of their properties. All of these properties are elementary. The student should read through this list once to see that each of these equations makes sense, and then refer to it again later if needed.

- $\max\{x, y\} + z = \max\{x + z, y + z\}$.
- $\min\{x, y\} + z = \min\{x + z, y + z\}$.
- $k \max\{x, y\} = \max\{kx, ky\}$ if $k > 0$.
- $k \min\{x, y\} = \min\{kx, ky\}$ if $k > 0$.
- $\max\{a, b\} + \max\{c, d\} = \max\{a + c, a + d, b + c, b + d\}$.
- $\min\{a, b\} + \min\{c, d\} = \min\{a + c, a + d, b + c, b + d\}$.
- $-\max\{x, y\} = \min\{-x, -y\}$.
- $-\min\{x, y\} = \max\{-x, -y\}$.
- $\max\{x, y\} + \min\{x, y\} = x + y$.
- $\max\{x, y\} - \min\{x, y\} = |x - y|$.
- $\max\{x, -x\} = |x|; \quad \min\{x, -x\} = -|x|$.

3.30. Let X be a set. A *unary operator* on X is simply a function from X into X. A *binary operator* on X is a function from X^2 into X — that is, a function that takes an ordered

Functions

pair of members of X into X. (See 3.7.g for the definition of X^2.) Commonly used binary operators are written with an infix notation instead of a prefix notation; for instance, we write $3+5$ rather than $+(3,5)$.

Here are some examples of unary and binary operators:

- $+$ and $\times$ are binary operators on $\mathbb{R}$.
- min and max, when applied to a pair of numbers, are binary operators on $\mathbb{R}$. (However, we can also take the min or max of more than two numbers.)
- $f(x) = e^x$ is a unary operator on $\mathbb{R}$. It is sometimes written as $\exp(x)$.
- Subtraction $(-)$ is a binary operator on $\mathbb{R}$, as in $6 - 2 = 4$. However, the same symbol $(-)$ is also used to denote a unary operator, for instance in the expression $-3 < 5$; here the minus sign means "multiply by -1." Most computer keyboards enter these two operators with the same keystroke, but many handheld calculators have separate keys for the two operators, distinguished by different colors and/or by a pair of parentheses around one of the minus signs.
- Let Ω be some nonempty set, treated as the universal set (see 3.19). We will see later in this chapter that $\cup$ and $\cap$ are binary operators on $\mathcal{P}(\Omega)$, and $\complement$ is a unary operator on $\mathcal{P}(\Omega)$.
- In later chapters we will see that $\vee, \wedge, \rightarrow$ are binary operators on formulas, and $\neg$ is a unary operator on formulas — i.e., these operators transform strings of symbols into strings of symbols. Likewise, $\ovee, \owedge, \ominus$ are binary operators on the semantic values of formulas, and $\ominus$ is a unary operator on those values — i.e., these operators transform numbers or sets into other numbers or sets.

3.31. An *involution* on a set X is a unary operator $f : X \to X$ which, if repeated, gets you back to where you started; that is, $f(f(x)) = x$. Some examples:

- $f(x) = -x$ is an involution of the real numbers;
- $g(x) = 1/x$ is an involution of the nonzero real numbers;

- $g(x) = 1/x$ is also an involution of the positive reals;
- $h(x) = 4 + \dfrac{7}{x-4}$ is an involution of $\mathbb{R} \setminus \{4\}$.

We mention this concept as motivation for both complementation 3.44.a and negation 5.17.

3.32. Let $\square$ and $\circ$ be binary operators on a set X. We say that $\square$ *distributes* over $\circ$ (from the left or from the right, respectively) if

$$a \square (b \circ c) = (a \square b) \circ (a \square c) \quad \text{or} \quad (b \circ c) \square a = (b \square a) \circ (c \square a)$$

for all $a, b, c \in X$. One familiar example is that multiplication distributes over addition: $a \times (b + c) = (a \times b) + (a \times c)$; but addition does not distribute over multiplication. We will see in 3.49 and 3.54 that intersection and union distribute over each other. In many logics, $\vee$ and $\wedge$ distribute over each other; this is discussed further in 9.5 and 14.20. Other results in logic that are related to distribution can be found in 14.1.h and 15.4.a.

The Axiom of Choice (optional)

3.33. The Axiom of Choice (AC) is an abstract principle used in many different parts of higher mathematics. It is not essential for this book's treatment of logic, but it does serve as one more illustration of the difference between the constructive and nonconstructive viewpoints. AC takes many forms; here is one of the simplest:

> *The Axiom of Choice (AC).* Suppose that $\mathcal{C} = \{T_x : x \in I\}$ is a given nonempty set whose members T_x are nonempty sets. Then it is possible to *choose* some member u_x in each T_x. That is, there exists a function f defined on the set I, with the property that $f(x) \in T_x$ for each $x \in I$.

(The function f is sometimes called a *choice function*.) To see what AC means, we should consider some examples:

- If $\mathcal{C}$ is the collection of all nonempty subsets of $\mathbb{N}$, then one selection procedure is to let $f(x)$ be the lowest member of T_x. That's an explicit algorithm for choosing $f(x)$.

The Axiom of Choice (optional)   93

- On the other hand, if $\mathcal{C}$ is the collection of all nonempty subsets of $\mathbb{R}$, then there is no explicit algorithm. By this I don't just mean that we don't know of an explicit algorithm; I mean that it is possible to prove that no explicit algorithm can ever be formulated. (Proving the unavailability of an algorithm, or even stating it more precisely, is quite difficult. An outline of the proof can be found in Schechter [1996].) But here is a *nonexplicit* procedure: For each nonempty set $T_x \subseteq \mathbb{R}$, just pick any number in that set, and call that number $f(x)$. If we allow "functions" of this sort, then there are infinitely many choice functions f available.

As we remarked in 2.9, the *real world* may include three apples or three airplanes, but the abstract concept of the number 3 exists only in our minds. Certainly we are in agreement about 3, but our mental pictures may differ on more sophisticated concepts such as the choice function for the reals. Thus, whether a choice function f "exists" or not depends on our philosophy about the existence of mathematical objects.

The Axiom of Choice was somewhat controversial when it was first formulated near the beginning of the 20th century, and it is still unacceptable to constructivists because it does not produce explicit examples; it simply postulates objects into existence. Moreover, some of its consequences are even more counterintuitive than the choice function for the reals.

Probably the most bizarre consequence of AC is the *Banach-Tarski paradoxical decomposition of the sphere*: Let

$$B = \{(x, y, z) \in \mathbb{R}^3 : x^2 + y^2 + z^2 \leq 1\}.$$

That's just the usual closed ball of unit radius in three dimensions. The Banach-Tarski Theorem says that it is possible to partition this set into finitely many disjoint sets which can then be moved through rigid motions (preserving size and shape; pieces are rotated, translated, and may also pass through one another) and reassembled to form two copies of B.

This seems to violate the physicists' principle of "conservation of mass," but in fact it doesn't. Think of mass as being proportional to volume. We know how to define the volumes of simple objects — cubes, spheres, icosahedrons, etc. — and considerably more complicated objects; but it turns out that the notion of "volume" cannot be extended to *all* subsets of $\mathbb{R}^3$ while preserving its most important properties (additivity, and invariance under rigid motion). In addition to being specified rather inexplicitly, the pieces in the Banach-Tarski decomposition are so irregular that their "volume" cannot be defined.

Neither AC nor its negation is implied by the more elementary axioms of set theory; the two halves of that fact were proved by Gödel in 1939 and Cohen in 1963. Thus we may accept or reject AC. If we accept it, we are not agreeing to a description of the real world. Rather, we are agreeing to the *convention* that we are permitted to use the choice function f in the writing of mathematical proofs, as though it "exists" in some sense, even though we have no particular example in mind for f. Most mathematicians today accept this convention; they feel that bizarre consequences like the Banach-Tarski Paradox are outweighed by the simplifications that the Axiom of Choice allows in many explanations in higher mathematics.

OPERATIONS ON SETS

3.34. Let Ω be any set. The *powerset* of Ω, denoted by $\mathcal{P}(\Omega)$ in this book, is the collection of all subsets of Ω. That is, $\mathcal{P}(\Omega) = \{S : S \subseteq \Omega\}$. For example, $\mathcal{P}(\{4, 9, 11\})$ is the set

$$\Big\{\varnothing, \{4\}, \{9\}, \{11\}, \{4,9\}, \{4,11\}, \{9,11\}, \{4,9,11\}\Big\}$$

which has eight elements. Note that $\varnothing$ and Ω itself are among the members of $\mathcal{P}(\Omega)$, since they are subsets of Ω.

Of course, our counting does not depend on any special properties of 4, 9, and 11. We could replace those numbers with any three different numbers, or any three different objects (not necessarily numbers). Just by relabeling 4, 9, and 11, we see that *any* set with three elements has a powerset with 8 elements.

More generally, if Ω is a finite set with n elements, then $\mathcal{P}(\Omega)$ has 2^n elements. To prove that, say the members of Ω are the numbers $1, 2, 3, \ldots, n$. To choose a set $S \subseteq \Omega$, we have to decide yes or no to each of these n questions: Should 1 be a member of S? Should 2 be a member of S? etc. Each question has two possible answers, so the number of ways that we can answer *all* the questions is $2 \times 2 \times \cdots \times 2 = 2^n$.

3.35. This further example may require more careful thought: the powerset of $\Big\{\varnothing, \{\varnothing\}, \{\varnothing, \{\varnothing\}\}\Big\}$ is

$$\Big\{\ \varnothing,\ \{\varnothing\},\ \{\{\varnothing\}\},\ \{\varnothing, \{\varnothing\}\},\ \{\{\varnothing, \{\varnothing\}\}\},$$
$$\{\varnothing, \{\varnothing, \{\varnothing\}\}\},\ \{\{\varnothing\}, \{\varnothing, \{\varnothing\}\}\},\ \{\varnothing, \{\varnothing\}, \{\varnothing, \{\varnothing\}\}\}\ \Big\}.$$

Such deeply nested sets will not be needed later in this book, but they are important in axiomatic set theory.

Operations on sets

3.36. *Binary operators.* Let S and T be any sets. We define

$S \cup T = \{x :$ at least one of $x \in S$ or $x \in T$ holds$\}$; this is the *union* of S and T.

$S \cap T = \{x :$ both conditions $x \in S$ and $x \in T$ hold$\}$; this is the *intersection* of S and T.

$S \setminus T = \{x : x \in S$ and $x \notin T\}$; this is the *relative complement* of T in S, also known as S *minus* T. (This set is also written as $S - T$ by some mathematicians, but that notation has other meanings as well.)

The definition of $S \setminus T$ does *not* require that T be a subset of S, contrary to the belief of some beginning students.

A mnemonic trick: $\cup$ and $\cap$, which is which? You might remember it this way: the symbols $\cup$ and $\cap$ resemble letters in the words "**u**nion" and "i**n**tersection"; also, if we allow misspellings, "**u**r" and "a**n**d." This is not very mathematical, but perhaps it will help some students. See also the related remarks in 5.12.

3.37. *Examples*

a. If $S = \{2, 3, 4\}$ and $T = \{3, 4, 5\}$, then

$$S \cup T = \{2, 3, 4, 5\}, \quad S \cap T = \{3, 4\}, \quad S \setminus T = \{2\}.$$

b. Suppose that

$$\begin{aligned} A &= \{\text{multiples of } 2\} = \{2, 4, 6, 8, 10, 12, \ldots\}, \\ B &= \{\text{multiples of } 3\} = \{3, 6, 9, 12, 15, 18, \ldots\}, \\ C &= \{\text{multiples of } 5\} = \{5, 10, 15, 20, 25, \ldots\}. \end{aligned}$$

Then

$$\begin{aligned} A \cap B &= \{\text{multiples of } 6\} = \{6, 12, 18, 24, \ldots\}, \\ A \cap B \cap C &= \{\text{multiples of } 30\} = \{30, 60, 90, 120, \ldots\}, \\ B \setminus A &= \{\text{odd mult's of } 3\} = \{3, 9, 15, 21, 27, \ldots\}, \\ A \cup B &= (\text{hard to describe}) = \{2, 3, 4, 6, 8, 9, \ldots\}. \end{aligned}$$

c. Let $\mathbb{R}^{[0,1]}$ represent the set of all real-valued functions defined on $[0,1]$. For each number $r \in [0,1]$, let

$$A_r = \{f \in \mathbb{R}^{[0,1]} : f(r) = 0\}.$$

Then (see definitions in 3.40)

$$A_{1/3} \cap A_{2/3} = \{f \in \mathbb{R}^{[0,1]} : f(1/3) = f(2/3) = 0\},$$
$$A_{1/3} \cup A_{2/3} = \{f \in \mathbb{R}^{[0,1]} : f(1/3)f(2/3) = 0\},$$
$$\bigcap_{r \in \mathbb{Q}} A_r = \left\{f \in \mathbb{R}^{[0,1]} : \begin{array}{l} f(r) = 0 \text{ for every} \\ \text{rational number } r \end{array}\right\},$$
$$\bigcup_{r \in \mathbb{Q}} A_r = \left\{f \in \mathbb{R}^{[0,1]} : \begin{array}{l} f(r) = 0 \text{ for at least} \\ \text{one rational number } r \end{array}\right\},$$
$$\bigcap_{r \in \mathbb{R}} A_r = \{0\}.$$

3.38. *Observations*

a. The operators $\cup, \cap$ are *commutative* — that is,

$$S \cup T = T \cup S, \qquad S \cap T = T \cap S.$$

b. The relative complement operator generally is not commutative — that is, the equation $S \setminus T = T \setminus S$ generally does not hold. *Exercise.* Give an example in which $S \setminus T = T \setminus S$, and another example in which $S \setminus T \neq T \setminus S$. Under what conditions does equality hold?

c. The operators $\cup$ and $\cap$ are *associative* — that is,

$$R \cup (S \cup T) = (R \cup S) \cup T, \qquad R \cap (S \cap T) = (R \cap S) \cap T.$$

Consequently, we can omit the parentheses when writing either of those sets. In fact, we can omit the parentheses when we take the union or intersection of more sets. Regardless of where we introduce the parentheses, for any positive integer *n* we get

$$\bigcup_{j=1}^{n} S_j = S_1 \cup S_2 \cup \cdots \cup S_n$$
$$= \{x : x \text{ belongs to } \textit{at least one} \text{ of the } S_j\text{'s}\},$$

$$\bigcap_{j=1}^{n} S_j = S_1 \cap S_2 \cap \cdots \cap S_n$$
$$= \{x : x \text{ belongs to } \textit{all} \text{ of the } S_j\text{'s}\}.$$

Note that taking the union of more sets yields a bigger result. On the other hand, taking the intersection of more sets yields a *smaller* result, because there are fewer x's that can meet the increased number of simultaneous restrictions.

(The relative complement operator is not associative; we will investigate that in 3.54.g and 3.54.h.)

d. $S \cup \varnothing = S$ and $S \cap \varnothing = \varnothing$.

e. $S \cap \Omega = S$ and $S \cup \Omega = \Omega$, if Ω is the universal set (defined in 3.19).

f. $\mathcal{P}(S) \cap \mathcal{P}(T) = \mathcal{P}(S \cap T)$.

g. $\mathcal{P}(S) \cup \mathcal{P}(T) \subseteq \mathcal{P}(S \cup T)$.
  Exercise. Give one example in which $\mathcal{P}(S) \cup \mathcal{P}(T) = \mathcal{P}(S \cup T)$, and another example for $\mathcal{P}(S) \cup \mathcal{P}(T) \subsetneq \mathcal{P}(S \cup T)$.

h. $S \subseteq T \Leftrightarrow S \cup T = T \Leftrightarrow S \cap T = S \Leftrightarrow S \setminus T = \varnothing$.

i. Relative complementation is *order reversing*; i.e.,

$$\text{If } Q \subseteq P, \text{ then } R \setminus Q \supseteq R \setminus P.$$

(Compare with 3.45 and 3.53.)

j. $(A \times B) \cap (X \times Y) = (A \cap X) \times (B \cap Y)$. (Recall from 3.7.g the definition of the product of sets.)

k. $(A \times B) \cup (X \times Y) \subseteq (A \cup X) \times (B \cup Y)$. Exercises:
  (i) Give an example in which $(A \times B) \cup (X \times Y) = (A \cup X) \times (B \cup Y)$.
  (ii) Give an example in which $(A \times B) \cup (X \times Y) \subsetneq (A \cup X) \times (B \cup Y)$.

3.39. *Exercises*

a. Let $A = \{2, 3, 5, 7, 11\}$ and $B = \{3, 6, 9, 12, 15\}$. Write $A \cup B$, $A \cap B$, $A \setminus B$, $A \times B$.

b. Let $A = [0, 5)$ and $B = (3, 7]$. Write $A \cup B$, $A \cap B$, $A \setminus B$.

c. Recall that $|A|$ is the number of members of A. If A and B are finite sets, prove that

  (i) $|A \cap B| + |A \cup B| = |A| + |B|$,

  (ii) $|A \times B| = |A| \, |B|$.

  (iii) A^B denotes the set of all functions from B into A. Prove that $|A^B| = |A|^{|B|}$. This equation even works with the empty set, provided that we define $0^0 = 1$; discuss.

3.40. We also have notations for the union or intersection of an infinite sequence of sets $S_1, S_2, S_3, \ldots$, as follows:

$$\bigcup_{j=1}^{\infty} S_j = S_1 \cup S_2 \cup S_3 \cup \cdots$$
$$= \{x : x \in S_j \text{ for at least one positive integer } j\},$$

$$\bigcap_{j=1}^{\infty} S_j = S_1 \cap S_2 \cap S_3 \cap \cdots$$
$$= \{x : x \in S_j \text{ for every positive integer } j\}.$$

More generally, suppose that $\{S_\lambda : \lambda \in \Lambda\}$ is any collection of sets, indexed by a set Λ, and suppose that $L \subseteq \Lambda$. We define

$$\bigcup_{\lambda \in L} S_\lambda = \{x : x \in S_\lambda \text{ for at least one } \lambda \in L\},$$

$$\bigcap_{\lambda \in L} S_\lambda = \{x : x \in S_\lambda \text{ for every } \lambda \in L\}.$$

Note that the former definition yields $\bigcup_{\lambda \in \varnothing} S_\lambda = \varnothing$. That is, the union of *no* sets is the empty set, since no x can satisfy "$x \in S_\lambda$ for at least one $\lambda \in \varnothing$."

If the S_λ's are subsets of some universal set Ω, then $\bigcap_{\lambda \in \varnothing} S_\lambda$ is sometimes understood to mean Ω. That is, the intersection of *no* sets is the whole of the universe. This is simply a new definition, not a consequence of previous definitions. The idea is that if we don't impose *any* restrictions on x, then we get the collection of "all" points x.

Operations on sets

3.41. Two sets A and B are said to *meet* if $A \cap B \neq \emptyset$. Two sets A and B are called *disjoint* if $A \cap B = \emptyset$.

A collection of sets $\{S_\lambda : \lambda \in \Lambda\}$ is called *pairwise disjoint* if each pair of sets in that collection is either equal or disjoint — that is, if $\mu \neq \nu \Rightarrow S_\mu \cap S_\nu = \emptyset$. For a collection of sets, most mathematicians use the term "disjoint" to mean the same thing as "pairwise disjoint"; thus the word "pairwise" is included mainly for emphasis. (However, a few mathematicians use the term "disjoint" for a collection $\{S_\lambda : \lambda \in \Lambda\}$ of sets such that $\bigcap_{\lambda \in \Lambda} S_\lambda = \emptyset$.)

A *partition* of a nonempty set Ω is a collection of pairwise disjoint, nonempty sets whose union is Ω. (Thus, each point of Ω is a member of one and only one of those sets.)

3.42. *(Postponable.)* An *equivalence relation* on a nonempty set Ω is a relation $\approx$ that holds between some pairs of members of Ω, and that satisfies these three conditions:

> *reflexive:* $\quad x \approx x$,
> *symmetric:* $\quad$ if $x \approx y$ then $y \approx x$,
> *transitive:* $\quad$ if $x \approx y$ and $y \approx z$ then $x \approx z$,

for all $x, y, z \in \Omega$. If $\approx$ is such a relation, then the *equivalence class* determined by any $x \in \Omega$ is the set $[x] = \{u \in \Omega : u \approx x\}$. It is easy to see that the equivalence classes form a partition of Ω.

Equivalence relations are used widely in higher mathematics. Their main purpose is this: If F and G are two objects that are identical in all aspects that are currently of interest to us, then we should be able to disregard the ways in which those objects differ, and use those objects as interchangeable. In effect, we focus on the equivalence classes rather than the objects themselves.

Here are some examples:

a. Equality ($=$) is an equivalence relation on any set. Each equivalence class is a singleton.

b. "Differs by a multiple of 5" is an equivalence relation on $\mathbb{Z}$. With this relation, for instance, $3 \approx 8$ and $3 \approx -2$. There are five equivalence classes; one of them is the set $\{\ldots, -2, 3, 8, \ldots\}$.

c. "First name begins with the same letter" is an equivalence relation on any collection of people. There are 26 equivalence classes.

d. *Any* partition of a set Ω can be viewed as the collection of equivalence classes of an equivalence relation $\approx$ on Ω. Indeed, just define $x \approx y$ to mean that x and y belong to the same member of the partition.

e. Any function $f : \Omega \to X$ determines an equivalence relation on the domain Ω of that function, by this rule: $s \approx t$ if and only if $f(s) = f(t)$. Conversely, any equivalence relation can be represented by a function: Just let X be the collection of equivalence classes, and let $f(t) = [t]$.

f. In later chapters we shall say that two formulas F and G are *semantically equivalent* in some interpretation, if $[\![F]\!] = [\![G]\!]$ for every valuation $[\![\ ]\!]$ in that interpretation. This is easily seen to be an equivalence relation on the set of all formulas. *Example:* $\neg(\pi_1 \vee \pi_2)$ and $(\neg\pi_1) \wedge (\neg\pi_2)$ are not the same formula; for instance, one includes the symbol $\vee$ and the other includes the symbol $\wedge$. But they are semantically equivalent in most of the logics that we shall study.

g. In later chapters we shall say that two formulas F and G are *syntactically equivalent* in some axiomatic system, if $\vdash F \to G$ and $\vdash G \to F$ in that system — i.e., if the formulas $F \to G$ and $G \to F$ are both theorems of that system. An example is the pair of formulas $\neg(\pi_1 \vee \pi_2)$ and $(\neg\pi_1) \wedge (\neg\pi_2)$; see 14.12.

It is perhaps not obvious that the term "syntactic equivalence" is justified — i.e., that the relation $\approx$ defined in 3.42.g actually *is* an equivalence relation on the set of all formulas. That is true for most axiomatic systems studied in this book; we can see that as follows:

- Reflexivity of the relation $\approx$ follows from $\vdash A \to A$, an axiom or theorem in the logics studied in this book. See 13.2.b.
- Symmetry of $\approx$ simply says that if

$$\vdash F \to G \quad \text{and} \quad \vdash G \to F,$$

then

$$\vdash G \to F \quad \text{and} \quad \vdash F \to G.$$

Operations on sets

- Transitivity of $\approx$ follows from the transitivity inference rule, studied in 13.5.b.

3.43. *Definition.* If some universal set Ω (defined as in 3.19) has been specified or is understood, then the *complement* of a set S is the set

$$\complement S \;=\; \Omega \setminus S \;=\; \{x : x \notin S\}.$$

Example. If the universal set is $\mathbb{N}$, then

$$\complement\{2,4,6,8,\ldots\} = \{1,3,5,7,\ldots\}.$$

Remarks. The set $\complement S$ is also written by some mathematicians as S^c or S' or $\overline{S}$. Some of those notations also have other meanings in mathematics. I prefer the notation $\complement S$ for several reasons: it is seldom used for other purposes in mathematics; it permits the use of exponents for iteration, as in $\complement^3 S = \complement\complement\complement S$; and it is not so small a mark that it can be confused with a smudge when written on a chalkboard.

3.44. *Observations*

a. For any set T, we have $\complement\complement T = T$. (Thus complementation is an *involution*; see 3.31.)
b. $S \setminus (\complement T) = T \setminus (\complement S)$. (Explain.)
c. $S \setminus T = S \cap \complement T$. (An illustration will be given at the end of 3.48.)
d. $S \cap \complement S = \varnothing$.
e. $S \cup \complement S = \Omega$, where Ω is the universal set.

3.45. Analogously to 3.28, suppose Γ is a function from one collection of sets to another. We shall say that Γ is

- *order preserving* if $S \subseteq T \;\Rightarrow\; \Gamma(S) \subseteq \Gamma(T)$, or
- *order reversing* if $S \subseteq T \;\Rightarrow\; \Gamma(S) \supseteq \Gamma(T)$.

A function with either of those properties is *monotone*. The results for composing monotone functions are just as in 3.28: The result is order preserving or order reversing, respectively, if there is an even or odd number of order reversing factors in the composition (regardless of the number of order preserving factors). We may also describe a function of several variables as monotone in one of its arguments when the others are held fixed. For instance,

- $S \cap T$ and $S \cup T$ are order preserving functions of S and T.
- $\complement S$ is an order reversing function of S. That is, if $S \subseteq T$, then $\complement T \subseteq \complement S$.
- $S \setminus (\complement T) = T \setminus (\complement S)$.
- $S \cap \complement T$ and $S \cup \complement T$ are order preserving functions of S and order reversing functions of T.

These notions will be used starting in 11.2.

3.46. *(Optional).* The *symmetric difference* of two sets S and T is
$$S \triangle T = (S \setminus T) \cup (T \setminus S) = \{x : x \in S \text{ or } x \in T \text{ but not both}\}.$$
For example, if $S = \{1, 2, 3, 4\}$ and $T = \{3, 4, 5, 6\}$, then $S \triangle T = \{1, 2, 5, 6\}$. Verify these properties of the symmetric difference:

  a. It is commutative and associative — that is,
  $$S \triangle T = T \triangle S \quad \text{and} \quad R \triangle (S \triangle T) = (R \triangle S) \triangle T.$$

  b. Consequently we can write expressions such as $S_1 \triangle S_2 \triangle \cdots \triangle S_n$; the meaning of that expression is not affected by where we put the parentheses, nor is it altered if we change the order of the S_j's.

  c. For any positive integer n,
  $$\triangle_{j=1}^{n} S_j = S_1 \triangle S_2 \triangle \cdots \triangle S_n = \left\{ x : \begin{array}{l} x \text{ belongs to an odd} \\ \text{number of the } S_j\text{'s} \end{array} \right\}.$$
  (There is no evident way to generalize this to infinitely many sets.)

  d. $S \triangle S = \emptyset$, $S \triangle \emptyset = S$, $S \triangle \Omega = \complement S$, $S \triangle \complement S = \Omega$.

  e. $\cap$ distributes over $\triangle$. That is, $R \cap (S \triangle T) = (R \cap S) \triangle (R \cap T)$.

  f. *Extra for algebraists:* $\mathcal{P}(\Omega)$ forms a commutative ring with unit, if we use $\triangle$ for addition, $\cap$ for multiplication, $\emptyset$ for additive identity, and Ω for multiplicative identity.

Venn diagrams

3.47. *Definition.* A *Venn diagram* is a rectangular diagram containing a few circles or other simply shaped regions. The

Venn diagrams

rectangle is not optional; it represents the universal set. (Omitting the rectangle is one of the most common errors made by beginning students.)

See the first Venn diagram, at right. The rectangle is Ω, and the three circles are labeled P, Q, R. It must be understood (for instance) that P is the entire upper circle, not just the upper subregion of that circle.

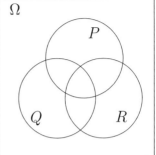

The overlap of two (or more) circles represents the intersection of two (or more) sets. Unless some relationship between the sets is known, we draw the n circles or similarly shaped regions in *general position*, so that all 2^n intersections of regions and their complements are represented. For instance, if there are three regions

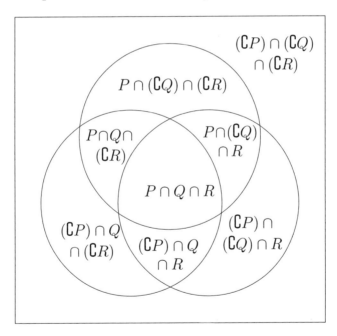

P, Q, R, then there will be eight subregions as a result. I've labeled all eight in this one example, but usually we only label the circles. Note in particular that the region outside all three circles is one of the eight subregions; that is the subregion

representing $(\complement P) \cap (\complement Q) \cap (\complement R)$.

3.48. Here is an elementary example: Most of the students in my logic class know a foreign language (i.e., another language in addition to English). As you can see from the three-circle diagram below,

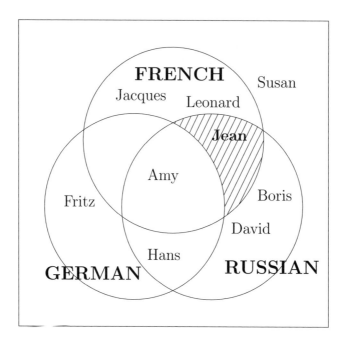

- Susan is the only student who does not know a foreign language.
- David and Boris both know Russian, but not French or German.
- Amy knows French, German, and Russian.
- I have shaded in the region that represents the set of students who know French and Russian but not German; you can see that Jean is the only student in that set.

Shading can be used to call attention to one particular portion of a Venn diagram; that technique will be used in other Venn

Venn diagrams 105

diagrams in the next few pages. *Example.* Use the two-circle diagram given here to convince yourself that $P \setminus Q = P \cap \complement Q$.

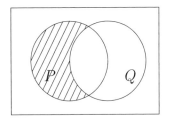

Venn diagrams have certain limitations, of course. They are not very useful for illustrating an infinite collection of sets, or for sets-within-sets such as $\{\{1,2\}, \{1, \{2,3,4\}\}\}$. But for illustrating finitely many non-nested sets, Venn diagrams are very helpful, particularly for beginners in set theory.

3.49. Distribution of operations was introduced in 3.32. We will now show that union distributes over intersection — that is,

$$(P \cap Q) \cup R \;=\; (P \cup R) \cap (Q \cup R).$$

This rule is not at all obvious to beginners, but we can make it more evident by drawing pictures of it.

First make a picture of $(P \cap Q)$. Then take the union of that set with R, to get a picture of $(P \cap Q) \cup R$.

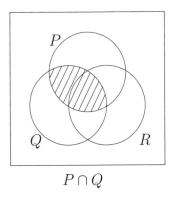
$P \cap Q$

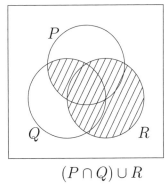
$(P \cap Q) \cup R$

On the other hand, make pictures of $(P \cup R)$ and $(Q \cup R)$; then it's easy to see what their intersection is — but I'm not drawing another picture of that, because you can see that $(P \cup R) \cap (Q \cup R)$ turns out to be the same as the picture we already obtained for $(P \cap Q) \cup R$. That completes our pictorial "proof" of the distributive law for union over intersection.

In arithmetic, distribution is not symmetric: Multiplication

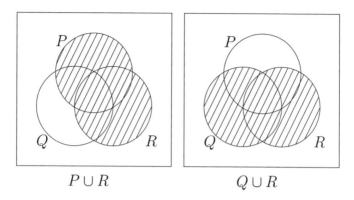

$P \cup R$ $Q \cup R$

distributes over addition, but addition does not distribute over multiplication. For instance, $2+(5\times 3)$ is not equal to $(2+5)\times(2+3)$. However,

> distribution in set theory works both ways: union and intersection each distribute over the other.

We've just finished proving one of the distributive laws. The other, which will be proved in exercises 3.54.a through 3.54.d later in this chapter, is

$$(P \cup Q) \cap R \;=\; (P \cap R) \cup (Q \cap R).$$

3.50. If used carefully, Venn diagrams can be relied upon for correct simple statements such as the distributive laws. However, it is difficult to formalize pictures in a fashion that makes them as reliable and certain as algebraic proofs. For this reason, most mathematicians prefer algebraic proofs most of the time. Let us see how to prove one of the distributive laws by a purely algebraic approach — i.e., without any use of pictures.

To prove that the sets $(P \cap Q) \cup R$ and $(P \cup R) \cap (Q \cup R)$ are equal, we must prove that they contain the same members — i.e., we must prove that $x \in (P \cap Q) \cup R$ if and only if $x \in (P \cup R) \cap (Q \cup R)$.

The proof is subdivided into many cases and subcases.

1. First, we must show that if $x \in (P \cap Q) \cup R$ then $x \in (P \cup R) \cap (Q \cup R)$.

    Assume, then, that $x \in (P \cap Q) \cup R$. This means that

Venn diagrams

at least one of the conditions $x \in P \cap Q$ or $x \in R$ must hold. We must show that, in either of those cases, we have $x \in (P \cup R) \cap (Q \cup R)$. It suffices to show that, in either of those cases, we have both $x \in P \cup R$ and $x \in Q \cup R$. We prove that conclusion separately in those two cases.

(a) Suppose $x \in P \cap Q$. In that case we have both $x \in P$ and $x \in Q$. Now

- from $x \in P$ we can deduce $x \in P \cup R$,
- from $x \in Q$ we can deduce $x \in Q \cup R$.

(b) Suppose $x \in R$. In that case

- from $x \in R$ we can deduce $x \in P \cup R$,
- from $x \in R$ we can deduce $x \in Q \cup R$.

Thus, in either case we have both $x \in P \cup R$ and $x \in Q \cup R$, so we conclude $x \in (P \cup R) \cap (Q \cup R)$.

2. Second, we must show that if $x \in (P \cup R) \cap (Q \cup R)$ then $x \in (P \cap Q) \cup R$. This part can be proved by reasoning in a style similar to that above; the details are left as an exercise for the student.

3.51. Three sets is enough for most applications with Venn diagrams, but occasionally it is useful to draw a Venn diagram with more than three sets.

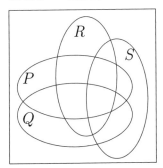

Actually, it is *not possible* to draw four or more *circles* "in general position" (i.e., with all 2^4 possible combinations of intersections represented). But four *ellipses* can be drawn in general position, as shown in the diagram. In fact, it is possible to draw *any* number of ellipses in general position, though the drawing is difficult for $n > 4$. For a proof, see Grünbaum [1975].

Drawing four ellipses in general position takes a bit of practice. But drawing four *rectangles* in general position may be easier. See the Venn diagrams in 4.26.

3.52. At right is a diagram in which the circles are *not* in general position. Consequently the circles satisfy some unusual conditions. For instance, they satisfy the equation $(P \cup Q) \cap (R \cup S) = (P \cap R) \cup (Q \cap S)$; that condition is not always satisfied by sets in general.

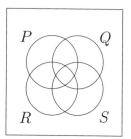

Exercise. Give an example of some sets P, Q, R, S for which $(P \cup Q) \cap (R \cup S) \neq (P \cap R) \cup (Q \cap S)$. You can do this in at least a couple of ways:

  a. Draw two Venn diagrams, each of which has four regions in general position (as in 3.51). In one diagram, shade $(P \cup Q) \cap (R \cup S)$; in the other, shade $(P \cap R) \cup (Q \cap S)$.
  b. Let P, Q, R, and S be different subsets of $\{1, 2, 3, \ldots, 15, 16\}$; choose them so that $(P \cup Q) \cap (R \cup S) \neq (P \cap R) \cup (Q \cap S)$. Actually, you could use a smaller number than 16; how small a number can you use?

3.53. Sometimes it is useful to draw a Venn diagram with the regions *not* in general position. For instance:
  a. We know that $\mathbb{N} \subseteq \mathbb{Z} \subseteq \mathbb{Q} \subseteq \mathbb{R}$; see the diagram in 3.7.d.
  b. If $Q \subseteq P$, then $R \setminus Q \supseteq R \setminus P$, as shown below.

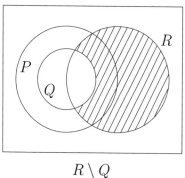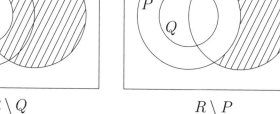

$R \setminus Q$ \qquad\qquad $R \setminus P$

Venn diagrams

3.54. *Exercise.* Represent each of the following expressions as the shaded region in a Venn diagram. Each exercise requires a separate diagram; do not try to combine several exercises into one diagram.

Each diagram should have three circles labeled Q, R, S, even if only two of those labels are mentioned in the expression you are depicting. (That's because we want to compare the diagrams later.) To reduce the paper-grader's work, let's all use the same arrangement of labels: Q should be the top circle; R should be the bottom left circle; S should be the bottom right circle.

a. $Q \cup R$.
b. $(Q \cup R) \cap S$.
c. $Q \cap S$.
d. $R \cap S$.
e. $(Q \cup R) \cap (R \cup S) \cap (S \cup Q)$.
f. $(Q \cap R) \cup (R \cap S) \cup (S \cap Q)$.
g. $(Q \setminus R) \setminus S$.
h. $Q \setminus (R \setminus S)$.

3.55. *Exercises*

a. Draw one Venn diagram of $Q \cap (R \cup S)$ and another of $(Q \cap R) \cup S$. Use these diagrams to explain that $Q \cap (R \cup S) \subseteq (Q \cap R) \cup S$. (That fact will be used in 11.6.)

b. Draw Venn diagrams of the sets $R \cap S \cap \complement T$, $R \cap S$, and $T \cup \complement S$. By considering which of the eight subregions of the diagram must be empty, explain that the following three statements are equivalent — i.e., that if one of these statements is true, then the other two are also.

$$R \cap S \cap \complement T = \varnothing, \qquad R \cap S \subseteq T, \qquad R \subseteq T \cup \complement S.$$

3.56. *Exercises.* Represent each of the following expressions as the shaded region in a Venn diagram. Use just two circles, labeled Q and R. (The circle on the left should be Q; the circle on the right should be R.)

a. $\complement Q$.
b. $\complement R$.
c. $\complement(Q \cup R)$.
d. $\complement(Q \cap R)$.

3.57. *Remarks.* Exercises 3.56 demonstrate *De Morgan's laws*:

$$\complement(P \cup Q) = (\complement P) \cap (\complement Q), \qquad \complement(P \cap Q) = (\complement P) \cup (\complement Q).$$

A corollary of De Morgan's laws is the

Duality principle. (Let Ω denote the universal set.) Any true statement about sets remains true if we simultaneously swap the roles of

$$\cup \text{ and } \cap, \qquad \subseteq \text{ and } \supseteq, \qquad \varnothing \text{ and } \Omega.$$

Here are some examples:

$$\varnothing \subseteq S \subseteq S \cup T, \qquad S \subseteq T \Leftrightarrow S = S \cap T \Leftrightarrow S \cap \complement T = \varnothing,$$
$$\Omega \supseteq S \supseteq S \cap T, \qquad S \supseteq T \Leftrightarrow S = S \cup T \Leftrightarrow S \cup \complement T = \Omega,$$
$$S \cup \Omega = \Omega = S \cup \complement S, \qquad S \cup \varnothing = S = S \cup S,$$
$$S \cap \varnothing = \varnothing = S \cap \complement S, \qquad S \cap \Omega = S = S \cap S,$$
$$R \cap (S \cup T) = (R \cap S) \cup (R \cap T), \qquad \complement(S \cup T) = (\complement S) \cap (\complement T),$$
$$R \cup (S \cap T) = (R \cup S) \cap (R \cup T), \qquad \complement(S \cap T) = (\complement S) \cup (\complement T).$$

3.58. The distributive laws generalize to infinite collections of sets:

$$\left(\bigcap_{\lambda \in \Lambda} P_\lambda\right) \cup \left(\bigcap_{\mu \in M} Q_\mu\right) = \bigcap_{\substack{\lambda \in \Lambda \\ \mu \in M}} (P_\lambda \cup Q_\mu),$$

$$\left(\bigcup_{\lambda \in \Lambda} P_\lambda\right) \cap \left(\bigcup_{\mu \in M} Q_\mu\right) = \bigcup_{\substack{\lambda \in \Lambda \\ \mu \in M}} (P_\lambda \cap Q_\mu),$$

and so do De Morgan's laws:

$$\complement\left(\bigcap_{\lambda \in \Lambda} P_\lambda\right) = \bigcup_{\lambda \in \Lambda} (\complement P_\lambda), \qquad \complement\left(\bigcup_{\lambda \in \Lambda} P_\lambda\right) = \bigcap_{\lambda \in \Lambda} (\complement P_\lambda).$$

These formulas cannot be illustrated with Venn diagrams, but we can prove these formulas by considering a typical member x of any of these sets. For instance,

x is a member of $\complement\left(\cap_{\lambda \in \Lambda} P_\lambda\right)$
$\iff$ it is not true that x is a member of $\cap_{\lambda \in \Lambda} P_\lambda$
$\iff$ it is not true that x is a member of every P_λ
$\iff$ x is outside at least one of the P_λ's
$\iff$ x is a member of at least one of the $\complement P_\lambda$'s
$\iff$ x is a member of $\cup_{\lambda \in \Lambda}(\complement P_\lambda)$.

The middle step of this proof actually uses a property of quantifiers discussed in 5.49.

Syllogisms (optional)

3.59. Syllogistic reasoning was developed mainly by the Greek philosopher Aristotle (384–322 B.C.). Two typical syllogisms are

- All humans are mortal; Socrates is a human; hence Socrates is mortal.
- All humans are mortal; all philosophers are humans; hence all philosophers are mortal.

Aristotle classified reasonings of these and similar types in abstract terms, almost like what we would today call an algebraic syntax. Aristotle was so influential that his theory of syllogisms became the main theory of logic. For over two millennia, it was the framework for almost all research on formal logic.

But a whole new approach to logic appeared in 1847, with the publication of Augustus De Morgan's *Formal Logic* and George Boole's *Mathematical Analysis of Logic*. De Morgan and Boole began with Aristotle's syntactic style, but added to it the analyses of connectives such as "and," "or," and "implies," which became the basis of modern logic.

Aristotelian syllogism still can be found in the introductory chapters of many logic textbooks, but its placement there may confuse some beginning students. It no longer plays a central role in logic. Indeed, today syllogisms are mostly seen as an old-fashioned way of describing some very elementary parts of set theory — a subject which is undeniably important, but which is now viewed as *separate* from logic.

More precisely, Aristotle's syntactic reasoning corresponds to a semantics of sets. Aristotle's "is" can be translated as "is a member of" or "is a subset of." For instance, the two examples about mortality given a few paragraphs ago could be restated, with obvious abbreviations, as

- $H \subseteq M$ and $s \in H$; therefore $s \in M$.

- $H \subseteq M$ and $P \subseteq H$; therefore $P \subseteq M$.

For some additional examples of syllogisms, we turn to the work of Charles Lutwidge Dodgson (1832–1898). He is known among mathematicians for research in several areas of mathematics, particularly in matrices and determinants. But none of his mathematical discoveries are nearly as famous as the books that he wrote under the pseudonym of "Lewis Carroll," particularly *Alice's Adventures in Wonderland* (1865) and *Through the Looking Glass* (1872). Like those books, *Symbolic Logic* (1896) was written for the amusement of children, but this book was also intended to teach the children mathematical logic. It contains many silly syllogisms, intended to teach not-so-silly mathematical principles. Dodgson accompanied these syllogisms with diagrams of a type slightly different from Venn diagrams; we will not consider his diagrams here.

Below are some of Dodgson's simplest syllogisms (with a few minor changes to bring the language up to date).

(1) No ducks waltz.

(2) No officers ever decline to waltz.

(3) All my poultry are ducks.

Abbreviating $D = \{\text{ducks}\}$, $F = \{\text{officers}\}$, $M = \{\text{my poultry}\}$, $W = \{\text{willing to waltz}\}$; we can restate the given information as

(1') $D \cap W = \emptyset$.

(2') $F \subseteq W$.

(3') $M \subseteq D$.

From this information, one of the conclusions we can draw is $M \cap F = \emptyset$; that is, "none of my poultry are officers."

Here is a final example:

(1) Every eagle can fly.

(2) Some pigs cannot fly.

Let $E = \{\text{eagles}\}$, $F = \{\text{creatures that can fly}\}$, $P = \{\text{pigs}\}$. Then the two given pieces of information can be restated:

Syllogisms (optional)

(1′) $E \subseteq F$.

(2′) $P \setminus F$ is not empty — that is, $P \cap \complement F \neq \varnothing$. That can also be restated as $(\exists x)(x \in P\setminus F)$, if we use quantifiers (see 5.41).

That information can be represented by a Venn diagram, as shown. Assumption (1) can be restated as $\complement F \subseteq \complement E$. Combining that with (2), we conclude that there is some $x \in P \cap \complement E$. That is, the shaded region (at least) is known to be nonempty, so "there are some pigs that are not eagles."

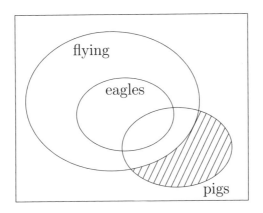

The rules of the game are that we seek conclusions that can be drawn *from the given assertions*. We may use English grammar, but not common sense. For instance, in the last example, we cannot conclude that no pigs can fly, nor can we conclude that any eagles exist! It may help to imagine all of this taking place on some other planet, where the animals are different from those here on earth.

When we say "every eagle can fly," this really means "if there are any eagles, then they can all fly." On the other hand, we're given that "some pigs cannot fly," which means "there exists at least one pig that cannot fly" — which implies, as a corollary, that there does exist at least one pig.

3.60. Some clarification should be given regarding the word "some." In everyday nonmathematical English, "some" has two different meanings:

- "Some" may mean "at least one, but not all." For instance, "I took some of the peanuts" usually means "I took at least one peanut, perhaps most of the peanuts, but I didn't take all of them." This is *not* the meaning of "some" used customarily in mathematics.

- "Some" may mean "at least one, and possibly all." For instance, if "I know that some French people like wine," then I don't know about the other French people; it might or might not be true that all French people like wine. This *is* the meaning of "some" used customarily in mathematics, and it should be used in all the exercises below.

3.61. *Exercises.* In each of the following problems (taken from Lewis Carroll), several pieces of information are given, and then several possible conclusions are given. Do not concern yourself about whether the statements are actually *true* in the world where we live. List all the given conclusions that *can be reached from the given information*. Some, all, or none of the listed conclusions may be reachable (or, in mathematicians' language, some or none of the listed conclusions may be reachable).

a. Given these two pieces of information:

> The only birds that are proud of their tails are peacocks.
> Some birds that are proud of their tails cannot sing.

Which of these conclusions can be reached?

(i) There are some peacocks that cannot sing.

(ii) There are some singing birds that are not peacocks.

(iii) Every peacock is proud of its tail.

(iv) Every bird that is proud of its tail is a peacock.

(v) There are some peacocks that can sing.

(vi) There are some singing birds that are peacocks.

b. Given information:

> All cats understand French.
> Some chickens are cats.

Syllogisms (optional)

Which of these conclusions can be reached?
- (i) Some cats are chickens.
- (ii) Some cats understand French.
- (iii) All chickens understand French.
- (iv) Some chickens understand French.
- (v) All chickens are cats.

c. Given information:

> Babies are illogical people.
> No person who can manage a crocodile is despised.
> Illogical people are despised.

Which of these conclusions can be reached?
- (i) Babies are despised.
- (ii) Babies cannot manage crocodiles.
- (iii) There is at least one illogical person.
- (iv) Some illogical persons can manage crocodiles.
- (v) Despised people are illogical.

d. Given information:

> Some oysters are silent.
> No silent creatures are amusing.

Which of these conclusions can be reached?
- (i) Some oysters are not silent.
- (ii) Some silent creatures are not amusing.
- (iii) Some oysters are not amusing.
- (iv) Some amusing creatures are oysters.

e. Given information:

> No kitten that loves fish is unteachable.
> No tailless kitten will play with a gorilla.
> All kittens with whiskers love fish.
> No unteachable kitten has green eyes.
> Every kitten with a tail has whiskers.

Which of these conclusions can be reached?

(i) Every kitten with whiskers will play with a gorilla.

(ii) No kitten with whiskers will play with a gorilla.

(iii) No green-eyed kitten will play with a gorilla.

(iv) No green-eyed kitten loves fish.

(v) Every green-eyed kitten loves fish.

(vi) Every fish-loving kitten has green eyes.

INFINITE SETS (POSTPONABLE)

Results of this subchapter will not be needed until 29.2.

3.62. In the late 19th century, Georg Cantor (1845–1918) published several papers that greatly changed the way mathematicians view infinity and infinite sets; he is now considered to be the "father of modern set theory."

Prior to Cantor's time, the only infinity that was commonly accepted in mathematics was a so-called *potential infinity* — an infinity that could never be reached. For instance, the sequence

$$1, 2, 3, \ldots$$

gets higher and higher, but it never actually gets to infinity. Here is another example: In calculus, the equation $\lim_{x \to 0} x^{-2} = \infty$ means that when x gets near 0, but not quite equal to 0, then x^{-2} gets higher and higher without bound. But x is not actually allowed to become 0, and x^{-2} never actually gets to infinity. (Contrast that with the semicircle diagram in 3.64.)

A similar conceptual difficulty may have been what directed the ancient Greek mathematicians toward geometry instead of algebra, over two millennia ago. (They did not use decimal numbers, but we will use them to simplify this explanation.) The sequence

$$3, \quad 3.1, \quad 3.14, \quad 3.141, \quad 3.1415, \quad 3.14159, \quad \ldots$$

Infinite sets (postponable)

gets closer and closer to π, but in some sense it never actually gets to π. Thus π is not seen as an ordinary "number." However if two circles have circumferences C_1 and C_2 and diameters D_1 and D_2, then $C_1/D_1 = C_2/D_2$. The common value of the two ratios is π, but we may discuss π as a relationship between geometric lengths without ever mentioning that it is a number. That is largely the attitude taken by the ancient Greek mathematicians.

An *actual infinity*, or *completed infinity*, is obtained if the end of the infinite procedure somehow is actually reached. Such a notion had been suggested many times throughout history, but Cantor was the first mathematician to really make sense of it, and to show how it could be used in a systematic way. The simplest example probably is just the set of natural numbers:

$$\mathbb{N} = \{1, 2, 3, \ldots\}.$$

If we remove the braces, we see a sequence of infinitely many numbers. But if we enclose it in braces, we have just one object. It is *a* set that has infinitely many members. We emphasize that conceptually, the set $\mathbb{N}$ is just *one* object, though it contains *all* the natural numbers.

Once they have grasped this concept, modern mathematicians generally take it for granted, and they may forget how difficult it was at first. But the mathematical community took more than two millennia to prepare for this concept. It should not be surprising that many nonmathematicians and many students initially have difficulty with the idea. How is it "possible" to reach the end of the infinite process?

Perhaps this mental picture will be helpful: Imagine counting "1" in 1 second, then counting "2" in 1/2 second, then counting "3" in 1/4 second, then "4" in 1/8 second, and so on. It's not clear what happens at the very end of the two seconds, but after that we have all the natural numbers.

Of course, we can't count that fast, so that explanation is fiction; it exists only as "what if." But that is true of infinity too. Present-day astronomers are mostly in agreement that the physical universe is finite (though enormous) in size, and contains only a finite (though enormous) amount of matter and energy.

There is no infinity in the real world.

Still, we may use unreal concepts in our intermediate steps, to get from a real-world problem to a real-world solution.[7] Infinity is a concept that does help us to do some of our mathematics. It is not a vague, amorphous, or dreamy concept; it is a concept that has well-defined rules and structures. We will look at some of those rules and structures in the next few pages.

3.63. The *cardinality* of a set is the "size" of that set — i.e., the "number of members" that the set has. The meaning of this definition is very clear for finite sets, but much harder for infinite sets. We shall see that there are different kinds of infinity — i.e., not all infinite sets have the same size. Defining the "size" of an infinite set turns out to be difficult[8]; it is beyond the scope of this book. Surprisingly, it turns out to be fairly easy (and adequate for our needs) to define whether two infinite sets have the same "size," even without saying what that "size" *is*.

3.64. A *one-to-one correspondence* between two sets S and T is a function with domain S and range (not just codomain) T, such that different members of S get mapped by the function to different members of T.

---

[7]A more blatant example of the usefulness of unreal intermediate steps is given by the "imaginary" number $i = \sqrt{-1}$. That number was useful in intermediate steps for many decades before it was recognized to represent a very real *rotation*, through 90°, of the Euclidean plane. The name has persisted, however; $\sqrt{-1}$ is still called an "imaginary number."

[8]Having the same "size" is an equivalence relation, so it is tempting to say that the sizes are just the equivalence classes. But the resulting classes are too big to be sets; they have the same kind of difficulties as the sets in Russell's paradox.

So, in a more advanced set theory than that used in this book, here is the most common method for defining cardinality: First we define the *ordinals*, a complicated collection of sets within sets within sets that have the remarkable property that they can be ordered so that any collection of ordinals has a smallest member. (The ordinals play a role similar to the counting numbers $1, 2, 3, \ldots$, except that there are far more ordinals.) Then we use the Axiom of Choice to prove that every set has the same cardinality as some ordinal. Finally, the *cardinal number* of a set X can be defined to be the smallest ordinal having the same cardinality as X.

Infinite sets (postponable)

Two sets S and T are said to be *equipollent*, or to *have the same cardinality*, written $|S| = |T|$, if there exists[9] a one-to-one correspondence between them.

For instance, the set $\{1, 2, 3, 4, \ldots\}$ of natural numbers and the set $\{1, 4, 9, 16, \ldots\}$ of squares of natural numbers are equipollent, even though the latter is a proper subset of the former. Indeed, the one-to-one correspondence is shown below:

$$\{ \quad 1, \quad 2, \quad 3, \quad 4, \quad \ldots \quad \}$$
$$\updownarrow \quad \updownarrow \quad \updownarrow \quad \updownarrow$$
$$\{ \quad 1^2, \quad 2^2, \quad 3^2, \quad 4^2, \quad \ldots \quad \}.$$

Two more examples are given by the two illustrations below. The first picture shows that any two line segments AB and $A'B'$ have the same cardinality, even if they have different lengths — i.e., they have the same "number" of points, since any point x in one line segment corresponds to a point x' in the other line segment. (Think of one segment as a stretched copy of the other.)

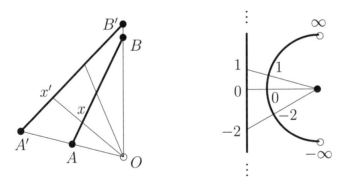

The second picture shows, in a similar fashion, that a line (with infinite length) has the same cardinality as a semicircle (with finite length) with endpoints of the semicircle omitted — that is, there are no points on the line that correspond to the points labeled ∞ and $-\infty$ on the semicircle. If we add those two

---

[9] That we should actually be able to *find* such a correspondence is an additional requirement, imposed by constructivists but not by most mathematicians. In some abstract cases one may be able to prove the *existence* without actually finding the correspondence.

endpoints to the semicircle, we obtain a geometric representation of the "actual infinity" on the straight line — i.e., points that correspond to the "numbers" $\lim_{x\to 0} x^{-2}$ and $-\lim_{x\to 0} x^{-2}$.

3.65. It is customary to write

- $|S| \leq |T|$ if S is equipollent with some subset of T, and
- $|S| < |T|$ if we have $|S| \leq |T|$ but not $|S| = |T|$.

The $\leq$ and $<$ notation are suggestive, but they should not be taken for granted. They are used because the ordering of cardinalities shares some properties with the familiar ordering of the real numbers, but those properties are not simple to prove. Here are two such properties:

> *The Schroeder-Bernstein theorem.* If $|S| \leq |T|$ and $|T| \leq |S|$ then $|S| = |T|$.
>
> *Hartogs's equivalent of the Axiom of Choice.* If S and T are any two sets, then $|S| \leq |T|$ or $|T| \leq |S|$.

Those properties might appear to be "obviously true," but only if one assumes (erroneously) that cardinalities are just like numbers. To see how nontrivial those two theorems are, let us restate them without the simplifying notation. They actually say:

> *Schroeder-Bernstein.* If there exists a one-to-one correspondence between S and a subset of T, and there exists a one-to-one correspondence between T and a subset of S, then there exists a one-to-one correspondence between S and T.
>
> *Hartogs.* For any two sets, there exists a one-to-one correspondence between one and some subset of the other.

The proofs can be found in books on set theory.

3.66. *Definitions.* A set is *finite* if it has the same cardinality as one of the sets

$$\varnothing, \quad \{1\}, \quad \{1,2\}, \quad \{1,2,3,\}, \quad \{1,2,3,4\}, \quad \ldots \quad .$$

Otherwise it is *infinite*.

A set is called *countably infinite* if has the same cardinality as the set of positive integers — i.e., if its members can be rearranged into an infinite sequence, indexed by $\mathbb{N}$. For instance, the

Infinite sets (postponable)

set of *all* integers is countably infinite, since it can be reordered by this "alternating signs trick":

$$0, \ 1, \ -1, \ 2, \ -2, \ 3, \ -3, \ 4, \ -4, \ \ldots \ .$$

The set of all rational numbers is also countably infinite, by a proof that will be given in 3.69.b.

A set is *countable* if it is either finite or countably infinite — i.e., if its members can be arranged in a finite or infinite list. Any subset of a countable set is countable (see 3.68); thus any "sufficiently small" set is countable.

A set is *uncountable* if it is not countable — i.e., if it is infinite and it cannot be arranged into a sequence. Any superset of an uncountable set is uncountable; thus any "sufficiently large" set is uncountable.

In this book we will mainly be concerned with countable sets. But to make proper use of their countability, we will need to have some further understanding (beyond mere definitions) of the properties of countable sets, and of the differences between countable and uncountable sets. Those are the topics of the next few pages.

3.67. The set of all finite sequences of positive integers is countable.

Proof. We list the finite sequences in a chart. The nth row of the chart contains 2^{n-1} of the finite sequences — namely, those finite sequences whose terms add up to n. For instance, in the third row, $1 + 1 + 1 = 1 + 2 = 2 + 1 = 3$.

(1),

(1,1), (2),

(1,1,1), (1,2), (2,1), (3),

(1,1,1,1), (1,1,2), (1,2,1), (1,3), (2,1,1), (2,2), (3,1), (4),

(1,1,1,1,1), (1,1,1,2), (1,1,2,1), (1,1,3), (1,2,1,1),
(1,2,2), (1,3,1), (1,4), (2,1,1,1), (2,1,2), (2,2,1), (2,3),
(3,1,1), (3,2), (4,1), (5),

... .

Of course, the members of each row must be ordered in some fashion within that row. For that purpose, we have used lexicographical order — i.e., the analogue of dictionary order, with $a, b, c, \ldots$ replaced by $1, 2, 3, \ldots$, but where (like Dr. Seuss) we don't stop at z. For instance, row 3 consists of (1,1,1), (1,2), (2,1), (3), which translates to aaa, ab, ba, c. Those four "words" are in dictionary order.

We emphasize that each row is ordered by dictionary order *separately*. If we put the entire set of *all* finite sequences into one dictionary order together, it would not be in one-to-one correspondence with the positive integers. Indeed, such a "list" would never get to (2), because there would be infinitely many items before it: (1), (1,1), (1,1,1), (1,1,1,1), ..., as well as (1,2) and (1,2,1) and many others.

3.68. If X is a countable set and $Y \subseteq X$, then Y is countable too.

Proof. Write the members of X in a list. Then go through the list, crossing out any terms that aren't members of Y. What remains is a list of all the members of Y.

3.69. *Corollaries*
  a. The set of all ordered pairs of positive integers is countable. (*Hint:* Combine the two previous theorems.)
  b. The set of all rational numbers is countable. (*Hint:* Combine 3.69.a, 3.68, and the alternating signs trick in 3.66.)
  c. The union of countably many countable sets is countable. Hence

  > any uncountable set is much, much, much bigger than any countable set!

d. The set of all finite sequences of keystrokes on a conventional computer keyboard is countable. (*Hint.* Put different numbers on the keys; use 3.67.)

e. The set of all paragraphs that can be written in English is countable.

(Here we must consider English as though it were a fixed language; we must disregard the ways that English is always changing. Also, we take it as understood that each paragraph by itself is finite in length, though there are infinitely many different paragraphs and there is no finite upper bound for the lengths of all of the paragraphs. Now view each paragraph as a sequence of keystrokes on a conventional keyboard.)

f. The collection of all mathematical objects that can be described in English (by a finite sequence of paragraphs) is countable.

(We do not view that collection itself as a "mathematical object." Our terminology is not precise, nor is our conclusion; this "result" will not be used in later proofs. We're just trying to convey some intuition at this point, leading up to the surprising conclusion in 3.71. To make any sense of this statement, we have to use a language that prohibits self-referencing; but ordinary English actually permits self-referencing.)

3.70. The set of all real numbers is *not* countable.

Hints. Suppose it were countable; then the interval $(0,1)$ would also be countable. Arrange the members of $(0,1)$ into a sequence $s_1, s_2, s_3, \ldots$. Then form a number $r \in (0,1)$ whose kth digit after the decimal point is different from the kth digit in s_k. Choose r also so that its digits are not 9's or 0's (thus avoiding things like $0.379999\cdots = 0.3840000\cdots$). Thus r is not in the sequence, a contradiction.

Remarks. Actually, it can be shown that $|\mathcal{P}(\mathbb{N})| = |[0,1]| = |\mathbb{R}|$, but we will not need that fact in this book. In one of the proofs of that fact, one of the interesting steps is the mapping

$$f(S) = \sum_{j \in S} 2^{-j},$$

which maps each set $S \subseteq \mathbb{N}$ to a number $f(S) \in [0,1]$. That mapping is "almost" a one-to-one correspondence: Its range is all of $[0,1]$, and there are only countably many members of $[0,1]$ that correspond to two S's.

3.71. *Remarks.* The set of all real numbers, and indeed many other mathematical sets of interest, are uncountable. Nevertheless, as noted above, only countably many of them can be described. Thus,

"most" mathematical objects cannot be described.

We permit "descriptions" to be quite complicated. For instance, the number $1.634 + e^{6/7} + \frac{1}{2}\sqrt{2}$ can be described; we merely have to translate the expression "$1.634 + e^{6/7} + \frac{1}{2}\sqrt{2}$" into paragraph form. Such a number has an individual personality of its own. But there are only countably many such descriptions. Most real numbers cannot be described even in such a complicated fashion. They are like unfamiliar and unrecognized faces in a crowd, or like extras in a cinema background.

All concrete mathematical examples use describable numbers, but most abstract mathematical principles (such as $x+y = y+x$) apply equally well to describable numbers and to indescribable numbers; they do not mention the distinction. That general observation applies to this book as well; describability will not be investigated further in this book. However, we briefly remark that it is possible to reason about describability. If we assume that we have fixed some version of English — i.e., that we are no longer permitting our language to change — then it is meaningful to start a discussion by saying something like "suppose x is one of the indescribable real numbers." We can also give a partial description of a number — for instance, "suppose x is an indescribable number that is less than 5" narrows things down a bit, but still leaves uncountably many candidates.

Infinite sets (postponable)

3.72. *Cantor's powerset theorem.* $|\mathcal{P}(X)| > |X|$, for any set X.

Proof. It is easy to see that $|\mathcal{P}(X)| \geq |X|$, since the map $f(x) = \{x\}$ sends different members of X to different subsets of X. We will show that no function can give a one-to-one correspondence between X and $\mathcal{P}(X)$ — in fact, we shall show that no function defined on X can have range equal to $\mathcal{P}(X)$. Indeed, let $b : X \to \mathcal{P}(X)$ be any function. Define the set

$$S_0 = \{x \in X : x \notin b(x)\}.$$

It suffices to show that $S_0 \notin \text{Range}(b)$ — i.e., it suffices to show that for each $x \in X$, we have $b(x) \neq S_0$. We consider two cases:

- if $x \in S_0$ then $x \notin b(x)$,
- if $x \notin S_0$ then $x \in b(x)$,

in each case using just the definition of S_0. Hence, in each of the two cases, x is a member of one of the sets $b(x)$ and S_0 but not a member of the other set, and therefore those two sets are different.

Remark. Russell's Paradox (in 1901) was inspired by Cantor's proof (published in 1891). Note the resemblance between the set S_0 above and the set N in 3.11.

3.73. *Corollary.* $\mathcal{P}(\mathbb{N})$, the set of all subsets of $\mathbb{N}$, is uncountable. However, (*exercise*) the set of all finite subsets of $\mathbb{N}$ is countable.

3.74. *Remarks.* By Cantor's powerset theorem, we have $|\mathbb{N}| < |\mathcal{P}(\mathbb{N})| < |\mathcal{P}(\mathcal{P}(\mathbb{N}))| < |\mathcal{P}(\mathcal{P}(\mathcal{P}(\mathbb{N})))| < \cdots$. Hence

there are infinitely many different kinds of infinities.

That begs the question: Which kind of infinity is the kind we use when we talk about *how many* kinds of infinities there are? Trying to make sense out that question is one of the things set theorists do. However, most other mathematicians have few uses for any sets larger than $\mathcal{P}(\mathcal{P}(\mathbb{N}))$.

Chapter 4

Topologies and interiors (postponable)

4.1. *Preview.* In classical, two-valued logic, either a thing is or it is not. Thus $P \vee \overline{P}$ is a tautology of classical logic, known as the *Law of the Excluded Middle.* This viewpoint is reflected very simply by set theory: For any set S and any point x, either x is a member of S or it is a member of the complement of S. Thus $S \cup \complement S = \Omega$.

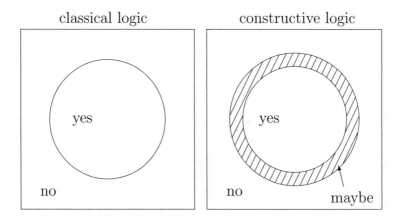

But in constructive logic, $P \vee \overline{P}$ is *not* a theorem. Between "yes" and "no" there is a *boundary* region of "maybe" (or "we don't know yet"). The middle case is *not* excluded. The middle case is not used as a semantic value, but the cases of "yes" and "no" do not make up everything. This philosophical or logical idea can be represented readily by topology, which is like set theory with boundaries. We represent "yes" (a proposition S) and "no" (its negation $\neg S$) as the sets that remain after the boundary is subtracted. The negation $\neg S$ may be strictly smaller than the complement $\complement S$, and so $S \cup \neg S$ might not be equal to Ω.

Topology, introduced in this chapter, is a natural continuation of set theory. However, the results in this chapter are more specialized and can be postponed; these results will not be needed until Chapter 10. In fact, this chapter and Chapter 10 could both be postponed further; their combined results will not be needed until 22.14.

Remarks. Topology is a major branch of mathematics in its own right. Students majoring in mathematics generally take at least one course devoted entirely to topology, covering far more than what is in this chapter. Logic is not the only use for topology, nor even its most common use. Topology is better known for notions related to the following.

- *Limits and continuity.* A function $f : \Omega_1 \to \Omega_2$, from one topological space (Ω_1, Σ_1) into another topological space (Ω_2, Σ_2), is *continuous* if for each $S \in \Sigma_2$ we have $f^{-1}(S) \in \Sigma_1$. (That may not look much like the definition of continuity that you learned in calculus, but it turns out to be equivalent to that definition, if you use the topology given in 4.7.)

- *Shapes.* A topological space (Ω, Σ) is *connected* if $\varnothing$ and Ω are the only clopen subsets of Ω (see definition in 4.4); otherwise the space is *disconnected*. This rather abstract definition does give the right pictures, at least in the case of topological spaces that are simple enough to have meaningful pictures.

TOPOLOGIES

4.2. The following definition is rather abstract. Students should glance ahead to the examples, which begin in 4.6.

Definition. Let Ω be any nonempty set. A *topology* on Ω is a collection $\Sigma \subseteq \mathcal{P}(\Omega)$ (that is, a collection Σ of subsets of Ω) satisfying these requirements:

a. $\varnothing$ and Ω are members of Σ;

b. any intersection of two members of Σ is also a member of Σ; and

c. any union of arbitrarily many members of Σ is also a member of Σ.

In principle, any letters could be used for a set and a topology on that set. However, in this book I will generally use Ω and Σ (Greek uppercase omega and sigma) for those purposes.

4.3. The preceding definition often will be used in this fashion: We will be given some particular set Ω and some particular collection Σ of subsets of Ω, and we'll need to verify that Σ is a topology on Ω, so that we can apply some general principles about topologies. In some cases that verification will be hard, and in other cases it will be easy.

Characterization of finite topologies. Checking whether Σ is a topology turns out to be particularly easy in those cases where Σ is a *finite* set, because under that additional assumption we can simplify condition 4.2.c to

>Any union of two members of Σ is also a member of Σ.

The justification for this simplification is left as an exercise for the reader. Here is a hint: $S_1 \cup S_2 \cup S_3$ is the union of the *two* sets $S_1 \cup S_2$ and S_3.

4.4. *Further remarks and further definitions.* When a set Ω is equipped with a topology Σ, then:

- The pair (Ω, Σ) is called a *topological space*. The set Ω itself may also be referred to as the "topological space," if the choice of Σ is understood.

- The members of Σ are said to be the *open* subsets of Ω.

- In the theory of topology, a set S is said to be *closed* if its complement is open — that is, if the set $\Omega \setminus S$ is a member of Σ. A set is *clopen* if it is both open and closed. Note that $\varnothing$ and Ω are both clopen subsets of any topological space. In some topological spaces there are also other clopen subsets.

  (We won't actually do anything with closed sets in this book, but I mention the definition of "closed" to avoid students' misusing that word.

Unless I mention it, students tend to use "closed" to mean "not open," and that grates terribly on my ears. Among topologists, "closed" does *not* mean "not open." In fact, in most topological spaces of interest, most subsets are neither open nor closed.)

It should be emphasized that this terminology depends on our having already selected some particular topology Σ. Usually we will study only one topology at a time. However, occasionally we may study the relationships between several different topologies $\Sigma_1, \Sigma_2, \Sigma_3, \ldots$ on the same set Ω — that is, several different collections Σ of subsets of Ω, each one of which satisfies the axioms in 4.2 when considered by itself. Then we may have to qualify the term "open set" in some fashion. For instance, if $\Sigma_1 \subseteq \Sigma_2$, we might say that "every Σ_1-open set is also a Σ_2-open set."

4.5. The term "closed" has another meaning in mathematics, unrelated to topology. We say that a set is closed under some operation(s) if applying the operation(s) to members of that set yields only members of that set. This meaning of "closed" will be used in 10.9 and 29.3.

4.6. *Elementary examples of topologies*
  a. Let $\Omega = \{4, 7, 9\}$. Then

$$\Sigma = \Big\{ \varnothing, \{7\}, \{9\}, \{4,7\}, \{7,9\}, \{4,7,9\} \Big\}$$

is a typical topology on Ω — indeed, one picked almost at random, from among the many topologies possible on Ω. If we use this topology, then the four sets $\varnothing, \{9\}, \{4,7\}, \{4,7,9\}$ are clopen; the two sets $\{7\}, \{7,9\}$ are open but not closed; and the two sets $\{4\}, \{4,9\}$ are closed but not open.

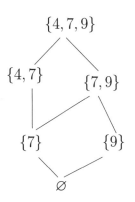

The accompanying diagram shows the inclusion relations in this topology — i.e., the diagram shows all the open sets, and a downward path goes from any open set to any open subset of that set. Compare with 4.6.i.

b. Let $\Omega = \{1, 2, 3, 4\}$. There are many different topologies with which we can equip Ω. A typical one is

$$\Sigma = \Big\{ \varnothing, \{1\}, \{2\}, \{1,2\}, \{2,3\}, \{1,2,3\}, \{1,2,3,4\} \Big\}.$$

Again, there is no evident pattern; the subsets were chosen almost at random. If we use this topology, then the seven sets $\varnothing, \{1\}, \{2\}, \{1,2\}, \{2,3\}, \{1,2,3\}, \{1,2,3,4\}$ are the open sets, and the other nine subsets of Ω are not open. In this example, $\{2,4\}$ is neither open nor closed.

c. Again let $\Omega = \{1, 2, 3, 4\}$. An arbitrarily chosen collection of subsets of Ω might or might not be a topology. For instance, the collection of sets

$$\mathcal{C} = \Big\{ \varnothing, \{1\}, \{2\}, \{1,2\}, \{2,3\}, \{3,4\}, \{1,2,3\}, \Omega \Big\}$$

is *not* a topology on Ω. It violates rule 4.2.c. Indeed, $\{2,3\}$ and $\{3,4\}$ are members of $\mathcal{C}$, but their union, $\{2,3,4\}$, is not a member of $\mathcal{C}$.

d. *Exercise.* Show that $\Big\{ \varnothing, \{1\}, \{2\}, \{1,3\}, \{1,2,3\} \Big\}$ is not a topology on the set $\{1, 2, 3\}$.

e. If Ω is any set, then $\{\varnothing, \Omega\}$ is a topology on Ω. It is called the *indiscrete topology*. It is the smallest possible topology on Ω. If we use this topology, then the empty set and the whole space are the only open sets.

Example. If $\Omega = \{1, 2, 3, 4\}$, then the indiscrete topology on Ω is the doubleton $\{\varnothing, \{1,2,3,4\}\}$.

f. If Ω is any set, then $\mathcal{P}(\Omega) = \{\text{subsets of } \Omega\}$ is also a topology on Ω. It is called the *discrete topology*. It is the largest possible topology on Ω. If we use this topology, then every subset of Ω is an open set.

g. A set $S \subseteq \mathbb{N}$ is a *lower set* if it has the property that whenever a number x is a member of S, then any number lower than x is also in S. The lower sets form a topology on $\mathbb{N}$, which we may refer to as the *lower set topology*:

$$\Sigma = \Big\{ \varnothing, \{1\}, \{1,2\}, \{1,2,3\}, \{1,2,3,4\}, \ldots, \mathbb{N} \Big\}.$$

Topologies

h. Similarly, a set $S \subseteq \mathbb{N}$ is an *upper set* if it has the property that whenever a number x is a member of S, then any number higher than x is also in S. The upper sets form a topology on $\mathbb{N}$, which we shall call the *upper set topology*:

$$\Sigma = \Big\{ \mathbb{N}, \{2,3,4,\ldots\}, \{3,4,5,\ldots\}, \{4,5,6,\ldots\}, \ldots, \varnothing \Big\}.$$

i. Let $\Omega = \mathbb{N} = \{\text{natural numbers}\}$, and let

$$\Sigma = \Big\{ \mathbb{N}, \varnothing, \{\text{even numbers}\}, \{\text{multiples of 4}\},$$
$$\{\text{odd numbers}\}, \{\text{multiples of 4}\} \cup \{\text{odd numbers}\} \Big\}.$$

Note that some of the members of Σ are infinite sets. Nevertheless, we shall call Σ a *finite topology*, since the Σ itself has only finitely many members — i.e., there are only finitely many open sets. The inclusion diagram is shown below.

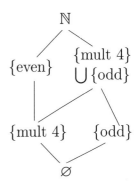

Remarks. Note that the inclusion diagrams in 4.6.a and 4.6.i have the same arrangement of lines. Actually, those two topologies determine the same logic (via a procedure studied in Chapter 10). The logic depends only on the arrangement of inclusions, not on the particular sets and their points.

Consequently, researchers in logic prefer to work with just the arrangement of lines, or equivalently with the abstract algebraic structure represented by those lines. That structure is a special kind of poset (see 3.20); it is known as a *frame* or a *complete Heyting algebra*.

However, the "pointless" approach is more abstract and harder to define; it requires greater knowledge of advanced algebra. Consequently, topologies may be more suitable for beginners in logic, despite the superfluous baggage of the particular sets and their points.

4.7. The *usual topology on the reals* is the collection of those sets $S \subseteq \mathbb{R}$ that satisfy this condition:

S is a union of some (or all, or none) of the intervals of the form (a, b) — i.e., intervals that do not contain their own endpoints.

The real line is always be understood to be equipped with this topology, except on those rather infrequent occasions when some other topology on $\mathbb{R}$ is specified.

How do we prove that the collection of all such unions *is* a topology? *Sketch of proof.* It is easy to see that the collection satisfies 4.2.a and 4.2.c; we just need to verify 4.2.b. Let S_1 and S_2 be two such unions of boundaryless intervals; we must show that $S = S_1 \cap S_2$ is another such union. Temporarily fix any $x \in S$. By the given properties of S_1 and S_2, we know that there exist intervals (a_1, b_1) and (a_2, b_2) such that $x \in (a_1, b_1) \subseteq S_1$ and $x \in (a_2, b_2) \subseteq S_2$. Let $a = \max\{a_1, a_2\}$ and $b = \min\{b_1, b_2\}$; then we can verify that $x \in (a, b) \subseteq S$. Now let x vary over all points in S; it is easy to see that S is the union of the resulting intervals (a, b).

Examples. If $\mathbb{R}$ has its usual topology, then these sets are open:

$$\emptyset, \quad \mathbb{R}, \quad (0,1) \cup (3,5), \quad (-\infty, 5), \quad (0, \infty),$$

$$(-\infty, 0) \cup (0, \infty), \quad (0,1) \cup (2,3) \cup (4,5) \cup \cdots,$$

but these sets are not open:

$$[0, 1), \quad [0, 1], \quad [0, \infty).$$

4.8. The *usual topology on the plane* is similar to that of the line, but it has the advantage that we can draw pictures in $\mathbb{R}^2$ that are easier to understand.

Recall that the disk centered at the point (a, b), with radius r, and without the boundary of the disk, is the set

$$\{(x, y) \in \mathbb{R}^2 \;:\; (x-a)^2 + (y-b)^2 < r^2\}.$$

The usual topology on the plane is the collection of those sets $S \subseteq \mathbb{R}^2$ that satisfy this condition:

S is a union of (some/all/none) boundaryless disks.

This collection can be proved to be a topology, by an argument similar to that given in 4.7.

Though we shall not prove this characterization, it can be shown that

Interiors

among subsets of $\mathbb{R}^2$ that are simple to draw, a set is open if and only if it includes none of its boundary points.

That is illustrated by the pictures below. (That criterion also applies to figures that are *not* simple enough to draw, if we use the complicated but precise definition of "boundary" given in 4.18.)

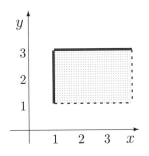

$\{(x,y) : 1 \leq x < 4,$
$1 < y \leq 3\}$ is *not* open

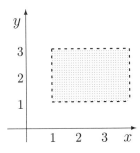

$\{(x,y) : 1 < x < 4,$
$1 < y < 3\}$ *is* open

4.9. *Exercise.* How many different topologies are there on the set $\{1,2\}$? List them.

Harder exercise. How many different topologies are there on the set $\{1,2,3\}$? List them.

INTERIORS

4.10. *Definition.* Let Σ be a topology on a set Ω. Suppose that S is some subset of Ω, not necessarily open. Then S has some open subsets (for instance, $\emptyset$ is an open subset of S). The *interior* of S, denoted by $\operatorname{int}(S)$, is defined to be

the union of all the open subsets of S.

By 4.2.c, we see that int(S) is itself an open subset of S. Hence int(S) is also equal to

$$\text{the largest of all the open subsets of } S$$

(where "larger" is understood to mean "a superset of," not just "having more members" — see 3.14).

4.11. *Examples*

a. Suppose Ω and Σ are as in 4.6.b.

  (i) What is the interior of the set $\{2,3,4\}$? — That set is not open, but some of its subsets are open. Specifically, it has $\varnothing, \{2\}, \{2,3\}$ as open subsets. Among these, $\{2,3\}$ is largest. Thus $\text{int}(\{2,3,4\}) = \{2,3\}$.

  (ii) On the other hand, what is the interior of $\{2,3\}$? — The set $\{2,3\}$ is open, so among the open subsets of $\{2,3\}$, the largest is $\{2,3\}$. Therefore $\text{int}(\{2,3\}) = \{2,3\}$.

  Exercise. List all 16 subsets of Ω, and give the interior of each. I've already done 2 of the 16 for you, in the paragraphs above.

b. To get the idea across, let's cheat a little: Let's use Ω and $\mathcal{C}$ as in 4.6.c; let's pretend that $\mathcal{C}$ is a topology on that Ω. Let $S = \{2,3,4\}$. What is then the "interior" of S? The open subsets of S are the sets $\varnothing, \{2\}, \{2,3\}, \{3,4\}$. There is no "largest" among these sets — i.e., no one of these sets is a superset of all the others. Thus our definition of int(S) doesn't work. That's because — as we noted in 4.6.c — the collection $\mathcal{C}$ that we're using is not a topology.

c. Let $\mathbb{R}$ be equipped with its usual topology. Let $S \subseteq \mathbb{R}$. Then it can be shown that $x \in \text{int}(S)$ if and only if there exists an interval (a,b) such that $x \in (a,b) \subseteq S$. Using that criterion, it is easy to see that:

  (i) The numbers $\frac{1}{2}, \frac{2}{3}, \frac{1}{4}\sqrt{2}, 0.9998$ are in the interior of the set $[0,1)$, but the numbers 0, 1, 5 are not in the interior of that set. In fact, $\text{int}\big([0,1)\big) = (0,1)$.

(ii) The interior of the set $\{0, 1, 2\}$ is the empty set. That is, $\text{int}(\{0, 1, 2\}) = \emptyset$.

(iii) $\text{int}(\mathbb{Q}) = \emptyset$.

(iv) $\text{int}(\mathbb{R}) = \mathbb{R}$.

(v) The interior of the set $[3,7] \cup [11, 17]$ is the set $(3,7) \cup (11, 17)$.

d. Again, the topology of the plane (introduced in 4.8) provides good pictures. In the case of regions that are simple to draw, the *interior* of a region in the plane is the *inside* of that region. For instance, refer to the picture at the end of 4.8. The region in the right-hand graph is equal to the interior of the region of the left-hand graph.

4.12. Exercise. Let $\mathbb{R}$ have its usual topology.

a. Give the interior of each of the following sets:

$$(1,5), \quad \{1,5\}, \quad [1,5], \quad \mathbb{Z}, \quad [1,5] \cup (7,11), \quad \mathbb{R} \setminus \mathbb{Q}.$$

b. Let $S = \mathbb{Q}$ and $T = \mathbb{R} \setminus \mathbb{Q}$. Find both of the sets $\text{int}(S \cup T)$ and $\text{int}(S) \cup \text{int}(T)$.

c. Find both of the sets

$$\bigcap_{q \in \mathbb{Q}} \text{int}(\mathbb{R} \setminus \{q\}) \quad \text{and} \quad \text{int}\left(\bigcap_{q \in \mathbb{Q}} (\mathbb{R} \setminus \{q\})\right).$$

Some students may find that confusing; here is some clarification.

Start with, say, 3/4. It is a rational number. The set $\mathbb{R} \setminus \{3/4\}$ is the set of all real numbers other than 3/4. You should find the interior of that set. Likewise, find the interior of $\mathbb{R} \setminus \{-7/12\}$ and the interior of $\mathbb{R} \setminus \{100/3\}$. Continue in this fashion, finding the interior of each set of the form $\mathbb{R} \setminus \{q\}$, where q is any rational number. After you've found all those interiors, take the intersection of those interiors. That's the first of the two sets you're being asked to find.

On the other hand: Find the intersection of the sets $\mathbb{R} \setminus \{3/4\}$ and $\mathbb{R} \setminus \{-7/12\}$ and $\mathbb{R} \setminus \{100/3\}$ and $\cdots$. Then find the interior of that intersection.

4.13. *Further properties of interiors.* Let Ω be a space equipped with some topology Σ. Then:

a. $S \supseteq \mathrm{int}(S)$. That is, every set is a superset of its interior.

b. $\mathrm{int}(S) = S$ if and only if the set S is open.

c. $\mathrm{int}(\Omega) = \Omega$. More generally, suppose $S \subseteq \Omega$; then $\mathrm{int}(S) = \Omega$ if and only if $S = \Omega$.

d. $\mathrm{int}(\complement S) = \Omega$ if and only if $S = \emptyset$.

e. $\mathrm{int}\bigl(\mathrm{int}(S)\bigr) = \mathrm{int}(S)$. That is, repeating the interior operator has no further effect.

f. The interior operator is *order preserving*. That is, whenever S and T are two subsets of Ω,

$$\text{if} \quad S \subseteq T, \quad \text{then} \quad \mathrm{int}(S) \subseteq \mathrm{int}(T).$$

g. $\mathrm{int}\bigl(\complement(S \setminus T)\bigr) = \Omega$ if and only if $S \subseteq T$.

h. $\mathrm{int}(S) \subseteq T$ if and only if $\mathrm{int}(S) \subseteq \mathrm{int}(T)$.

i. $\mathrm{int}(S) \subseteq \mathrm{int}(T)$ does *not* imply $S \subseteq T$.

For instance, let $\mathbb{R}$ have its usual topology. When $S = \mathbb{Q}$ and $T = \mathbb{Z}$, what happens?

j. In any topological space: $x \in \mathrm{int}(S)$ if and only if there is some open set G such that $x \in G \subseteq S$.

4.14. If S and T are any two subsets of a topological space Ω, then

$$\mathrm{int}(S \cup T) \supseteq \mathrm{int}(S) \cup \mathrm{int}(T), \qquad \mathrm{int}(S \cap T) = \mathrm{int}(S) \cap \mathrm{int}(T).$$

Sketch of proof. By 4.13.f we know that $\mathrm{int}(S \cup T) \supseteq \mathrm{int}(S)$. The same reasoning also shows $\mathrm{int}(S \cup T) \supseteq \mathrm{int}(T)$. Combining those two facts proves $\mathrm{int}(S \cup T) \supseteq \mathrm{int}(S) \cup \mathrm{int}(T)$.

By 4.13.f we know that $\mathrm{int}(S \cap T) \subseteq \mathrm{int}(S)$. The same reasoning also shows $\mathrm{int}(S \cap T) \subseteq \mathrm{int}(T)$. Combining those two facts proves $\mathrm{int}(S \cap T) \subseteq \mathrm{int}(S) \cap \mathrm{int}(T)$.

On the other hand, we know that $\mathrm{int}(S) \subseteq S$ and $\mathrm{int}(T) \subseteq T$ by 4.13.a. From those facts it follows that $\mathrm{int}(S) \cap \mathrm{int}(T) \subseteq S \cap T$. Now, $\mathrm{int}(S) \cap \mathrm{int}(T)$ is the intersection of two open sets, so it is

Interiors

an open set by 4.2.b. Thus it is an open subset of $S \cap T$. But $\mathrm{int}(S \cap T)$ is the *largest* open subset of $S \cap T$, so $\mathrm{int}(S) \cap \mathrm{int}(T) \subseteq \mathrm{int}(S \cap T)$.

4.15. Let Λ be some index set. For each $\lambda \in \Lambda$, suppose that S_λ is some subset of a topological space Ω. Then

$$\mathrm{int}\left(\bigcap_{\lambda \in \Lambda} S_\lambda\right) \subseteq \bigcap_{\lambda \in \Lambda} \mathrm{int}(S_\lambda), \qquad \bigcup_{\lambda \in \Lambda} \mathrm{int}(S_\lambda) \subseteq \mathrm{int}\left(\bigcup_{\lambda \in \Lambda} S_\lambda\right).$$

Remarks. We omit the proof, which is similar to that given in 4.14. However, note that the intersection result in 4.14 actually was equality; for the intersection of infinitely many sets we do not assert equality. Why doesn't the equality proof in 4.14 work for infinitely many sets?

4.16. *Exercises on comparing interiors*

a. Consider $\Omega = \mathbb{R}$ as a topological space; then $[0, 1)$ is a subset of that space. Find $\mathrm{int}\big([0, 1)\big)$ if the topology Σ is

(i) the discrete topology (4.6.f),

(ii) the indiscrete topology (4.6.e),

(iii) the usual topology on $\mathbb{R}$ (4.7).

(Those three questions have different answers.)

b. Consider $\Omega = \mathbb{N}$ as a topological space; then $\{1, 2, 5\}$ is a subset of that space. Find $\mathrm{int}(\{1, 2, 5\})$ if the topology Σ is

(i) the discrete topology (4.6.f),

(ii) the indiscrete topology (4.6.e),

(iii) the lower set topology (4.6.g).

(Those three questions have different answers.)

4.17. Technical lemma. (This will be needed in 10.7.c.) Let G and H be subsets of a topological space Ω. Assume the set G is open. Then

$$\mathrm{int}\bigl(\mathrm{int}(H) \cup \complement G\bigr) \;=\; \mathrm{int}\bigl(H \cup \complement G\bigr).$$

Remark. This lemma fails if we drop the assumption that G is open — for instance, if Ω is $\mathbb{R}$ with its usual topology, and $G = H = \mathbb{Q}$.

Proof of lemma. Proving an inclusion in one direction is fairly easy: We have $\mathrm{int}(H) \subseteq H$, hence $\mathrm{int}(H) \cup \complement G \subseteq H \cup \complement G$. Taking interiors on both sides of that inclusion (using 4.13.f) yields

$$\mathrm{int}\bigl(\mathrm{int}(H) \cup \complement G\bigr) \;\subseteq\; \mathrm{int}(H \cup \complement G).$$

The inclusion in the other direction is much harder. Reason as follows: Since the set G is open,

(1) $\qquad\qquad\qquad \mathrm{int}(G) \;=\; G.$

Also, by elementary set theory (without regard to topology),

(2) $\qquad\qquad\qquad (H \cup \complement G) \cap G \;\subseteq\; H.$

Then

(3) $\quad \mathrm{int}(H \cup \complement G) \cap G \;\stackrel{(1)}{=\!=}\; \mathrm{int}(H \cup \complement G) \cap \mathrm{int}(G)$
$\qquad\qquad\qquad\qquad \stackrel{4.14}{=\!=}\; \mathrm{int}\bigl((H \cup \complement G) \cap G\bigr) \;\stackrel{(2),\,4.13.f}{\subseteq}\; \mathrm{int}(H).$

From (3) and 3.55.b we obtain $\mathrm{int}(H \cup \complement G) \subseteq \mathrm{int}(H) \cup \complement G$. Take the interior of both sides of that inclusion; also use 4.13.e. Thus we obtain

$$\mathrm{int}\bigl(H \cup \complement G\bigr) \;\subseteq\; \mathrm{int}\bigl(\mathrm{int}(H) \cup \complement G\bigr),$$

completing the proof of the lemma.

4.18. Optional. We won't actually use exteriors or boundaries in the rest of this book, but boundaries were used for motivation in 4.1. Just in case you're interested, here is the precise definition: Let Ω be a topological space, and let $\complement$ denote complementation in Ω. Let S be a subset of Ω (not necessarily open). Then the *exterior* and *boundary* of S are the sets

$$\mathrm{ext}(S) = \mathrm{int}(\complement S), \qquad \mathrm{bdry}(S) = \complement\bigl(\mathrm{int}(S) \cup \mathrm{ext}(S)\bigr).$$

The exterior is always an open set; the boundary is always a closed set (i.e., the complement of an open set). In the case of regions in $\mathbb{R}^2$ that are simple to draw, the mathematical meanings of these terms is essentially the same as the English, nonmathematical meaning — i.e., the "exterior" is the "outside," and the "boundary" is everything "between" the interior and exterior.

GENERATED TOPOLOGIES AND FINITE TOPOLOGIES (OPTIONAL)

This subchapter is a bit technical. It can be postponed, or skipped altogether if one is willing to skip the proof of the optional result 10.11.

4.19. If $\{\Sigma_\lambda : \lambda \in \Lambda\}$ is a collection of topologies on a set Ω, then $\bigcap_{\lambda \in \Lambda} \Sigma_\lambda$ is also a topology on the set Ω. It is called the *inf topology* or *infimum topology* — that is, the infimum of the Σ_λ's. It is the largest topology that is smaller than all the Σ_λ's. Note that a set is open for the infimum topology if and only if it is open for every one of the Σ_λ's.

4.20. Let Ω be a set, and let $\mathcal{C}$ be some collection of subsets of Ω. Then $\mathcal{C}$ is not necessarily a topology, but there do exist some topologies Σ that contain $\mathcal{C}$ (i.e., that satisfy $\Sigma \supseteq \mathcal{C}$), and among all such topologies there is a smallest. It is called the *topology generated* by $\mathcal{C}$. One easy way to specify the generated topology (and also prove its existence) is as follows: Just take the infimum of all the topologies that contain $\mathcal{C}$ (as in 4.19); it is easy to verify that this topology has the right properties.

That existence proof, though mathematically sound, is somewhat indirect and hard to visualize. Following is a second existence proof, which is more constructive and yields more information. It is in two steps:

1. Let $\mathcal{D} = \{S : S \text{ is the intersection of finitely many members of } \mathcal{C}\}$. Here we adopt the convention that "finitely many" might be just one. The "intersection" of one member of $\mathcal{C}$ is just that member; hence $\mathcal{C} \subseteq \mathcal{D}$. Also we adopt the convention that the intersection of *no* subsets of Ω is just the set Ω (as explained at the end of 3.40), so we have $\Omega \in \mathcal{D}$.

2. Let $\Sigma_g = \{T : T \text{ is the union of arbitrarily many members of } \mathcal{D}\}$. Here it is understood that the "union" of one member of $\mathcal{D}$ is just that member of $\mathcal{D}$, so we have $\mathcal{D} \subseteq \Sigma_g$. Also it is understood that the union of *no* sets is $\varnothing$, so we have $\varnothing \in \Sigma_g$.

Then it can be proved that the set Σ_g constructed in this fashion must be a topology on Ω, and in fact Σ_g is the generated topology. This "construction" also makes it evident that

> if the given collection $\mathcal{C}$ is finite, then the intermediate collection $\mathcal{D}$ is finite, hence the generated topology Σ_g also finite.

4.21. *Example.* The collection $\mathcal{C}$ in 4.6.c is not a topology. What topology does it generate?

Follow the construction outlined above, with $\Omega = \{1, 2, 3, 4\}$. We start from

$$\mathcal{C} = \Big\{\varnothing, \{1\}, \{2\}, \{1,2\}, \{2,3\}, \{3,4\}, \{1,2,3\}, \Omega\Big\}.$$

Taking all intersections of those sets yields those same sets plus just one more set, $\{3\}$; thus we have

$$\mathcal{D} = \Big\{\varnothing, \{1\}, \{2\}, \{3\}, \{1,2\}, \{2,3\}, \{3,4\}, \{1,2,3\}, \Omega\Big\}.$$

Taking all unions of those sets yields those same sets plus $\{1,3\}$, $\{1,3,4\}$, $\{2,3,4\}$. Thus we have

$$\Sigma_g = \left\{\begin{array}{l} \varnothing, \{1\}, \{2\}, \{3\}, \{1,2\}, \{1,3\}, \{2,3\}, \{3,4\}, \\ \{1,2,3\}, \{1,3,4\}, \{2,3,4\}, \{1,2,3,4\} \end{array}\right\}.$$

That's all the subsets of $\{1, 2, 3, 4\}$ except for $\{4\}$, $\{1, 4\}$, $\{2, 4\}$, $\{1, 2, 4\}$.

4.22. *Another example.* The topology on $\mathbb{R}$ that is generated by the two sets

$$(-\infty, 0), \qquad (0, +\infty)$$

is this collection of five sets:

$$\Sigma = \Big\{\varnothing, (-\infty, 0), (0, +\infty), (-\infty, 0) \cup (0, +\infty), \mathbb{R}\Big\}.$$

Generated topologies and finite topologies (optional) 141

4.23. *A more complicated example.* Find the topology on $\mathbb{R}$ that is generated by the five sets

$$S_1 = (0,1), \qquad S_2 = (1,2), \qquad S_3 = \mathbb{R} \setminus [0,1],$$
$$S_4 = \mathbb{R} \setminus [1,2], \qquad S_5 = \mathbb{R} \setminus \{1\}.$$

Solution. First we list all the sets that can be obtained by taking the intersection of some of the S_j's; thus we obtain

$$\mathcal{D} \;=\; \Big\{ \mathbb{R}, \quad S_1, \quad S_2, \quad S_3, \quad S_4, \quad S_5, \quad \varnothing \quad S_3 \cap S_4 \Big\}.$$

(For reference we note that $S_3 \cap S_4 = \mathbb{R} \setminus [0,2]$.) Then we list all the sets that can be obtained as unions of some of the members of $\mathcal{D}$; those sets are the members of $\mathcal{D}$ plus the following sets:

$$(0,1) \cup (1,2), \qquad \mathbb{R} \setminus \{0,1\}, \qquad \mathbb{R} \setminus \{1,2\},$$
$$\mathbb{R} \setminus \{0,1,2\}, \qquad \mathbb{R} \setminus \big([0,1] \cup \{2\}\big), \qquad \mathbb{R} \setminus \big(\{0\} \cup [1,2]\big).$$

Thus, the resulting topology on $\mathbb{R}$ is this collection of 14 subsets of $\mathbb{R}$:

$$\Sigma_g = \left\{ \begin{array}{lllll} \varnothing, & (0,1), & (1,2), & \mathbb{R}, & \mathbb{R} \setminus \{1\}, \\ \mathbb{R} \setminus [0,1], & \mathbb{R} \setminus [1,2], & \mathbb{R} \setminus [0,2], & (0,1) \cup (1,2), \\ \mathbb{R} \setminus \{0,1\}, & \mathbb{R} \setminus \{1,2\}, & \mathbb{R} \setminus \{0,1,2\}, \\ \mathbb{R} \setminus \big([0,1] \cup \{2\}\big), & \mathbb{R} \setminus \big(\{0\} \cup [1,2]\big) \end{array} \right\}.$$

4.24. We saw in 4.16 that one set may have different interiors when considered under different topologies, in general. But we shall now show that, under special circumstances, different topologies may yield the *same* interior.

Principle on comparing topologies. Suppose Σ_1 and Σ_2 are topologies on the same set Ω, with interior operators int_1 and int_2. Suppose that $\Sigma_1 \subseteq \Sigma_2$. Let S be some particular subset of Ω (not necessarily open for either topology). Then:

a. $\text{int}_1(S) \subseteq \text{int}_2(S)$.

b. If the set $\text{int}_2(S)$ is a member of Σ_1, then $\text{int}_1(S) = \text{int}_2(S)$.

Proof of a. For $j = 1, 2$, recall that $\text{int}_j(S)$ is the largest Σ_j-open subset of S. Now, $\text{int}_1(S) \in \Sigma_1 \subseteq \Sigma_2$, so $\text{int}_1(S)$ is *one of* the Σ_2-open subsets of S. But $\text{int}_2(S)$ is the largest of those, so $\text{int}_1(S) \subseteq \text{int}_2(S)$.

Proof of b. If the set $\text{int}_2(S)$ happens to be a member of Σ_1, then it is one of the Σ_1-open subsets of S. But $\text{int}_1(S)$ is the largest of those, so $\text{int}_2(S) \subseteq \text{int}_1(S)$.

4.25. *Corollary.* Suppose Σ_i is some topology on a set Ω, and $\mathcal{C} \subseteq \Sigma_i$ is just *some* of the open sets from that topology. Let Σ_g be the topology on Ω generated by $\mathcal{C}$. (Hence $\Sigma_g \subseteq \Sigma_i$.) Let int_i and int_g be the interior operators for those two topologies. Let S be a subset of Ω, not necessarily open for either topology. If $\text{int}_i(S) \in \mathcal{C}$, then $\text{int}_i(S) = \text{int}_g(S)$.

Proof. This is a special case of 4.24.b; we have restated it separately merely because that will make the application a little simpler in 10.11.

4.26. *Finite representation principle.* Let (Ω, Σ) be a topological space, and suppose Σ is a finite set — i.e., the topology has only finitely many open sets. (We do not assume that the set Ω is finite.)

Then there exists a topological space $(\widehat{\Omega}, \widehat{\Sigma})$ in which the underlying set $\widehat{\Omega}$ is finite, and having the same inclusion diagram of open sets as (Ω, Σ) has.

(In other words, from any topology like that in 4.6.i we can go to a corresponding topology like that in 4.6.a.)

Abbreviated proof for algebraists. For any points $x, y \in \Omega$, say that x and y are "equivalent" if the topology Σ cannot distinguish between them — i.e., if there does not exist an open set containing one of x, y and not the other. Let $\widehat{\Omega}$ be the set of equivalence classes. Let $\widehat{\Sigma}$ be the topology obtained from Σ by collapsing each equivalence class down to a point.

That's essentially the whole proof, but it may be unsuitable for some intended readers of this book, as it is rather abstract and relies on a strong familiarity with equivalence relations. Accordingly, the longer proof given

Generated topologies and finite topologies (optional)

below gives a concrete representation $varphi(x)$ for the equivalence class containing x.

Proof. Suppose $\Sigma = \{G_1, G_2, \ldots, G_p\}$. That is, the open sets of the given topological space are $G_1, G_2, \ldots, G_p$. The order of this list does not matter, but the number p and the order of the list will remain fixed throughout the remainder of this proof. We emphasize that $\varnothing$ and Ω are two of the G_i's.

For each i, let $\mathbb{1}_{G_i}$ be the characteristic function of the set G_i (defined as in 3.27). Define a function $\varphi : \Omega \to \{0,1\}^p$ by

$$\varphi(x) = \Big(\mathbb{1}_{G_1}(x),\ \mathbb{1}_{G_2}(x),\ \mathbb{1}_{G_3}(x),\ \ldots,\ \mathbb{1}_{G_p}(x) \Big).$$

Thus, $\varphi(x)$ is a sequence of p zeros and ones. There are 2^p different such sequences, but not all those sequences are necessarily in the range of φ; thus the range of φ has 2^p or fewer members. Say it has M members, so

$$\widehat{\Omega} \overset{\text{def}}{=\!=} \operatorname{Range}(\varphi) = \{\lambda_1,\ \lambda_2,\ \lambda_3,\ \ldots,\ \lambda_M\}.$$

Now, for each i $(1 \leq i \leq p)$, define the set

$$\widehat{G_i} = \varphi(G_i) = \Big\{ \lambda_j \in \widehat{\Omega}\ :\ \lambda_j = \varphi(x) \text{ for some } x \in G_i \Big\}.$$

Then it can be verified that $\widehat{\Sigma} = \{\widehat{G_1}, \widehat{G_2}, \ldots, \widehat{G_p}\}$ is a topology on the set $\widehat{\Omega}$, with the same inclusion diagram as (Ω, Σ).

The preceding argument is rather abstract, but it may be clarified by a typical example. Start from the topology of 4.6.i. That is, $\Omega = \mathbb{N}$ has these six open subsets:

$G_1 = \mathbb{N}$, $\quad G_2 = \varnothing$, $\quad G_3 = \{\text{multiples of 4}\}$,
$G_4 = \{\text{even numbers}\}$, $\quad G_5 = \{\text{odd numbers}\}$,
$G_6 = \{\text{odd numbers}\} \cup \{\text{multiples of 4}\}$.

Thus $p = 6$. The Venn diagram below shows how these six sets are related to one another. G_2 is not represented in the Venn diagram because it is the empty set, and G_1 is the enclosing box since it is the universal set.

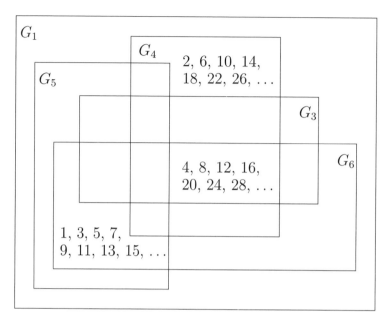

Though the set $\mathbb{N}$ has infinitely many members, they can only appear in $2^4 = 16$ different locations in the Venn diagram. Some of those locations are empty; for instance, $G_3 \cap G_4 \cap G_5 \cap (\complement G_6) = \varnothing$. On the other hand, some of those locations contain infinitely many members of $\mathbb{N}$; for instance,

$$2, 6, 10, 14, 18, 22, 26, \ldots \quad \in \quad (\complement G_3) \cap G_4 \cap (\complement G_5) \cap (\complement G_6).$$

The strategy of our proof is to replace all those numbers in one location with just one mathematical object. The resulting Venn diagram, shown below, will have the same shape; it simply replaces all the numbers in each location with a single mathematical object. Really, any mathematical object would do, but we want something that is easy to describe in the general case. One object that is well suited to this purpose is the sequence mapped by φ from all those numbers:

$$\varphi(2) = \varphi(6) = \varphi(10) = \varphi(14) = \cdots = (1, 0, 0, 0, 1, 1)$$

(or more briefly 100011). This equation holds because each number $x = 2, 6, 10, 14, \ldots$ satisfies

$\mathbb{1}_{G_1}(x) = 1$ since $x \in G_1$, $\quad \mathbb{1}_{G_2}(x) = 0$ since $x \in \complement G_2$,
$\mathbb{1}_{G_3}(x) = 0$ since $x \in \complement G_3$, $\quad \mathbb{1}_{G_4}(x) = 0$ since $x \in \complement G_4$,
$\mathbb{1}_{G_5}(x) = 1$ since $x \in G_5$, $\quad \mathbb{1}_{G_6}(x) = 1$ since $x \in G_6$.

Generated topologies and finite topologies (optional)

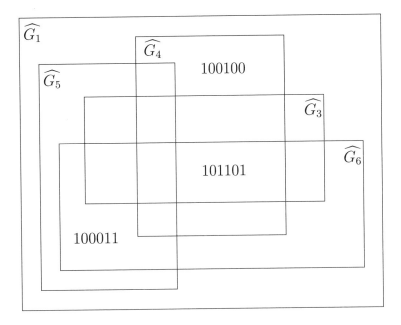

The sequence 100011 is different from the sequence obtained from any other location in the Venn diagram. For instance,

$$\varphi(1) = (\quad 1, \quad 0, \quad 0, \quad 0, \quad 1, \quad 1, \quad)$$
since $1 \in G_1$, $1 \in \complement G_2$, $1 \in \complement G_3$, $1 \in \complement G_4$, $1 \in G_5$, $1 \in G_6$.

There are $2^6 = 64$ different possible length-6 sequences of zeros and ones, but only three of those 64 sequences are in the range of φ, so $M = 3$ in this example. The resulting topology on $\widehat{\Omega}$ has these open sets:

$$\widehat{G_1} = \widehat{\Omega} = \{100100, 101101, 100011\}, \qquad \widehat{G_2} = \varnothing,$$
$$\widehat{G_3} = \{101101\}, \qquad \widehat{G_4} = \{100100, 101101\},$$
$$\widehat{G_5} = \{100011\}, \qquad \widehat{G_6} = \{101101, 100011\}.$$

Those sets are related as shown in the Venn diagram on the previous page.

Obviously, the three sequences 100100, 101101, and 100011 can be replaced by any three distinct objects. If we replace them with the numbers 4, 7, and 9 respectively, we get the example in 4.6.a.

Chapter 5

English and informal classical logic

5.1. Much of this chapter is intended merely as a preview of later chapters. Except where noted by exercises, material in this chapter is not intended to be mathematically precise. Assertions made in this preview should not be taken as proved, and will not be used as ingredients in later proofs.

Language and bias

5.2. As much as possible, the mathematical content of this book is independent of the language in which the material is explained. For instance, the mathematical and philosophical significance of the expression $P \vee \overline{P}$ do not depend on one's nationality or ethnic origin. However, our *method of explaining* the mathematics does depend on the language used. Some parts of this book, dealing specifically with the idiosyncrasies of English, will not translate directly to other languages. I apologize for my ignorance of those other languages.

5.3. *Is truth independent of language?* I have already said that *as much as possible*, the mathematical content of this book is independent of the language in which the material is explained. But is it possible to separate language entirely from mathematical truth?

Language and bias

For many years, linguists have studied the ways in which our perception and language shape each other. Benjamin Whorf was one of the pioneers in this research. In a foreword to some of Whorf's work (Whorf [1956]), Stuart Chase wrote:

> The Greeks took it for granted that back of language was a universal, uncontaminated essence of reason, shared by all men, at least by all thinkers. Words, they believed, were but the medium in which this deeper effulgence found expression. It followed that a line of thought expressed in any language could be translated without loss of meaning into any other language.
>
> This view has persisted for 2500 years, especially in academic groves. Whorf flatly challenges it in his second major hypothesis. "A change in language," he says, "can transform our appreciation of the Cosmos."

In the 1930s, Whorf was studying the language of the Hopi Indians. He found that they do not have nouns separate from verbs. Instead, their words describe space-time *events*, combining subject and verb, much as in Einsteinian relativity. Whorf concluded that the effects of language are not superficial.

To a large extent, we see what we expect to see, and our expectations are shaped by our language. For instance, for many years the terms "black" and "white" were used in the United States to describe the skin colors of humans. Objective examination shows that no humans are as black as the ink on this page, nor as white as the paper. All humans are shades of brown, if we include lights and darks, reddish browns and yellowish browns — which we certainly would, if we were discussing the colors of some less emotionally charged objects such as wallpaper or carpets. Using the terms "black" and "white" (e.g., on survey forms) perpetuates the myth that people fall naturally into two distinguishable groups.

Any human endeavor, even one so seemingly objective as mathematics, is affected by assumptions inherent in the language. For instance, in 3.10 and 3.33 we see that classicists and constructivists have different meanings and uses for the words "exist" and "or." In 2.41 we see that classicists and relevantists have different meanings for "implies." This leads them to different conclusions about what a sentence means or which sentences are acceptable.

5.4. *How can we communicate with one another at all?* With any human language, there is always a possibility for misunderstanding. Even mathematicians, with their precise definitions and notations, cannot be certain of perfect communication. When mathematicians talk, they are attempting to copy ideas from the inside of one mathematician's head to the inside of another mathematician's head, but there is no way to be absolutely certain that the two heads end up containing identical mathematics. Ultimately, each mathematician must construct a separate mathematics inside his or her own head. This admittedly solipsistic view was part of Brouwer's "intuitionistic" philosophy.

Perhaps Brouwer was excessively pessimistic. After all, most of us do succeed in communicating with each other most of the time. Brouwer's pessimism may be partly due to the fact that he developed ideas far from the mainstream, and was particularly prone to being misunderstood.

Some concepts are more easily understood than others. For instance, everyone knows how difficult it is to explain "love." Here is another example that I can present much more objectively:

- Differential equations is a fairly straightforward subject. Over the years, in teaching that subject I have found that I am seldom misunderstood. My students do not always instantly understand everything I say, but it is extremely seldom that they invent alternate interpretations different from the one I intended. When they don't understand something I have said, usually they have no interpretation at all, and so they ask.

- Logic is more subtle in some ways, but much of its vocabulary resembles the vocabulary of everyday English. Consequently, when I teach logic, my students sometimes find an alternate interpretation for something I say, quite different from the interpretation I had intended. This happens because I had

Language and bias 149

not anticipated some of the possible misinterpretations, and had not prevented those misinterpretations by choosing my wording carefully enough.

The latter situation is characterized by this popular[1] saying:

I know you believe you understand what you think I said, but I am not sure you realize that what you heard is not what I meant.

5.5. *How can we analyze our languages?* Mathematical logic attempts to write the concepts of mathematics in a bias-free form, by carefully analyzing the language. However, this analysis itself must be carried out in some language — i.e., the natural language of the humans who are discussing the mathematical logic. An understanding of mathematical logic cannot be built from nothing. It must be intuitively based on the student's informal but logical understanding of everyday experience, which in turn is shaped by the student's natural language.

This book is grounded in American English, my own native language. It is conceivable that this book is subject to some biases of which I am not aware — e.g., biases that are inherent in English, or that are inherent in all Indo-European natural languages or in all human languages. (As philosopher John Culkin said, "We don't know who discovered water, but we are certain it wasn't a fish.") In this book we attempt to present that "universal essence of reason" in whose existence the Greeks believed, but it is only an English presentation or a human presentation. Quite possibly, someday we will meet beings who evolved on a different planet, under circumstances very different from our own, and who have a different "universal essence of reason."

Though we cannot know whether we have found *all* our biases, some biases become apparent when two languages are contrasted. By contrasting nonmathematical English with corresponding notions in mathematical logic, in this chapter we attempt to clear

---

[1] Popularized in the 1960s with lapel buttons and wall posters. Attributed to various sources, including Irish folk sayings and S. I. Hayakawa (semanticist and senator, 1906–1992).

away some confusions and preconceptions of natural language, to prepare for the formal logic. Some of the insights given in these pages will be specific to English, and may not translate readily to other natural languages (Spanish, Russian, etc.); I apologize for my ignorance of those other languages. Students who are trying to learn English as a second language may have extra difficulty reading this chapter. On the other hand, those students may find that this chapter helps them to better understand a few of the trickier parts of English.

Parts of speech

5.6. The four main *connectives* that we shall study are

$$\text{or,} \quad \text{and,} \quad \text{not,} \quad \text{implies.}$$

How are they used in everyday nonmathematical English?

In mathematical logic, "conjunction" just means "and"; but in studies of English grammar, "and" and "or" are *both* called conjunctions. They combine two words or phrases of some type into a longer phrase of the same type. For instance,

- Abraham will fry the fish *or* count the tickets.
- Paul has an apple *and* a banana.

Similarly, "implies" is a transitive verb, and "not" is an adverb.

However, the examples above are really the wrong examples for a preview of this book. I mentioned them only in order to emphasize that they do *not* indicate what we're going to do. Those examples use "or" and "and" to connect together mere parts of a sentence. This book is about propositional logic, also known as sentential logic; the smallest units of meaning considered in this book are propositions (i.e., sentences). Here are examples.

- Abraham fries the fish *or* Abraham counts the tickets.
- Paul has an apple *and* Paul has a banana.
- *It is not true that* Abraham is frying the fish.
- *If* Abraham fries the fish, *then* Paul will count tickets.

This book is about "Abraham fries the fish," not about Abraham or about frying or about the fish.

5.7. *Another difference between mathematical and nonmathematical English.* In mathematical logic, nested implications can be used freely. For instance,

> If I were hungry, then if I had some bread, then if I had some turkey, then I would make a turkey sandwich.

In symbols, that's $H \to (B \to (T \to S))$. The "implies" operator combines two statements into one statement — for instance, it combines B and $T \to S$ into $B \to (T \to S)$. We can repeat this kind of operation as many times as we wish.

In nonmathematical English, nested implications are uncommon; we just don't usually speak that way. We would probably rephrase that sentence as

> If I were hungry and I had some bread and turkey, then I would make a a turkey sandwich.

In symbols, that's $(H \wedge B \wedge T) \to S$. This avoids the nested implications. It introduces a long list of and's, but that is not a difficulty for English.

The formulas $H \to (B \to (T \to S))$ and $(H \wedge B \wedge T) \to S$ are equivalent in some of this book's logics, but not in others. Consequently, we shall treat those formulas as different except where explicitly noted.

Semantic values

5.8. The classical two-valued semantic logic, introduced in 2.38, is computationally the simplest of all logics. It has only two semantic values: one kind of "true" and one kind of "false." But other logics studied in this book have more than two values. Sometimes the values can be explained in intuitive, nonmathematical terms, such as "not known yet," "plausible," or "mostly true." Other, less intuitive values are also possible.

For each semantic logic studied in this book, we divide the set Σ of semantic values into two subsets, Σ_+ and Σ_-, which (for lack of better names[2]) we shall call the set of *true values* and the set of *false values*. Each of those sets contains at least one member but may contain many members. Thus, there may be more than one kind of "true" and more than one kind of "false."

We caution the reader not to inadvertently assume properties of the classical "true" and "false" just based on this terminology. For instance, in classical logic, the negation of any true value is a false value, and vice versa, but that relation fails in most of the nonclassical logics studied in this book. In fact, a semantic value may even be equal to its own negation; examples of this phenomenon are given in 8.16, 8.28, and 9.8.

DISJUNCTION (OR)

5.9. *Two kinds of "or."* The English word "or" has at least two different meanings. Those meanings are represented by two different words in some other languages, such as Latin.

- The Latin word *vel* means "one or the other or both" — i.e., *at least one* of the alternatives holds. In English it is sometimes written as *and/or*. In mathematics we call this *disjunction*, *inclusive or*, or simply *or* (since it is the most commonly used type of "or"). A symbol commonly used for it is $\vee$.

- The Latin word *aut* means "one or the other but *not* both" — i.e., *exactly one* of the alternatives holds. In English it is sometimes written *either/or*. In mathematics we call this *exclusive or*, occasionally abbreviated in writing (but not in speech, please!) as *xor*. Several symbols are used for it in the literature; we shall represent it with $\oplus$.

The word "or" is ambiguous in nonmathematical English, but usually the correct meaning is evident from the context. For

---

[2] Some of the literature calls members of Σ_+ the *designated elements* or *positively designated elements*, but in my opinion that terminology conveys far too little intuition.

Disjunction (or)

instance, in the sentence

> I will go to the movies or I will go home,

it is clear that I can't do both actions simultaneously, so $\oplus$ makes sense. In the sentence

> I want a new computer or a big raise in salary,

it is clear that I would be pleased to get both, so $\vee$ makes sense.

But the contexts in mathematics are far less familiar than the contexts in real life, so in mathematics a correct interpretation often is not evident from the context. When the choice between *or*'s is not evident, then *or* usually means $\vee$, simply because that is the custom among mathematicians.

5.10. *Variants of "or."* In classical logic, the four formulas

$$A \vee B, \qquad (A \to B) \to B, \qquad \overline{A} \to B, \qquad \overline{\overline{A} \wedge \overline{B}}$$

are "equivalent," in the sense that they yield the same valuations, or in the sense that each can be proved to imply the others. Thus the four formulas can be used interchangeably. That is not the case in some nonclassical logics.

For instance, substituting $B = \overline{A}$ yields

$$A \vee \overline{A}, \qquad (A \to \overline{A}) \to \overline{A}, \qquad \overline{A} \to \overline{A}, \qquad \overline{\overline{A} \wedge \overline{A}}.$$

All four of those formulas are theorems of classical logic. But the first formula is not a theorem of constructive logic; the first two formulas are not tautological in fuzzy logics.

5.11. In constructive logic, if $A \vee B$ is a theorem, then at least one of the formulas A or B is a theorem. Thus the theorems of constructive logic form a *prime set of formulas* (see 29.5); this is called the *Disjunction Property* of constructive logic. It doesn't hold in classical logic. (For instance, in classical logic, $\pi_1 \vee \neg \pi_1$ is a theorem, but neither π_1 nor $\neg \pi_1$ is a theorem.)

To understand why constructive logic is prime, let us restate things a bit:

- In classical logic, $A \vee B$ means "at least one of the statements A or B is true."

- In constructive logic, $A \vee B$ means, roughly, "we know how to carry out at least one of the procedures A or B — and if we only know one of those, we know which one it is."

Why does constructive "or" involve this additional knowledge? That is because it refers to procedures. For instance, assume that

(∗) Given wood, nails, and tools, I know how to build a birdhouse or a bookend.

It is not likely that, while I'm hammering away, I'm wondering what the finished product will turn out to be. Thus, given statement (∗), at least one of these two statements undoubtedly is also true:

- Given wood, nails, and tools, I know how to build a birdhouse.
- Given wood, nails, and tools, I know how to build a bookend.

Moreover, if only one of those two statements is true, I know which one it is.

Still, if we know which one it is, why do we bother to mention the other? One reason is that the formula $A \vee B$ can be used as a *hypothesis*. For instance, the constructive interpretation of $(A \vee B) \to C$ is, roughly,

if I knew at least one of A or B (and I knew which one it was), then I would also know C.

That statement can make sense even before we establish at least one of A or B, and thus before we have determined which one. Reasoning of this type was used in Jarden's proof (2.45–2.46), which could be interpreted as a nonconstructive proof of Jarden's theorem or as a constructive proof of this seemingly weaker statement:

If we knew at least one of the statements "$\sqrt{2}^{\sqrt{2}}$ is rational" or "$\sqrt{2}^{\sqrt{2}}$ is irrational," then we would be able to find positive irrationals p and q such that q^p is rational.

The primeness of constructive logic may seem rather intuitive and elementary; perhaps it should be included as one of the initial assumptions of constructive logic. But instead it is an advanced theorem, proved quite late in our investigations (in 27.11), because we shall follow the custom of building constructive logic from the properties that it shares with classical logic.

CONJUNCTION (AND)

5.12. The use of "and" varies slightly from one formal, mathematical logic to another. For instance, $P \to [Q \to (P \wedge Q)]$ is a theorem in classical and constructive logic, but not in relevant logic. That fact is discussed further in 23.17.

However, there is little variation in the use of "and" in informal logic. The expression $x \wedge y$ means that "both of conditions x and y hold." This is not ambiguous in English, but it does cause difficulties for some beginning math students who become confused about its relation to intersections and unions. Following is a typical error from precalculus.

Problem. Find all x satisfying the inequality $x^2 - 4x > -3$.

Correct solution. Rewrite the inequality as $(x-1)(x-3) > 0$. That condition can be restated as: $x - 1$ and $x - 3$ are both positive or both negative. This leads to the correct answer, $\boxed{(-\infty, 1) \cup (3, \infty)}$, which can also be written correctly as

$$\boxed{\{x \in \mathbb{R} \ : \ x < 1 \text{ or } x > 3\}}.$$

Note the use of the word "or" here — it means that we want a number x that satisfies *at least one* of the two inequalities. It doesn't really matter whether the "or" is $\vee$ or $\oplus$, because it is not possible for one number x to satisfy *both* of the inequalities.

Erroneous solution. Some students arrive at the answer above but write it incorrectly, because they are thinking in these terms:

the numbers which satisfy $x < 1$, *and also* the numbers which satisfy $x > 3$.

Note the use of the word "and" in the preceding line. These students write the answer as "$x < 1$ and $x > 3$." Most math teachers will dislike that answer, and some will even mark it wrong, because it can be understood to mean

$$\{x \in \mathbb{R} : x \text{ is a number satisfying both } x < 1 \text{ and } x > 3\}.$$

That's the set $(-\infty, 1) \cap (3, \infty)$, or more simply $\varnothing$; it's not the right answer.

NEGATION (NOT)

5.13. *Notation.* In this book, the expressions $\neg P$ and $\overline{P}$ are used interchangeably; they can both be read aloud as "not-P." One expression is more convenient than the other in certain contexts:

- $\neg P$ is more convenient when we wish to emphasize that negation is an operator, or when we wish to calculate the length or rank of a formula.
- $\overline{P}$ is convenient in longer expressions, since it also acts as a delimiter — i.e., it reduces the need for parentheses. For instance, $\overline{A \wedge B} \to (\overline{A} \wedge \overline{B})$ is the first De Morgan's law (14.12.a). With the other notation for negation we would have to write $(\neg(A \vee B)) \to ((\neg A) \wedge \neg B)$, which is much longer and harder to read.

5.14. In classical logic, P and $\overline{P}$ are opposites, in the sense that (roughly speaking) we must have

at least one or the other, but not both.

That's really two properties of negation, which may be best understood separately:

5.15. *At least one of the cases P, $\overline{P}$ must hold.* That is the content of the formula
$$P \vee \overline{P},$$
known in this book as the *Law of the Excluded Middle*; it is also

Negation (not)

known as the "Law of the Excluded Third Case" or "Tertium Non Datur" in some of the literature.[3] A logic is called

- *complete for negation* if it satisfies $P \vee \overline{P}$, or
- *paracomplete* if it does not.

For instance, we shall see that classical logic is complete for negation, but constructive logic is paracomplete.

5.16. *But not both.* When some formula P and its negation $\neg P$ are both present in a logic, we say they form a *contradiction* or an *inconsistency*. Those words may also be applied to the formula $P \wedge \neg P$, which combines the two contradictory formulas. A logic that generates no contradictions is called *consistent*; a logic that generates at least one contradiction is called *inconsistent*.

The *Law of Noncontradiction* is $\overline{P \wedge \overline{P}}$. This is a theorem in most of our logics; see 15.3.d. It says, roughly, that any negation of a contradiction is true. This law prohibits some kinds of inconsistencies, but not all. Indeed, the Law of Noncontradiction is satisfied by the comparative interpretation (8.33), a logic that is somewhat useful despite being inconsistent (see 2.40 and 8.37.g).

A logic is called *explosive* if it includes the formula $(P \wedge \overline{P}) \to Q$, known among some mathematicians as the *explosion principle*. This formula says that a contradiction implies anything at all. Thus, if we were to add even one contradiction to our collection of formulas, we would suddenly be able to prove *all* formulas — an "explosive" growth in our collection, rendering it trivial and useless. A logic is *paraconsistent* if it lacks the explosion principle — i.e., if it is possible for us to add some contradictions without reducing the system to triviality. For instance, classical logic is explosive, but comparative logic is paraconsistent. A

---

[3] Unfortunately, all three of those traditional names are *wrong* — i.e., they describe something slightly different from $P \vee \overline{P}$. Those names all say that "no third case can hold," whereas $P \vee \overline{P}$ actually says "at least one of these two cases must hold." Such inaccurate terminology is perhaps a consequence of the dominance of classical logic in the early history of this subject. After all, it is difficult to accurately describe a principle when no plausible alternatives to that principle are known. See related remarks in 1.13 and 5.5.

paraconsistent logic might be more suitable for a juror who must hear conflicting testimonies.

5.17. Still another way that P and $\overline{P}$ are opposites is this: In most logics that we shall study, negation is an involution (defined in 3.31), so two negations cancel each other out. Thus $\overline{\overline{P}}$ is in some respects interchangeable with P (see 19.3).

However, $P \leftrightarrow \overline{\overline{P}}$ is really two properties which are best understood separately:

$P \rightarrow \overline{\overline{P}}$ (adding two bars) is called $\neg$-*introduction* or *negation-introduction* or *not-introduction* in this book. That formula is a theorem in all the major logics studied in this book.

$\overline{\overline{P}} \rightarrow P$ (subtracting two bars) is called $\neg$-*elimination* or *negation-elimination* or *not-elimination* in this book. That formula fails in constructive logic.

Translating very loosely, the constructive $\overline{\overline{P}}$ means that "we can prove that it is not possible to prove that P is false." This is a bit weaker than saying "we can prove P is true," because we might not be capable of proving many things.[4]

5.18. In nonmathematical English, a double negative is somewhat ambiguous:

- Among well-educated speakers of English, a second negation cancels out the first, just as in classical logic, though most often the two negations take different forms. For instance,

  I do not lack questions to investigate

  means the same thing as "I *have* questions to investigate." Here "not" and "lack" have canceled each other out. Another

---

[4]Indeed, a constructive proof of some formula X is stronger, more informative, perhaps more authoritative than a classical proof of that X, but fewer formulas X can be proved constructively than classically.

Negation (not)   159

example: Advertisements for Sara Lee brand foods use the slogan

Nobody doesn't like Sara Lee

which of course means everyone likes Sara Lee.

- On the other hand, among poorly educated speakers of English, a second negation may be intended for emphasis. For instance,

I did not see no one

probably is intended to mean "I saw no one."

English is unique in its cancellation of double negatives. The poorly educated English speaker is actually following the well-educated sentence patterns of most other Indo-European languages. For instance, translate the sentence "I saw no one" into Spanish, in a fashion that preserves the meaning. We obtain

no ví a nadie.

Translating that *word for word* back into English, without regard to the meaning, we obtain two negatives:

Not [I] did see (of) no one.

Why does English differ from all the other Indo-European languages? We must look at its history. Usually, languages evolve in a natural fashion, steered largely by popular culture; grammar books attempt to describe the best contemporary uses of a language. However, English went through an authoritarian period in the 18th century, when its first grammar textbooks were produced. Instead of descriptive books, the grammarians wrote *prescriptive* books; they tried to make English follow rules similar to those of Latin or of logic. Particularly influential was *A Short Introduction to English Grammar*, published by Robert Lowth in 1762. It is to Lowth that we owe many of the rules of modern English grammar. Grammar books like his were viewed as authoritative, and were used to teach students for many generations.

The English language had a rhythm and logic of its own, not

at all like those of Latin or logic, and not taken into account by Lowth and his contemporaries. Their imposition of Latin-like rules amputated organic parts of English and replaced those parts with artificial limbs, much to the regret of more recent students of literature. English grammar lessons have been made unnatural, and an earlier literature has been made less accessible to us. Even Shakespeare now sounds peculiar and appears to be riddled with grammatical errors.

However, English-speaking students of logic can thank Lowth for his influence on the language. Surely the classical theorem $\overline{\overline{P}} \leftrightarrow P$ is easier to understand if the student's native tongue reflects this theorem.

5.19. *Joke.* A linguistics professor was lecturing to his class one day. "In English," he said, "a double negative forms a positive. In some languages, such as Russian, a double negative is still a negative. However," he pointed out, "there is no language wherein a double positive can form a negative." A weary voice from somewhere in the back of the room commented, "Yeah. Right."

Further remarks. Increasingly often in recent years, negation has been indicated in spoken English by context or by tone of voice, as in the joke above. Of course, a tone of voice cannot be reliably conveyed on paper, and so it is not an acceptable method of communication in mathematics at present.

5.20. *Duality.* In classical and some other logics, $\vee$ and $\wedge$ are *dual* to each other, in this sense:

(i) $\qquad \overline{A \vee B}$ is equivalent to $\overline{A} \wedge \overline{B}$;

(ii) $\qquad \overline{A \wedge B}$ is equivalent to $\overline{A} \vee \overline{B}$.

These are the *De Morgan's laws* of logic. They are much like the De Morgan's laws of set theory (3.57). Statements about $\vee$ can be converted to statements about $\wedge$, and vice versa. For instance, as an example of (i), these two statements mean the same thing:

> It is not true that today is Monday or today is Tuesday.
>
> Today is not Monday and today is not Tuesday.

Material implication

Also, as an example of (ii), these two statements mean essentially the same thing:

> It is not true that both cars can be seen from here.
>
> At least one of the cars is not visible from here.

MATERIAL IMPLICATION

5.21. The expression $A \to B$ can be read aloud in any of these eight ways, all of which mean the same thing in this book:

- if A then B,
- A implies B,
- A only if B,
- A is a sufficient condition (for B to hold),
- whenever A then B,
- B is a consequence of A,
- B is a necessary condition (for A to hold),
- B if A.

We will use several different symbols for different kinds of implications:

- $\to$ in the formal system (replaced by "$\supset$" in some books),
- $\rightsquigarrow$ for a modified formal implication, used in fuzzy syntactics (see 24.10),
- $\Rightarrow$ in the metasystem,
- $\vdash$ for syntactic consequence,
- $\vDash$ for semantic consequence.

All of these symbols are explained in later chapters.

5.22. The definition of classical implication is given in the last column of the table in 2.38. It is called *material implication* when we need to distinguish it from other kinds of implications. It can be restated this way: $A \to B$ means that *at least in those cases where A is true, B is also true.*

Sometimes we can tell whether $A \to B$ is true, even without knowing whether A or B are themselves true. For instance, the decimal expansion $3.1415926535\cdots$ of the number pi is rather erratic; it does not have any simple patterns. Consequently, I don't know[5] whether one, both, or neither of these statements is true:

A = "The sequence 0123456789 occurs somewhere in the decimal expansion of pi."

B = "The sequence 123456789 occurs somewhere in the decimal expansion of pi."

But I do know that *if* A is true, *then* B must be true.

Material implication is really quite simple, but beginners find it confusing. That's because they're accustomed to the much more complicated "conventional" implications of everyday non-mathematical English. The real difficulty is not in learning mathematical implication, but in unlearning those "conventional" implications. The differences between the two are discussed in the next few pages.

A large part of the difference lies in the fact that, in the mathematical implication $A \to B$, the statements A and B are understood to be *mathematical* statements such as "$3+5 < 9$" or "x is an odd number." Such statements can be viewed in isolation, free of any context or culture. In everyday English, on the other hand, the statements A and B are laden with unspoken, understood meanings that affect how we read the implication $A \to B$.

---

[5] Actually, that's a little lie. The answer to this question was not known in the early 20th century when Brouwer and Heyting first proposed such questions, and the answer still was not known in the late 20th century when I first started writing this book. But in 1997, using powerful computers, Kanada and Takahashi computed over 50 billion digits of π. (See Borwein [1998].) The sequence 0123456789 appears a while after the 17 billionth digit. Thus, statement A is true. But I'm going to pretend that we don't know that — i.e., that we're still living in the pre-1997 world. The idea I'm trying to convey here isn't really about the number pi; it's about implication between any statements whose truth or falsehood has not yet been determined.

It keeps getting harder to find examples of unsolved mathematical problems that are simple to state. Another old favorite that was taken away recently was *Fermat's Last Theorem*, finally proved by Taylor and Wiles [1995] after going unsolved for about 300 years.

Material implication 163

5.23. The formula $A \leftrightarrow B$ is an abbreviation for
$$(A \to B) \land (B \to A);$$
thus it holds if and only if both $A \to B$ and $B \to A$ hold. This is the semantic or syntactic equivalence described in 3.42.f and 3.42.g. It can be read aloud as

- "*A if and only if B*," or as
- "*A* is a *necessary and sufficient condition* for *B* to hold."

Sometimes "if and only if" is abbreviated in writing as *iff*, but this abbreviation generally is not used in spoken discussions, since "iff" sounds too much like "if."

Beginners sometimes confuse the three expressions $A \to B$, $B \to A$, $A \leftrightarrow B$. The confusion probably arises from the fact that those expressions often are interchangeable in nonmathematical contexts, due to additional, implicitly available information.

For instance, I might say

If I get hungry, I'll go out and buy a snack.

But most people will guess (quite correctly) that I *won't* interrupt my work to go out for a snack if I'm *not* hungry. Thus the implication above is equivalent to the statement

I'll go out and buy a snack if and only if I get hungry.

In some contexts that additional bit of information is not available, and then $A \to B$ and $B \to A$ are *not* the same. For instance, let us assume that (i) anyone who has a pebble in his or her shoe will walk with a limp, but (ii) a limp may have other causes, not involving a pebble. Then this is a valid piece of reasoning:

If $\begin{pmatrix} \text{Jack has a pebble} \\ \text{in his shoe today} \end{pmatrix}$ then $\begin{pmatrix} \text{Jack walks with} \\ \text{a limp today} \end{pmatrix}$.

However, we can *not* conclude that this piece of reasoning is valid:

If $\begin{pmatrix} \text{Jack walks with} \\ \text{a limp today} \end{pmatrix}$ then $\begin{pmatrix} \text{Jack has a pebble} \\ \text{in his shoe today} \end{pmatrix}$.

(Indeed, I happen to know that Jack injured his knee in a football game several years ago. He has walked with a limp ever since, regardless of shoes or pebbles.) Thus, in this example $A \to B$ is true but $B \to A$ is false.

5.24. For mathematical statements, we cannot rely on familiarity with some body of contextual knowledge and common sense; mathematical statements deal with issues that are less familiar. Consequently, in mathematics, "if" generally is *not* used interchangeably with "if and only if."

There is one common exception. When we define a word or phrase, if we emphasize it with **boldface**, *italics*, an underline, "putting the word or phrase in quotes," or some other typographical device, then we routinely write "if" where we mean "if and only if." For instance, 3.13 includes this definition:

> We say that S is a *subset* of T, written $S \subseteq T$, if every member of S is also a member of T.

Because of the italics, that "if" is understood as an "if and only if" — i.e., the sentence really means this:

> We say that S is a subset of T, written $S \subseteq T$, if and only if every member of S is also a member of T.

5.25. The *converse* of the formula $P \to Q$ is the formula $Q \to P$. For instance, the two implications about the pebble and the limp, in 5.23, are converses of each other. In general, a formula and its converse are not equivalent.

The definition of "converse" is clear enough when we have a formula or sentence of the simple form $P \to Q$. But sentences in English often take a more complicated form, and then it isn't always clear just what is the converse. For that reason, the word "converse" should only be used cautiously.

For example, consider the old joke:

> If I had some bread, I could make a ham sandwich — if I had some ham.

Material implication

(It's supposed to be funny because ham, being more expensive than bread, should not be an afterthought.) What is the converse of that sentence? Well, by reading it in different ways we get different converses, and the different converses are *not* all equivalent to one another. To see that, abbreviate

$$b = \text{"I had some bread,"}$$
$$h = \text{"I had some ham,"}$$
$$s = \text{"I could make a ham sandwich."}$$

Then one interpretation of the given sentence is $(b \wedge h) \to s$, which has converse $s \to (b \wedge h)$. But a second interpretation of the given sentence is

(*) \hspace{2cm} Assume h. Then $b \to s$.

That is, $h \to (b \to s)$. With this interpretation, the converse is $(b \to s) \to h$.

However, a third interpretation is possible. In the expression (*), we could describe "Assume h" as an *underlying assumption* or *background assumption* for the implication $b \to s$. Usually, when we're looking for a converse, we don't change the underlying assumptions. The converse of $b \to s$ is $s \to b$, so the converse to (*) would be

Assume h. Then $s \to b$.

That is, $h \to (s \to b)$.

We have produced three different converse formulas. These are different in appearance, and they are different in values too: they produce different true/false values, for certain choices of s, b, h, as we shall show in the exercise below.

5.26. *Exercise.* Complete the following true/false table. (To reduce the paper grader's workload, everyone in the class should use the same ordering of rows and columns.)

s	b	h	$s \to (b \wedge h)$	$(b \to s) \to h$	$h \to (s \to b)$
F	F	F			
F	F	T			

F	T	F
F	T	T
T	F	F
T	F	T
T	T	F
T	T	T

5.27. For any implication $A \to B$, the *contrapositive*[6] is the implication $\overline{B} \to \overline{A}$. An implication and its contrapositive *are* equivalent, at least in classical logic. For instance, the implication

$$\text{if } \left(\begin{array}{c} \text{Jack has a pebble} \\ \text{in his shoe today} \end{array} \right) \text{ then } \left(\begin{array}{c} \text{Jack walks with} \\ \text{a limp today} \end{array} \right)$$

is just as true, or just as untrue, as the implication

$$\text{if } \left(\begin{array}{c} \text{Jack is not} \\ \text{limping today} \end{array} \right) \text{ then } \left(\begin{array}{c} \text{Jack has no pebble} \\ \text{in his shoe today} \end{array} \right).$$

There are several contrapositive laws, which say in various ways that $A \to B$ and $\overline{B} \to \overline{A}$ are the same. See 14.1.i, 14.10, 19.2.

5.28. Another reason that beginners in logic become confused by material implication is that they are accustomed to seeing implications with at least some ingredient of *causality*. An example of a causal implication is

$$\text{if } \left(\begin{array}{c} \text{I fall out of an airplane} \\ \text{without a parachute} \end{array} \right) \text{ then } \left(\begin{array}{c} \text{I get} \\ \text{hurt} \end{array} \right).$$

Here we have a formula of the form $A \to B$ in which A causes B. This is also a material implication, since

$$\begin{array}{c} \text{in each} \\ \text{case} \\ \text{where} \end{array} \left(\begin{array}{c} \text{I fall out of an} \\ \text{airplane without} \\ \text{a parachute} \end{array} \right) \begin{array}{c} \text{it also} \\ \text{is true} \\ \text{that} \end{array} \left(\begin{array}{c} \text{I get} \\ \text{hurt} \end{array} \right).$$

Some students naively expect all implications to be like that. But

---

[6]The student is cautioned not to confuse converse (5.25), contrapositive (5.27), contradiction (5.35), contraction (Chapter 15), and constructive (2.42 and Chapter 22).

Material implication

not every material implication is causal. Indeed, some material implications seem to go in the direction opposite to causality.

For instance, Mr. Simon brings an umbrella to work if it is raining at the time when he leaves his house; otherwise he leaves the umbrella at home. Mr. Simon and Ms. Lido work in an interior office that has no windows. Nevertheless, Ms. Lido — whose work shift begins two hours earlier than Mr. Simon's — can get some idea of the weather conditions by noting whether Mr. Simon is carrying an umbrella when he arrives. She knows his habits, and so she correctly reasons that

$$\text{if } \begin{pmatrix} \text{Mr. Simon arrives} \\ \text{with his umbrella} \end{pmatrix}, \text{ then } \begin{pmatrix} \text{it has recently} \\ \text{been raining} \end{pmatrix}.$$

This is a material implication: In all the cases where Mr. Simon arrives with his umbrella, it is also true that it has recently been raining. But this is not a causal implication — Mr. Simon's bringing an umbrella does not *cause* the rain.[7] In fact, it is opposite to causality: the presence or absence of rain determine Mr. Simon's decision about the umbrella.

Classical logic is indifferent to causality. It is only concerned with *what* happens, not *why*. That is part of why classical logic is so simple to use in mathematics; but that is also what some people find objectionable in classical logic.

On the other hand, all material implications show a direct "causality of knowledge." For instance,

$$\text{if } \begin{pmatrix} \text{Ms. Lido sees} \\ \text{Mr. Simon arrive} \\ \text{with his umbrella} \end{pmatrix}, \text{ then } \begin{pmatrix} \text{Ms. Lido knows} \\ \text{that it has recently} \\ \text{been raining} \end{pmatrix},$$

despite the lack of windows. This can be restated as a cause-and-effect relationship:

$$\begin{pmatrix} \text{Ms. Lido's seeing} \\ \text{Mr. Simon arrive} \\ \text{with his umbrella} \end{pmatrix} \text{ causes } \begin{pmatrix} \text{Ms. Lido's knowing} \\ \text{that it has recently} \\ \text{been raining} \end{pmatrix}.$$

---

[7]Though we sometimes joke that carrying an umbrella can prevent rain, just as starting a picnic can cause rain.

5.29. More extreme than the lack of causality, and more bewildering to most students, is a lack of *relevance* in material implication. In everyday nonmathematical experience, when we say that "A implies B," we expect A and B to be related in some way, no matter how tenuously. But for material implication, A and B don't need to have any relation at all. For instance, consider:

(∗)   If it is raining now, then red is a color.

That implication is *nonsense* in everyday nonmathematical conversation, because rain and red are unrelated; the two clauses are irrelevant to each other. But that implication is *true* in classical logic, since red *is* a color. In classical logic, if B is true, then $A \to B$ is true, regardless of what A and B are.

Students are confused by an implication like (∗) because they look to the meanings of A and B for clues about whether A implies B. But the truth or falseness of $A \to B$ is determined by the table in 2.38, and depends only on the truth or falseness of the formulas A and B; it does not depend on the *meanings* of A and B. Classical logic is indifferent to meanings.

This indifference to meanings is quite natural in the writing of mathematical proofs, where we have less contextual information available. We are very familiar with rain and red, and we know what behavior to expect from them. But a mathematician investigates objects that are far less familiar, and searches for unexpected behaviors of those objects.

For example, consider the following theorem from functional analysis. (Don't worry about understanding what this theorem *means*; just concentrate on its grammatical structure.)

> Let X be a normed vector space. Then the following conditions on X are equivalent (i.e., either condition implies the other):
>
>   (A)   Every lower semicontinuous seminorm on X is continuous.
>
>   (B)   Whenever S is a subset of the dual of X such

Material implication 169

> that S is bounded on each point of X, then S is bounded in norm.

In this theorem, $A \to B$ and $B \to A$ are material implications. But A and B seem nearly irrelevant to each other: A makes no mention of boundedness; B makes no mention of seminorms or lower semicontinuity or continuity. The only apparent connection between the two conditions is the background assumption that X is a normed vector space. That assumption is rather mild (take my word for it); it's not the main point of the theorem.

Indeed, the main point of the theorem is that conditions A and B actually *are* related. Some of the most interesting and important discoveries in mathematics involving the uncovering of hidden connections — i.e., showing that two seemingly unrelated concepts are in fact closely related, or even equivalent.

Most mathematicians feel that the peculiarities and nonsense of classical logic are outweighed by the convenience of its simplicity. But a few mathematicians, the "relevantists," have investigated and developed some alternative logics that avoid some of the nonsense inherent in classical logic. We will look at a couple of these relevant logics in later chapters.

5.30. *More vocabulary.* An implication $A \to B$ is

> *trivially true* if we know by some easy observation either that A is false or that B is true;
>
> *vacuously true* if we know by some easy observation that A is always false.

(Vacuously true is a special case of trivially true.) In either of these cases, the material implication $A \to B$ is then *true*, as evaluated in classical logic; but it isn't a very informative or helpful statement, and it may be unacceptable in relevant logic. Following are four examples of trivially true statements; the first two are vacuously true.

- If $2 + 2 = 5$, then a woman was elected president in the year 1900.

- If 2+2 = 5, then a man was elected president in the year 1900.
- If it rains next Tuesday, then a man was elected president in the year 1900.
- If it doesn't rain next Tuesday, then a man was elected president in the year 1900.

The term "vacuously" refers to the fact that, when I try to verify that "in all cases where A holds, B also holds," I find that there *aren't* any cases where A holds, so there is nothing to verify. The set of cases to be checked is an empty set — i.e., a vacuum.

The terms "trivially true" and "vacuously true" will only be used in our informal explanations, not in our formal logic. Indeed, our definitions of those terms are imprecise, since they use the subjective word "easy."

COTENABILITY, FUSION, AND CONSTANTS (POSTPONABLE)

5.31. The symbols $\vee, \wedge, \rightarrow, \neg$ correspond to basic notions of our everyday nonmathematical reasoning. The symbols $\circ, \&, \bot$ are less intuitive, and consequently they play only a secondary role in this book. They do not appear in our lists of axioms. Still, they will be useful in our studies of $\vee, \wedge, \rightarrow, \neg$.

5.32. The expression $A \circ B$ will be an abbreviation for $\overline{A \rightarrow \overline{B}}$. We shall call that the *cotenability* of A and B, for reasons discussed below.

In English, "tenable" means "can be held." For instance,

bananas are the best fruit

is an opinion — it can't really be proved right or wrong — but someone might *hold* that opinion. Similarly, "cotenable" should refer to two statements that can be held together. For instance, these two statements are *not* cotenable:

Cotenability, fusion, and constants (postponable)   171

- The Beatles were the greatest rock and roll band that ever lived.
- The Rolling Stones were the greatest rock and roll band that ever lived.

But these two statements *are* cotenable:

(A) Bananas are the best fruit.

(B) The Beatles were the greatest rock and roll band that ever lived.

Let us now restate the notion of cotenability several times, transforming it gradually into a logical formula:

- A and B are cotenable;
- the two statements A and B are "compatible" — they do not preclude one another;
- that one of the statements A, B holds does not rule out the possibility that the other statement might hold;
- A does not imply the negation of B;
- it is not true that A implies not-B;
- $\overline{A \to \overline{B}}$.

Hence our definition.

The phrase "A and B do not preclude one another" sounds as if it ought to be symmetric in A and B. Well, it *is* symmetric: In most logics of interest we shall prove $(A \circ B) \leftrightarrow (B \circ A)$. Thus cotenability is commutative. Cotenability also turns out to be associative: $[A \circ (B \circ C)] \leftrightarrow [(A \circ B) \circ C]$; consequently we can write $A \circ B \circ C$ without parentheses. These properties will be proved in 18.3 and 18.4.

We have defined $A \circ B$ to mean $\overline{A \to \overline{B}}$. In most (not all!) logics of interest, it turns out that $\overline{A \to B}$ is the same thing as $A \circ \overline{B}$. Thus we can translate back and forth between statements about $\to$ (which are of greater interest intuitively) and statements about $\circ$ (which are easier to analyze computationally).

5.33. Implication is not associative; i.e., $A \to (B \to C)$ is not generally equivalent to $(A \to B) \to C$. However, in some computations it would be convenient to move the parentheses to the other position, as in

$$A \to (B \to C) \quad \text{is equivalent to} \quad \begin{pmatrix} \text{something} \\ \text{involving} \\ A \text{ and } B \end{pmatrix} \to C.$$

The "something," in those logics where it exists, generally is called the *fusion* of A and B. It will be denoted by $A \mathbin{\&} B$ in this book. Thus, the main property of fusion is that

$$A \to (B \to C) \quad \text{is equivalent to} \quad (A \mathbin{\&} B) \to C,$$

where "equivalent" is defined as in 3.42.f or 3.42.g.

Actually, this "fusion property" does not determine the formula $A \mathbin{\&} B$ uniquely from A and B — there may be many formulas that can be used for the "something." But those formulas for $A \mathbin{\&} B$ all turn out to be equivalent to one another, as shown in 18.2. Moreover, often we are able to single out a convenient representative from that collection of equivalent formulas. Indeed, in most logics of interest, the requirements for & are satisfied by either $\wedge$ or $\circ$. (Consequently, the symbol & is dispensed with in most articles or books that deal with only one logic.)

In most logics of interest, fusion is commutative. Also, it is associative; thus an expression such as $A \mathbin{\&} B \mathbin{\&} C \mathbin{\&} D$ does not require parentheses. Consequently, a string of implications that are nested from the right, such as $A \to (B \to (C \to (D \to E)))$, can be written more briefly and intuitively as $(A \mathbin{\&} B \mathbin{\&} C \mathbin{\&} D) \to E$. See the example in 5.25.

5.34. *Ackermann constants.* We will use the symbols $\top$ and $\bot$, respectively, for *Ackermann truth* and *Ackermann falsehood* (denoted by t and f in some of the literature). These objects are present in most, but not all,[8] of the logics studied in this book.

---

[8]*Exercise for advanced students.* Use the Belnap Relevance Property 23.9.b to show that the syntactic logic R (23.1) does not have either Ackermann constant. On the other hand, the crystal semantics, which has an analogous relevance property (9.12), has both Ackermann constants (9.9); explain.

Cotenability, fusion, and constants (postponable) 173

For our purposes, it is simplest to define them as any objects ⊤ and ⊥ having the following properties (with "equivalent" defined as in 3.42.f or 3.42.g):

$$P \text{ is equivalent to } \top \to P$$
$$\text{and } \neg P \text{ is equivalent to } P \to \bot$$

for all formulas P. The latter property gives us a way of describing negation in terms of implication, and thereby reducing by one the number of primitive operations that we need to consider. It will be particularly useful in 11.6 and Chapters 20 and 26. Ackermann falsehood takes the value 0 or ∅ in most of our semantic logics, but it takes the value $\{+1\}$ in Church's diamond 9.14, and the value $\{-1, +1, +2\}$ in the crystal interpretation 9.9. For Ackermann falsehood in syntactic logics, see 16.5, 17.3, and 20.6.c. Ackermann truth is less crucial for the purposes of this book, and is mentioned here partly just for symmetry.

The student is cautioned not to confuse Ackermann's constants with *Church's constants*, T and F, which he or she may run across elsewhere in the literature:

T is the *top*, or *most true*, or *weakest*, value — i.e., it is implied by any other formula. The existence of a top element is much like the validity of the positive paradox formula $A \to (B \to A)$; see the remark in 16.3.

F is the *bottom*, or *most false*, or *strongest*, value — it implies any other formula. The existence of a bottom element is much like the validity of the conjunctive explosion formula $(A \wedge \overline{A}) \to B$; see 17.2.

In some logics, some of the four constants may coincide. If all four constants are present and distinct

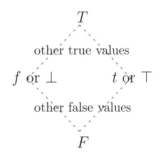

in a logic, they are related as shown in the diagram at right. The diagram is ordered by implication — i.e., if $X \to Y$ is valid, then there is a path upward from $X \to Y$. There may or may not exist a path between t and f.

The symbols ⊤ and ⊥ are used for the Ackermann constants

in some of the literature, and for the Church constants elsewhere in the literature. We emphasize that, in this book, $\top$ and $\bot$ are used for the Ackermann constants.

Optional. Though this book will define the Ackermann constants by the equivalences at the beginning of this section, we should mention a more algebraic definition used elsewhere in the literature:

- t is the least upper bound (the "or") of all the false values, and
- f is the greatest lower bound (the "and") of all the true values.

Here semantic values are lattice ordered by the rule that $x \preccurlyeq y$ when $x \ominus y$ is a true value.

METHODS OF PROOF

5.35. We consider some general methods for proving an implication of the form $P \to Q$.

- A *direct proof*, or *forward proof*, of $P \to Q$ generally goes like this: We start by assuming P is true. Then we figure out what are some of the consequences of P — i.e., we find more true statements. Then we figure out what are some of the consequences of those consequences — i.e., still more true statements. And so on. We continue this procedure, generating more consequences, until we discover Q to be among the consequences. (Thus Q is our "target" or "goal.") When we do find it among the consequences, that proves $P \to Q$.

    Technical remarks. The preceding informal description, if translated into the more precise language of our later chapters, says that we establish $\vdash P \Rightarrow \vdash Q$ and then use that to conclude $\vdash P{\to}Q$. Actually, $\vdash P \Rightarrow \vdash Q$ does imply $\vdash P{\to}Q$ in classical logic, but not in some other logics; see 27.13–27.18.

Besides direct proofs, we mention two kinds of *indirect proofs*.

- A *proof by contrapositive* is based on the fact that, in classical and most other logics of interest (though not all), the implications $P \to Q$ and $\overline{Q} \to \overline{P}$ are equivalent. Thus, to

Methods of proof 175

prove $P \to Q$, we start with the statement $\overline{Q}$, and find consequences of it, and find consequences of those consequences, until we find $\overline{P}$ among our conclusions. At that point, we're done. This procedure sometimes is easier than a direct proof simply because $\overline{Q}$ sometimes is easier to work with than P. Proof by contrapositive may seem to resemble "working backwards," but that is substantially different and not really a method of proof at all; see the discussion in 5.36.

- A *proof by contradiction* is more complicated to describe, but often easier to use. It is based on the fact that, in classical logic and in some other logics, the implication $P \to Q$ is equivalent to the implication $(P \wedge \overline{Q}) \to (R \wedge \overline{R})$, for *any* formula R. Our procedure is as follows: Start by assuming both P and $\overline{Q}$. Figure out what are some of the consequences of those two assumptions; then find some of the consequences of those consequences; and so on. Continue this procedure until a contradiction turns up among the consequences — i.e., until both R and $\overline{R}$ have turned up, for some formula R. (Thus the "target" or "goal" is $R \wedge \overline{R}$.) At that point, the proof is complete.

Proof by contradiction has these advantages:

- At first glance it might appear that we have a harder goal — we have to reach two targets (R and $\overline{R}$), rather than just one (Q). But actually we have more flexibility, because we can choose R to be whatever is convenient. We can choose R so that at least one of the two targets is very easy to reach.

- We work from two assumptions (both P and $\overline{Q}$) rather than just the one assumption of P; thus we have more statements on which to build. Thus it is easier to generate consequences.

Consequently, proofs by contradiction are often easier to *discover* than direct proofs. On the other hand, proofs by contradiction also have a couple of disadvantages:

- Proofs by contradiction are often harder to *read* than direct proofs because they are conceptually more complicated. A direct proof gradually increases the collection of statements that we consider to be "true"; this is a one-directional approach. But a proof by contradiction works simultaneously in two directions, mixing together statements (such as P and its consequences) that we take to be true with statements (such as $\overline{Q}$) that we temporarily pretend are true but shall eventually decide are false. This scheme must seem diabolical, or at least amoral, to beginners: It is not concerned so much with "what is true," but rather with "what implies what." Reading indirect proofs can be difficult not only for beginners, but even for experienced mathematicians; we all must work harder to follow the convoluted reasoning of an indirect proof.

- Either kind of indirect proof — by contrapositive or by contradiction — is sometimes *nonconstructive*: It may prove the existence of some mathematical object without producing any explicit example of that object, and thus without producing much intuitive understanding. In our development of formal logic in later chapters, we will see that $(\overline{Q} \to \overline{P}) \to (P \to Q)$ is not a theorem of constructive logic, and

$$[(P \wedge \overline{Q}) \to (R \wedge \overline{R})] \to (P \to Q)$$

is not a theorem of relevant or constructive logic; see 8.39.c(ii) and 10.5.

Some theorems are inherently nonconstructive, and can only be established by indirect proofs. But other theorems can be proved in either fashion — directly or indirectly. In such cases, the indirect proof may be easier to discover, but in many cases the author of the proof *should then rewrite their presentation as a direct proof if possible*. The word "should" here refers to good taste and readability, not to an actual requirement for certification of truth.

More experienced mathematicians may sometimes combine ingredients of direct and indirect proof in their presentation. This can be accomplished with appropriately chosen phrases such as "suppose the contrary" and "but then we would have," etc. In some cases this makes a proof easier to read, because the proof then avoids surprises and nonintuitive tricks, and follows a line of discovery that appears natural and spontaneous to most readers. (But the appearance of spontaneity generally requires much preparation.)

Working backwards

5.36. *Working backwards* is not a method of proof at all — it is a method of discovery. However, beginners sometimes confuse it with proof, perhaps because it slightly resembles an indirect proof, or because they do not understand the difference between discovery and certification (discussed in 2.13), or because working backwards often succeeds in some lower-level math courses. *It does not work in most advanced parts of math.* That is because, though most of the computations in lower-level math courses are reversible, many of the reasoning steps in higher-level math courses are not reversible.

The underlying idea is something like this: Consider the statement that we're trying to prove. Suppose that it were true. What other statements then would also be true? Are any of those other statements easier to prove? Could we prove some of those statements first, and then use them to prove the statement that is our main goal?

The procedure is something like this: Start with the statement that you're trying to prove. Transform it in some fashion to get a new statement, perhaps simpler or closer to a known truth. Then transform that, to get another new statement. Continue in this fashion, generating more statements, until you finally arrive at a statement you know to be true.

Some beginners stop at that point, but we're only half done. We've now *discovered* the steps that we *hope* will make a proof. Now rewrite the steps, in the opposite order to the order in which you discovered them, and check whether each transformation is reversible. If all the transformations are reversible, you've got a proof.

Here is a suggestion that may improve your use of working backwards: During the discovery process, put a *question mark* over each equality or inequality or other statement that you *hope to prove but have not proved yet*, so that you will not confuse it with statements that you have already established. However, that's only recommended for the discovery process. When you rewrite your ideas into a proof (i.e., certification), you should present them in the correct order, so that each assertion is made only when you prove it.

Actually, it is possible to turn a "working backwards" argument into a rigorous proof without changing the order of steps, but only by working very carefully. The proof must be interspersed with frequent phrases like

- "what we want to establish is,"
- "it suffices to show that,"
- "it remains only to prove," and
- "equivalently"

in appropriate places, and the author of the proof must check that each step *is* reversible. Occasionally a proof like this is more readable, because it follows a more natural path of reasoning — i.e., closer to the path of discovery. However, beginning mathematicians are not likely to handle this sort of thing correctly; even experienced mathematicians sometimes make a mess of it. (And even if such a proof is written correctly, beginners may have difficulty *reading* it.) Beginners might be safer if they rewrite their "working backwards" proofs into forward proofs.

5.37. *Example.* Prove that $\sqrt[3]{3} > \sqrt{2}$.

Method of discovery

- Start with the inequality that we want to prove, $\sqrt[3]{3} \stackrel{?}{>} \sqrt{2}$.
- Rewrite that as $3^{1/3} \stackrel{?}{>} 2^{1/2}$.
- Raise both sides to the power 6. That yields $(3^{1/3})^6 \stackrel{?}{>} (2^{1/2})^6$. (Note that raising an inequality to a positive power is a reversible step, if both sides of the inequality are positive numbers.)

- Simplify both sides, using the rule of exponents that says $(a^b)^c = a^{(bc)}$. Thus we obtain $3^2 \overset{?}{>} 2^3$.

- Evaluate. Thus we obtain $9 \overset{?}{>} 8$. This is clearly *true*.

But that's just the discovery process. Now we have to reverse all the steps to get a proof. In this example, all the steps happen to be reversible. Below is the finished argument, written as a *three-column proof* to exhibit each step's justification explicitly.

label	statement	justification
(1)	$9 > 8$	elementary fact of arithmetic.
(2)	$3^2 > 2^3$	rewrite (1), using more arithmetic.
(3)	$(3^{1/3})^6 > (2^{1/2})^6$	rewrite (2), using $(a^b)^c = a^{(bc)}$.
(4)	$3^{1/3} > 2^{1/2}$	Take sixth root on both sides of (3). (That preserves the inequality since everything is positive.)
(5)	$\sqrt[3]{3} > \sqrt{2}$	Rewrite (4) in different notation.

5.38. *Another example.* Prove that $x > \sqrt{x^2 - 1}$ for all real numbers x. (That's actually not true; let's see where the proof goes awry.)

Erroneous method of discovery

- Start with what we want to prove: $x \overset{?}{>} \sqrt{x^2 - 1}$. Note that the right side is a nonnegative number, since the symbol $\sqrt{}$ means "the nonnegative square root of." Since $x > \sqrt{x^2 - 1}$, we know that x is positive.

- Square both sides; thus we obtain $x^2 \overset{?}{>} x^2 - 1$. Note that squaring both sides of an inequality is a reversible step, if both sides are nonnegative.

- Obviously $x^2 > x^2 - 1$ is true.

Exercise. What went wrong with the preceding proof? Find all real numbers x for which the inequality $x > \sqrt{x^2 - 1}$ is actually true. If you can't answer the question algebraically, try drawing a graph of $y = x$ and a graph of $y = \sqrt{x^2 - 1}$; see where one graph is higher than the other. Be careful about the domains of the functions — i.e., where is $\sqrt{x^2 - 1}$ actually defined?

5.39. *"Cleaning" a proof may erase the discovery process.* When we rewrite a proof to make its steps follow the order of reasoning, the method of discovery may be obscured; the proof may become *less* readable. Following is an example. We wish to prove that

$$\text{the function } f(t) = t^2 \text{ is continuous.}$$

The discovery process. Let's start by just restating what it is we want to prove. According to the definitions, what we want to prove is that

(∗) for each real number x and each positive number ε there exists some positive number δ (which may depend on both x and ε) with the property that

whenever t is a real number that satisfies $|x - t| < \delta$, then $|f(x) - f(t)| < \varepsilon$.

This definition is discussed further in 5.45.) Rewrite that last clause, in terms of our function $f(x) = x^2$:

whenever t is a real number that satisfies $|x - t| < \delta$, then $|x^2 - t^2| < \varepsilon$.

Let's concentrate on the final inequality, $|x^2 - t^2| < \varepsilon$. That can be rewritten as $|x - t||x + t| < \varepsilon$. Since we will be assuming $|x - t| < \delta$, the final inequality will follow if we can just show $\delta|x + t| < \varepsilon$. And, in fact, aside from some degenerate boundary cases considered below, it would suffice to get $\delta|x+t| \leq \varepsilon$, since we will have $|x - t|$ *strictly* less than δ.

So, at first glance, it would appear that we could answer the problem by choosing $\delta = \varepsilon/|x + t|$. But that doesn't work, for a couple of reasons. One minor reason is that $|x + t|$ might be zero, so we can't divide by it. Again, this is just a boundary problem that we will deal with later. A more serious reason is that we are not allowed to use t in our definition of δ. If you look at how (∗) is worded, you'll see that we are required to find *one* value of δ that works for *all* choices of t.

Actually, any choice of δ satisfying the following requirements would work:

Let $\delta = \varepsilon/\text{something}$,

where $\begin{cases} \text{the "something" doesn't depend on } t, \\ \text{the "something" isn't 0, and} \\ \text{the "something" is } \geq |x+t|. \end{cases}$

Now let's concentrate on the last condition, since it contains the largest amount of algebra and may be most amenable to further transformation. How can we make the "something" greater than or equal to $|x+t|$? Well, for starters we know the triangle inequality

$$|x+t| \leq |x| + |t|.$$

So it suffices to make the "something" greater than or equal to $|x| + |t|$. That's a little simpler: Just make it the sum of two secondary "somethings," one of them greater than or equal to $|x|$, the other greater than or equal to $|t|$. The first of these conditions is easy: For something that doesn't depend on t and is greater than or equal to $|x|$, we could actually use $|x|$ itself.

The other part is a little harder. How do we find something that is greater than $|t|$, when we don't know what $|t|$ is? Well, we can use the triangle inequality another way: We know that

$$|t| = |x - (x-t)| \leq |x| + |x-t| \leq |x| + \delta.$$

So it suffices to get the latest "something" to be greater than or equal to $|x| + \delta$. But now our reasoning seems to be circular — we need to know what δ is, in order to define δ??

Not really. The key to calculus proofs is to get the inequality going in the right direction. For most of the things we do with δ, we don't need to know a precise value of δ; we just need to know δ is *small enough* to accomplish some task; any smaller positive number δ then will also work. This is true, for instance, in our original condition (∗). So, for instance, whatever preliminary value of δ we may come up with that works for other purposes,

$$\delta = \min\{1, \text{ preliminary value of } \delta\}$$

will also work. (This is one of the grab-bag of "tricks" that analysts use. I guess the only way to learn these tricks is to look at lots of examples.) Consequently we can assume $\delta \leq 1$, and so our preceding computation becomes

$$|t| = \text{etc.} \leq |x| + \delta \leq |x| + 1.$$

Thus, we can take $|x| + 1$ for our second secondary something. Combine that with $|x|$, and we get $2|x|+1$ for our original "something." This also satisfies the other two conditions we were looking for, so we've now found all the ingredients we need for δ. We have finished working backwards. We have finished the discovery process; we found δ at the *end* of the discovery process.

The presentation. The reasoning was a bit convoluted, and it might lose some readers. So we may want to rewrite the proof in a more direct fashion. We're just trying to prove (∗), so we might begin our proof this way:

> Let any real number x and any positive number ε be given. Then we shall choose
>
> $$\delta = \min\left\{1, \frac{\varepsilon}{1+2|x|}\right\}.$$
>
> We shall show that this δ has the property that, whenever t is a real number satisfying $|x - t| < \delta$, then $|f(x) - f(t)| < \varepsilon$.

And the proof continues from there. (*Exercise.* Write the rest of the proof, going forwards instead of backwards.)

This rewritten proof involves reasoning that is less convoluted, but its second sentence (wherein we "choose" δ) *is unexpected and nonintuitive.* Anyone who wasn't in the audience during the discovery process will be surprised — the complicated formula

$\delta = \min\{1, \varepsilon/(1+2|x|)\}$ seems to leap out of nowhere, like a rabbit pulled from a magician's hat. That may be entertaining in a magic show, and experienced mathematicians generally enjoy surprising each other, but in a classroom some beginning students find suprises utterly bewildering. Some of the students expect each step in a proof to somehow follow from the step before it; thus they expect the first step in a proof to be an obvious consequence of the hypotheses of the theorem. But that expectation is not being met here. How was that formula for δ chosen? It is at the *beginning* of the proof. Is it supposed to be somehow *obvious*?

No, it is not. Even to an experienced mathematician, the first step of the proof may seem mysterious — i.e., we don't yet see where this proof is going; we don't see why it has begun in this fashion. The author of the proof is saying, in effect, "Bear with me — humor me — my reasons for going in this direction will become clear later."

What is crucial for an acceptable proof, however, is that the experienced mathematician can see immediately that the mysterious first step is *permitted* (as in 2.16). The author of the proof is only choosing quantities that he or she is permitted to choose. (What we can choose and what we can't will be discussed further, in 5.45–5.46.)

QUANTIFIERS

5.40. We will not make quantifiers a part of the formal logic developed later in this book; we leave that for more advanced courses. However, they are part of the informal logic that we shall use in our metamathematical system — i.e., in our discussions *about* our formal logic. Moreover, a knowledge of the most elementary facts about quantifiers is essential for any mathematician; thus it should be included in any introduction to logic.

We emphasize that the rules developed below are only for quantifiers in *classical logic*, which is the metalogic used in this book. The rules must be modified slightly for constructive predicate logic or relevant predicate logic, etc.

5.41. The two kinds of quantifiers are

$\forall$ the *universal quantifier* ("for all") and

$\exists$ the *existential quantifier* ("there exists at least one").

To use the symbols $\forall$ and $\exists$, you must first choose what set of objects you're discussing — i.e., in what set do we find the x, y, z discussed below. Thus, "$\forall x$" really means "for all x in the set Ω," where Ω is some chosen set. The set Ω is the *domain*, or *domain of quantification*, or *universe*. The need for a universe is evident from the explanation of Russell's Paradox, in 3.12. (In some discussions the domain Ω is not specified explicitly because it is understood; but then it has nevertheless been "chosen.")

The symbols $\forall$ and $\exists$ may be viewed as extensions or generalizations of $\wedge$ and $\vee$, respectively. Indeed, if x is ranging over the positive integers, then

$\forall x P(x)$ means something like $P(1) \wedge P(2) \wedge P(3) \wedge \ldots$;

$\exists x P(x)$ means something like $P(1) \vee P(2) \vee P(3) \vee \ldots$.

However, the ellipsis (...) is not usually a part of formal logic; formalizing it would be difficult. To some extent, the role of $\forall$ and $\exists$ is to replace ellipses with a notation that can be formalized and made rigorous more readily.

5.42. It's not hard to remember which quantifier is which: "**All**" ($\forall$) and "**Exists**" ($\exists$).

But actually, "for all" isn't quite right. The mathematical meaning of $\forall$ is slightly closer in meaning to "for each." And, unfortunately, that messes up the mnemonic scheme indicated in the preceding paragraph.

In nonmathematical English, "for all x" suggests uniformity, as though perhaps *all* the x's are treated by the *same* procedure, or they all satisfy the same condition for the same reason. In contrast, "for each" conveys more of the idea that *each* of the x's is treated separately, individually, perhaps differently — all the x's satisfy the same condition, but perhaps for different reasons. That's closer to the meaning that mathematicians actually intend for $\forall x$.

Quantifiers

5.43. *Examples with integers.* In these examples, the domain (or universe) is $\mathbb{Z}$, the set of all integers. The boxes are just added for emphasis, and are not part of the official notation.

a. $\boxed{\forall x \quad x^2 \geq 0}$ says that "for each integer x, the number x^2 is nonnegative." That's *true*.

b. $\boxed{\exists x \quad x^2 = 9}$ says that "there exists at least one integer x such that $x^2 = 9$." That's *true*. In fact, there are two such integers: 3 and -3.

c. $\boxed{\forall x \exists y \quad x + y = 0}$ means that "for each integer x there exists at least one integer y such that $x+y = 0$." This is *true*: Given any x, we can find a suitable y. Indeed, just choose y to be equal to $-x$. In fact, that's the only choice that will work, so in this case "at least one" turns out to be "exactly one." Note that we get to choose y after x has been chosen; we can choose y so that it depends on x. The symbol $\forall$ is replaced better by the word "each" than by the word "all" in this example.

The quantifiers are *nested*; that means one is inside another. That could be emphasized with parentheses:

$$\boxed{\forall x (\exists y \quad (x + y = 0))}.$$

We might also emphasize it by writing with indented paragraphs:

> *For each* integer x, it is true that
> > *there is* some integer y such that
> > > $x + y = 0.$

d. $\boxed{\exists y \forall x \quad x + y = 0}$ means that "there exists some integer y with the property that, for every integer x, we have $x + y = 0$." We've changed the order of a few symbols, but in so doing we've greatly changed the meaning; the parentheses go this way: $\boxed{\exists y \, (\forall x \quad (x + y = 0))}$.

The resulting statement is *false*. For this statement, we would have to choose y before we choose x; we have to choose *one* value of y that works with *all* choices of x. In the indented paragraph form, this would say:

> *There is* some integer y with the property that
> $\left\{ \begin{array}{l} \textit{for every} \text{ choice of } x, \text{ we will have:} \\ \quad \left\{ x + y = 0. \right. \end{array} \right.$

5.44. The order of quantifiers is crucial, and keeping track of the order takes some practice. Here are some rules to keep in mind:

- As the examples above show, we cannot switch the order of an $\exists$ and an $\forall$.
- However, we can switch the order of two adjacent $\exists$'s. For instance, $\exists x \exists y \ \ x + y = 12$ means exactly the same thing as $\exists y \exists x \ \ x + y = 12$.
- Likewise, we can switch two adjacent $\forall$'s. For instance, $\forall x \forall y \ \ x^2 + y^2 \geq 0$ means the same thing as $\forall y \forall x \ \ x^2 + y^2 \geq 0$.

In general, quantifiers should be read from left to right. However, in short and simple statements, when there is little possibility of misinterpretation, mathematicians sometimes change the order to make it fit better with conventional English. For instance, $\forall x \ (x^2 \geq 0)$ is a true statement; it says that

> whenever x is an integer, then $x^2 \geq 0$.

Some mathematicians might write that as

> $x^2 \geq 0$ whenever x is an integer.

That sounds a little better, and it uses one less word (we don't need "then"). This kind of twisting of order should only be used with caution, and only in very simple sentences that have just one quantifier. And you should *never, never* use that twisted order in a symbolic expression, such as $x^2 \geq 0 \ \forall x$; in my opinion that is in very bad taste. (A few experienced mathematicians do write things that way, but I wish they wouldn't.)

Quantifiers

5.45. *Example: continuity of functions.* Calculus books sometimes give two definitions of continuity. The first is fairly simple: "f is continuous" means that "when t is close to x, then $f(t)$ is close to $f(x)$." That definition, supplemented by many examples, is probably adequate for the needs of a calculus course. Indeed, Isaac Newton had not much more than that when he invented calculus and calculated the movements of the planets.

But that definition is not precise enough for *proving* the calculations or the theorems of calculus, because it doesn't explain how close is "close enough" — i.e., *how* is the closeness of x and t related to the closeness of $f(x)$ and $f(t)$. Consequently, calculus books go on to give the epsilon-delta definition of continuity, developed over a century later by Cauchy and Weierstrass. We have presented it in statement (∗) of 5.39. It can be restated in symbols as

$$(\forall x)\,(\forall \varepsilon > 0)\,(\exists \delta > 0)\,(\forall t)\,\left[|x - t| < \delta \;\Rightarrow\; |f(x) - f(t)| < \varepsilon\right]$$

where the variables $x, \varepsilon, \delta, t$ represent real numbers. The quantifiers must be kept in the right order, or else the meaning is changed greatly.

This definition has *four* nested quantifiers, far more than one encounters in any ordinary sentence in nonmathematical English. I'm sure that's why so many calculus students have so much difficulty understanding the definition of continuity; its grammatical structure is very unfamiliar. Even an experienced mathematician takes a long time to digest an unfamiliar combination of three or more quantifiers, though a familiar combination gets easier with practice — e.g., an experienced mathematician has no difficulty with the definition of continuity given above.

Actually, we used some abbreviations in our notation when we wrote "$\forall \varepsilon > 0$" and "$\exists \delta > 0$." Those are sometimes called *bounded quantifiers*, though I think "qualified quantifiers" or "restricted quantifiers" would be more descriptive. Most mathematicians prefer bounded quantifiers, because it shortens what they must read or write. But logicians generally use plain, unrestricted quantifiers. For them, this is the definition of continuity:

$$(\forall x)\,(\forall \varepsilon)\left[(\varepsilon > 0) \Rightarrow \left\langle (\exists \delta)\left\{(\delta > 0) \quad \text{and} \quad \left[(\forall t)\left(|x-t| < \delta \Rightarrow |f(x)-f(t)| < \varepsilon\right)\right]\right\}\right\rangle\right].$$

Translated into words, with indentations for nested clauses, that takes the form of the boxed expression below.

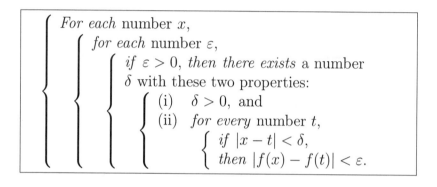

5.46. *The continuity game (optional).* The quantifiers $\forall$ and $\exists$ work in opposite directions, toward opposing goals. One way to understand quantifiers a little better is to describe the definition of continuity as a game played between two opposing persons or teams,

- the "$\exists \delta$ team," which tries to prove that some function $f(x)$ is continuous, and

- the "$\forall \varepsilon$ team," which tries to prove the function is discontinuous.

The game proceeds as follows.

(1) The $\forall \varepsilon$ team goes first. This team chooses a real number x and a positive number ε.

(2) The $\exists \delta$ must respond with some suitable positive number δ. Now x, ε, δ are held fixed, and we begin a subgame.

  (2-a) The $\forall \varepsilon$ team specifies some number t in the interval $(x-\delta, x+\delta)$.

(2-b) Then it's the $\exists\delta$ team's turn again. They must prove that the selected numbers satisfy $|f(x) - f(t)| < \varepsilon$. If they fail, they have lost the entire game, and f is declared discontinuous. But if they succeed, the game is not over. Rather, the number t is eliminated, and play goes back to step (2-a) again.

Steps (2-a) and (2-b) are repeated infinitely many times, eliminating numbers t one by one. Alternatively, if the $\exists\delta$ team includes some advanced players, they may attempt to produce an algebraic proof that can simultaneously eliminate all of the numbers in $(x - \delta, x + \delta)$.

(3) If the $\exists\delta$ team succeeds in eliminating all of $(x - \delta, x + \delta)$, then the game still is not over. Rather, it is declared that the pair (x, ε) has been defeated by the number δ. The pair (x, ε) is eliminated, and play goes back to step (1) again.

Steps (1)–(3) are repeated infinitely many times. If the $\exists\delta$ team succeeds in eliminating every ordered pair in $\{(x, \varepsilon) \in \mathbb{R}^2 : \varepsilon > 0\}$ — either one by one, or by an algebraic proof that covers infinitely many ordered pairs simultaneously — then the $\exists\delta$ team is declared victorious, and the function f is declared continuous.

5.47. *Exercises.* Run through the following statements twice — one time with the integers for your domain, and the second time using the positive real numbers for your domain. (*Caution:* 0 is an integer but not a positive number.) For each statement, say whether it is true or false.

a. $\forall x \quad x^2 = x$.
b. $\exists x \quad x^2 = x$.
c. $\exists x \quad x^2 = 2$.
d. $\forall x \forall y \quad (x > y \text{ or } y > x)$.
e. $\forall x \forall y \exists z \quad x + z = y$.

f. $\forall x \exists y \quad x^2 = y$.
g. $\exists y \forall x \quad x^2 = y$.
h. $\forall y \exists x \quad x^2 = y$.
i. $\exists x \forall y \quad x^2 = y$.

5.48. *Paradox of universality.* When we combine "implies" and "or" in one sentence, we run into some difficulties involving the distinction between "unvaryingly true" and "true at each instant." Mathematics makes that distinction, but English does not.

The formula $(P \to Q) \vee (Q \to P)$, called "chain ordering" elsewhere in this book, is a truth of classical logic.[9] It always evaluates to "true," regardless of whether we plug in "true" or "false" for P and Q; that was demonstrated in an exercise at the end of 2.38. But one instance of this formula is

(1) $\begin{cases} \text{if today is Sunday then today is Wednesday, or} \\ \text{if today is Wednesday then today is Sunday.} \end{cases}$

That sounds like two wrong statements; how could they make a right statement? What is going on here?

If we speak English in its usual fashion, statement (1) could be misinterpreted this way:

(2) $\begin{cases} \text{At least one of these is an always-true statement:} \\ \quad \text{(2a) If today is Sunday then today is Wednesday.} \\ \quad \text{(2b) If today is Wednesday then today is Sunday.} \end{cases}$

Clearly (2) is false. Indeed, (2a) is not an always-true statement, since it fails on Sundays; and (2b) is not an always-true statement, since it fails on Wednesdays.

However, the classical logician could read statement (1) this way:

(3) $\begin{cases} \text{On each day, at least one of the statements} \\ \text{(2a) or (2b) is true, though which is true} \\ \text{may vary depending on what day it is.} \end{cases}$

Statement (3) actually is true, though rather uninformative — it merely brings to our attention some facts that are trivially true. Indeed, (2a) is vacuously true on Wednesdays, since its "if" part is not satisfied on those days; likewise, (2b) is vacuously true on Sundays. On the other five days of the week, both (2a) and (2b) are vacuously true. The formula $(P \to Q) \vee (Q \to P)$ is true in classical logic regardless of what P and Q are; the statement doesn't tell us anything about P or Q.

All of this can be clarified with quantifiers. Abbreviate

$$S(x) = \text{``}x \text{ is Sunday,''} \qquad W(x) = \text{``}x \text{ is Wednesday,''}$$

---

[9]Not to be confused with $(P \to Q) \wedge (Q \to P)$, which is discussed in 11.13 for other purposes.

where x varies over the days of the week. Then (3) can be restated as

$$(\forall x) \left\{ [W(x) \rightarrow R(x)] \vee [R(x) \rightarrow W(x)] \right\}$$

(which is true), whereas (2) makes this rather different and false assertion:

$$\left\{ (\forall x) [W(x) \rightarrow R(x)] \right\} \vee \left\{ (\forall x) [R(x) \rightarrow W(x)] \right\}.$$

5.49. *Negated quantifiers.* In classical logic, the two quantifiers are dual to each other, in the sense that

$$\neg \begin{bmatrix} \text{one} \\ \text{quantifier} \end{bmatrix} \begin{bmatrix} \text{some} \\ \text{statement} \end{bmatrix} = \begin{bmatrix} \text{the other} \\ \text{quantifier} \end{bmatrix} \neg \begin{bmatrix} \text{the same} \\ \text{statement} \end{bmatrix}.$$

Following are some examples. These two statements mean the same thing:

$\neg \forall x\, S(x)$ It is *not* true that *every* student in this class is a senior.

$\exists x\, \neg S(x)$ There is *at least one* student in this class who is *not* a senior.

Also, these two statements mean the same thing:

$\neg \exists x\, P(x)$ It is *not* true that *at least one* of my pencils has a point sharp enough to use.

$\forall x\, \neg P(x)$ *Every one* of my pencils has a point so dull as to be *un*usable.

This means that we can move a negation sign past a quantifier, if we switch the type of quantifier.

5.50. *Ambiguity in English.* The word "all" is ambiguous in many uses of everyday English. For instance, the state song of the state of Kansas is "Home on the Range." The last line of the first verse and also of the chorus is

and the skies are not cloudy all day.

As written English, that line is ambiguous; it has two different but equally correct interpretations. Let t denote time of day, and let
$$C(t) = \text{"the sky is cloudy at time } t\text{."}$$
Then the two interpretations are

(i) $\quad (\forall t) \neg C(t) \quad$ (the sky is clear all day), or

(ii) $\quad \neg [(\forall t) C(t)] \quad$ (the sky is clear at least once a day).

However, in *spoken* or *sung* English, the line is not ambiguous. Oral English carries additional information of intonations. The song is arranged so that the stress is on the word "cloudy":

"and the skies are not <u>cloudy</u> all day"   means (i).

The meaning would be different if the stress were placed elsewhere:

"and the skies are <u>not</u> cloudy <u>all</u> day"   means (ii).

In written English and in mathematics, we must choose our words carefully to avoid such ambiguities.

5.51. *Example: The "stays away" theorem.* Let f be a real-valued function defined on all of $\mathbb{R}$. Then

$\quad f$ is not continuous

if and only if

> there exists a number x and a sequence (x_n) that converges to x (in fact, with $|x_n - x| < 1/n$), with the property that the sequence $(f(x_n))$ "stays away from" $f(x)$, in the sense that there exists a positive constant ε such that $|f(x_n) - f(x)| \geq \varepsilon$ for all n.

Remarks and proof. First we give an example. The function

$$f(x) = \begin{cases} \cos(1/x), & x \neq 0, \\ 0, & x = 0, \end{cases}$$

is discontinuous at 0. One of its "staying away" sequences is $x_n = 1/(2\pi n)$. Then $f(x_n) = 1$ for all n.

This theorem will not be used later in the book. Our goal here is not simply to prove the theorem, but to explain the use of quantifiers in the proof. Without such explanation, the proof would be a lot shorter. In fact, if the intended audience were experienced mathematicians, we probably wouldn't use the symbols ∀ and ∃ at all in this proof. Most mathematicians (other than logicians) generally prefer to use the words "for all" and "there exists" rather than the symbols — but most experienced mathematicians know how to use those words properly. A beginning student might be lost in such a proof. We shall use those symbols in the proof below, not to shorten the proof, but to emphasize and perhaps clarify the ways in which the words are being used.

We begin the proof by recalling from 5.45 the definition of continuity, and then putting a negation in front of it. Thus "f is not continuous" can be restated as

$$\neg (\forall x)(\forall \varepsilon > 0)(\exists \delta > 0)(\forall t) \left[|x - t| < \delta \Rightarrow |f(x) - f(t)| < \varepsilon \right].$$

Next, move the negation past the quantifiers, switching their types. Thus

$$(\exists x)(\exists \varepsilon > 0)(\forall \delta > 0)(\exists t) \neg \left[|x - t| < \delta \Rightarrow |f(x) - f(t)| < \varepsilon \right].$$

Move the negation inside the implication; thus

$$(\exists x)(\exists \varepsilon > 0)(\forall \delta > 0)(\exists t) \left[|x - t| < \delta \not\Rightarrow |f(x) - f(t)| < \varepsilon \right].$$

The logic we're using is classical, so $A \not\Rightarrow B$ is the same thing as $A \wedge \neg B$. Thus "f is not continuous" can be restated as

$$(\exists x)(\exists \varepsilon > 0)(\forall \delta > 0)(\exists t) \left[|x - t| < \delta \text{ and } |f(x) - f(t)| \geq \varepsilon \right].$$

Now comes the trickiest part of the proof: replacing the set of all positive numbers δ with the sequence $1, \frac{1}{2}, \frac{1}{3}, \ldots$. It will be helpful to temporarily adopt this notation: Think of x and ε as constants, and let

$$P(\delta) \quad = \quad \text{``} (\exists t) \left[|x - t| < \delta \text{ and } |f(x) - f(t)| \geq \varepsilon \right]. \text{''}$$

Thus $P(\delta)$ is a statement about the number δ. It might or might not be true, depending on what δ is. Whether or not it is true also depends on what x, ε, f are, but for now we're viewing those as fixed, and we're just going to look at what happens when δ changes; that's why we're writing $P(\delta)$ instead of $P(\delta, x, \varepsilon, f)$.

Also note that the expression "$P(\delta)$" does not mention t. That's because t is a bound variable, within the expression $P(\delta)$. That is, $P(\delta)$ is true if and only if *there exists* a t satisfying a certain condition.

Looking at the definition of $P(\delta)$ closely, we see that

$$\text{if} \quad \delta_1 \leq \delta_2 \quad \text{then} \quad P(\delta_1) \Rightarrow P(\delta_2).$$

Indeed, suppose that $\delta_1 \leq \delta_2$, and suppose that $P(\delta_1)$ is true. Then there is some t that satisfies $|x-t| < \delta_1$ and $|f(x)-f(t)| \geq \varepsilon$. Since $\delta_1 \leq \delta_2$, that same t also satisfies $|x-t| < \delta_2$, and so $P(\delta_2)$ is true.

Each member of either of the sets $\{1, \frac{1}{2}, \frac{1}{3}, \ldots\}$ and $(0, +\infty)$ is greater than or equal to some member of the other set. Consequently

$$\left[(\forall \delta > 0)\, P(\delta)\right] \quad \Longleftrightarrow \quad \left[P(1) \wedge P\left(\frac{1}{2}\right) \wedge P\left(\frac{1}{3}\right) \wedge \cdots\right].$$

Moreover, in the statement $P(1/n)$, we can relabel t as x_n. Thus $P(1/n)$ says

$$(\exists x_n)\left[|x - x_n| < \frac{1}{n} \ \text{ and } \ |f(x) - f(x_n)| \geq \varepsilon\right]$$

and so finally our statement "f is not continuous" can be restated as

$$(\exists x)(\exists \varepsilon > 0)(\forall n \in \mathbb{N})(\exists x_n)$$
$$\left[|x - x_n| < \frac{1}{n} \ \text{ and } \ |f(x) - f(x_n)| \geq \varepsilon\right].$$

That is just the "stays away" condition stated in the theorem.

INDUCTION

5.52. A definition or construction is *recursive*, or *by recursion*, if it is carried out by a sequence of steps, where each step after the first may depend on earlier steps. Examples can be found in 5.67, 6.7, 27.2, and 29.2.

A recursive *proof* generally is instead called *inductive*, or a *proof by induction*. Inductive proofs are explained below. We will use many inductive proofs in this book. Some of them are conceptually deeper than the simple examples given in the next few pages. See, for instance, 6.10, 21.1.a, and 22.5.

Remark. In nonmathematical English, an *inductive* argument is one that goes from the specific to the general, while a *deductive* argument is one that goes from the general to the specific. In mathematical English, the words "inductive" and "deductive" may be used interchangeably for any kind of reasoning, except that the principle listed in 5.55 is usually called the Induction Principle.

5.53. *Example of recursive definition.* The expression "$n!$" does not mean "shout n." It is called "n factorial," and it is defined recursively as follows: $0! = 1$, and

$$(n+1)! = n! \cdot (n+1) \qquad \text{for all} \quad n = 0, 1, 2, 3, \ldots.$$

(We customarily define $0! = 1$ just "because it works" — i.e., because many formulas come out more simply that way.)

For instance, $3! = 1 \cdot 2 \cdot 3 = 6$ and $4! = 1 \cdot 2 \cdot 3 \cdot 4 = 24$. In general, for positive integers n, we have

$$n! \;=\; 1 \cdot 2 \cdot 3 \cdot \cdots \cdot (n-1) \cdot n.$$

Note that $n!$ grows very rapidly when n gets large.

Exercise. Note that $10! = 3628800$ ends in two zeros. In how many zeros does $100!$ end?

5.54. We present two forms of the induction principle. The first one is only presented as a way of explaining the second one; the second one is the version we'll generally use.

Induction principle (obvious version). Assume that an infinite sequence of statements $P(1), P(2), P(3), \ldots$ is given, such that all of the statements

$$P(1), \quad P(1) \Rightarrow P(2), \quad P(2) \Rightarrow P(3), \quad P(3) \Rightarrow P(4), \quad \ldots$$

are true (where $\Rightarrow$ denotes "implies"). Then all of the statements $P(1), P(2), P(3), \ldots$ are true.

5.55. *Induction principle (customary formulation).* Assume an infinite sequence of statements $P(1), P(2), P(3), \ldots$ is given, satisfying both of these two conditions:
a. *Initialization step.* $P(1)$ is true.
b. *Transition step.* For each positive integer n,
$$\begin{cases} \text{if} & \text{(induction hypothesis)} \quad P(n) \text{ is true,} \\ \text{then} & \text{(induction conclusion)} \quad P(n+1) \text{ is too.} \end{cases}$$
Then all of the statements $P(1), P(2), P(3), \ldots$ are true.

5.56. *More remarks.* The induction principle can be restated in symbols:

$$\Big\{ P(1) \ \wedge \ \big[\forall n \ P(n) \Rightarrow P(n+1)\big] \Big\} \ \Rightarrow \ \big[\forall n \ P(n)\big],$$

where n varies over the positive integers.

Some variants of 5.55 are possible. For instance, instead of beginning with 1, we could use a sequence $P(0), P(1), P(2), \ldots$; then $P(0)$ would be the initialization step. Or, instead of verifying $P(n) \Rightarrow P(n+1)$ for all positive integers n, we could verify $P(n-1) \Rightarrow P(n)$ for all integers $n \geq 2$ (or for all positive integers, if we're initializing with 0). These alternate versions of the induction principle are proved and used in essentially the same way. Some students may be more familiar with one of these variants; experienced mathematicians use the variants interchangeably without even noticing the difference. But when I teach this subject, I usually ask all my students to follow the specific notation and formulation given in 5.55, just because grading homework papers will be easier if most of them look similar.

Induction

5.57. *Technical remarks (optional).* What is the difference between formulations 5.54 and 5.55? It is more formal and theoretical than practical. To apply 5.54, we must first verify *infinitely many different concrete assertions involving particular numbers* — i.e., we must separately verify each of the statements

$$P(1), \quad P(1) \Rightarrow P(2), \quad P(2) \Rightarrow P(3), \quad P(3) \Rightarrow P(4), \quad \ldots,$$

presumably by arithmetic or something like it; no variables are involved. The last three dots are an ellipsis, meaning "and so on." But we can't actually carry out that procedure, because writing out infinitely many verifications would take forever. To apply 5.55, we just need to verify the concrete statement $P(1)$ plus one *abstract* assertion involving the variable n, presumably by some algebraic method. Induction will be presented only informally in this book, but if the algebraic method is written carefully enough, it can be used in certain kinds of *formal* proofs. Induction is one of the tools used to replace ellipses, which cannot be used in formal proofs.

5.58. When applying the induction principle, beginners are urged to write out the first few statements in full detail — i.e., write out precisely what are the statements $P(1)$, $P(2)$, and $P(3)$. This is not a formal necessity in an induction proof, but I have found that it helps some beginners avoid misunderstanding just what $P(n)$ actually represents.

Students who are not yet mathematically sophisticated often have the idea that mathematics consists of just equations and inequalities. But a few *words* here and there are essential to mathematics; more words become necessary in higher mathematics. In an induction proof, it is essential to include a few words that make distinctions

- between what we have already established and what we intend to establish; and
- between what is asserted true for every n, and what is asserted true for some particular n.

5.59. To reduce confusion about the order and format of the steps, I sometimes recommend that beginners should follow a rigid format for induction proofs. Fill in the blanks:

    THEOREM: _____

PROOF: We will show, by induction on n, that the following statement $P(n)$ holds for all positive integers n.

$P(n)$: _____

(*Fill in the blank with a precise statement of $P(n)$. It might or might not be identical to the theorem you're trying to prove. If it is, repeat it anyway. If the statement $P(n)$ is an equation, you should understand that $P(n)$ is the name of the whole equation, not the value of the left side of the equation or the value of the right side of the equation. Therefore, at no point in your proof should you have any equations that say $P(n)$ is equal to something.*)

In particular, $P(1)$ says _____. That statement is true because _____. Thus, we've verified the INITIALIZATION STEP.

(*Optional but recommended:*) We also note, as examples, that $P(2)$ says _____ and $P(3)$ says _____. Those statements are true because _____.

Now for the TRANSITION STEP. Let n be some particular positive integer. We may assume the INDUCTION HYPOTHESIS, $P(n)$, which says _____. We want to use that to prove the INDUCTION CONCLUSION, $P(n+1)$, which says _____. That may be proved as follows:

(*Now you carry out some computational steps, generally involving some algebra. Make sure that each step is justified. It might be best to use a 3-column format, as in 5.36. This algebraic computation is the heart of your proof, but it is NOT the only part of the proof — the other steps are essential too, even if they merely seem like ritual incantations.*)

That proves $P(n+1)$, thus completing the transition step. That completes the proof of the theorem.

5.60. *Principle of strong induction (optional remarks, mainly for teachers).*

$$\Big\langle \forall n \, \{ \, [P(1) \wedge P(2) \wedge \cdots \wedge P(n-1)] \Rightarrow P(n) \, \} \Big\rangle \quad \Rightarrow \quad [\forall n \, P(n)].$$

Or, in words:

Suppose that $P(1), P(2), P(3), \ldots$ are statements, with the property that whenever n is a positive integer such that all the statements before $P(n)$ are true, then $P(n)$ is also true.

Then all the statements $P(1), P(2), P(3), \ldots$ are true.

This principle looks peculiar at first, and perhaps even incorrect — its transition step is more complicated than the one in 5.55.b, and its initialization step seems to be missing altogether. But actually the initialization step is *built into* the more complex transition step. (Indeed, what happens when $n = 1$? There *aren't any* statements before $P(1)$, so the condition that "all the statements before $P(1)$ are true" is vacuously satisfied. So the transition condition tells us that $P(1)$ must be true.)

Combining the initialization and transition steps in this fashion has this advantage: In some induction proofs, the initialization and transition conditions are verified by long but similar arguments; thus we save some work if we combine the two conditions and merge the two verifications. We will use this technique repeatedly; see, for instance, 23.6 and 22.5.

Strong induction has another, even more important advantage. In some arguments, proving $P(n+1)$ using just $P(n)$ would be difficult; proving $P(n+1)$ using all of $P(1), P(2), P(3), \ldots, P(n)$ is much easier.

However, presenting strong induction as a separate principle may be just an unnecessary complication. I have discussed strong induction in this book because some teachers will want to cover it, but I actually do not recommend it for beginners. Examples later in this book (for instance, 23.6) will make it clear that one can obtain all the power of strong induction just by using the ordinary induction principle 5.55 with a very carefully worded version of $P(n)$.

INDUCTION EXAMPLES (OPTIONAL)

5.61. The remainder of this chapter proves various properties about summations, binomial coefficients, the Fibonacci sequence, etc. None of those properties will be used in our study of logic in subsequent chapters. Those properties are proved in the following pages only because their proofs make good illustrations of induction. The goal of this subchapter is to make more understandable the induction proofs that will be covered in 21.1.a, 22.5, 27.8, etc.; but some students or teachers may feel that those proofs are already understandable. Teachers should cover as much or as little of this subchapter as they deem appropriate for their classes.

5.62. *Example: sums of squares.* $1^2 + 2^2 + 3^2 + \cdots + n^2 = \frac{1}{6}n(n+1)(2n+1)$ for all $n \in \mathbb{N}$.

We'll first prove this incorrectly, to illustrate a common error.

Incorrect proof. Let $P(n)$ be the equation $1^2+2^2+3^2+\cdots+n^2 = \frac{1}{6}n(n+1)(2n+1)$. The first three instances of this are as follows:

$P(1)$ is the equation $\quad 1^2 = \frac{1}{6} \cdot 1 \cdot 2 \cdot 3,$

$P(2)$ is the equation $\quad 1^2 + 2^2 = \frac{1}{6} \cdot 2 \cdot 3 \cdot 5,$

$P(3)$ is the equation $\quad 1^2 + 2^2 + 3^2 = \frac{1}{6} \cdot 3 \cdot 4 \cdot 7.$

All three of those are true, so we've probably understood the problem correctly. In particular, $P(1)$ is true; that's the initialization step.

Now assume $P(n)$ is true for some particular positive integer n; we want to prove $P(n+1)$. How can we get $1^2+2^2+3^2+\cdots+(n+1)^2$ on the left side? I've seen too many students do it this way: Start from what we already know, the induction hypothesis:

$$1^2+2^2+3^2+\cdots+n^2 = \frac{1}{6}n(n+1)(2n+1).$$

Now add $2n+1$ to both sides; thus

$$1^2+2^2+3^2+\cdots+n^2+2n+1 = \frac{1}{6}n(n+1)(2n+1)+2n+1.$$

Now, we know that $n^2+2n+1 = (n+1)^2$, so that last equation can be rewritten

$$1^2+2^2+3^2+\cdots+(n+1)^2 = \frac{1}{6}n(n+1)(2n+1)+2n+1.$$

At this point the student assumes that $\frac{1}{6}n(n+1)(2n+1)+2n+1$ must be equal to $\frac{1}{6}(n+1)(n+2)(2n+3)$ (for otherwise the thing we're trying to prove would not be true). So the student, not bothering to check it or unable to track down the error, writes that

$$\frac{n(n+1)(2n+1)}{6}+2n+1 = \frac{(n+1)(n+2)(2n+3)}{6} \quad \text{"by algebra"}$$

and then draws the conclusion

$$1^2+2^2+3^2+\cdots+(n+1)^2 = \frac{1}{6}(n+1)(n+2)(2n+3).$$

Induction examples (optional)

This is $P(n+1)$, as required.

Critique. The preceding proof is wrong. It has *two* serious errors. The last step looks correct, but that's only because the two errors canceled each other out.

The "by algebra" step is erroneous; actually

$$\frac{n(n+1)(2n+1)}{6} + 2n + 1 \ne \frac{(n+1)(n+2)(2n+3)}{6}.$$

That's easy enough to understand: We assumed equality because we expected it. The other error occurred earlier, and is more subtle. Can you find it before reading further?

The other error consists of losing some terms in the *ellipsis* — that is, in the three dots ($\cdots$). Remember, ellipses should only be used to drop terms that complete an evident pattern; enough terms must be retained to keep the pattern evident. Everything becomes clearer if we write out a few more terms in each line. Start with

$$1^2 + 2^2 + \cdots + (n-2)^2 + (n-1)^2 + n^2 = \frac{n(n+1)(2n+1)}{6}.$$

Now add $2n+1$ to both sides; thus

$$1^2 + 2^2 + \cdots + (n-2)^2 + (n-1)^2 + n^2 + 2n + 1$$
$$= \frac{n(n+1)(2n+1)}{6} + 2n+1.$$

Making use of the fact that $n^2 + 2n + 1 = (n+1)^2$, we can rewrite the last equation as

$$1^2 + 2^2 + \cdots + (n-2)^2 + (n-1)^2 + (n+1)^2$$
$$= \frac{n(n+1)(2n+1)}{6} + 2n+1.$$

But now it becomes evident that the n^2 term is missing from the left side, so the left side is *not* equal to $1^2 + 2^2 + 3^2 + \cdots + (n+1)^2$.

A corrected proof (sketched). Start from

$$1^2 + 2^2 + 3^2 + \cdots + n^2 = \frac{1}{6}n(n+1)(2n+1).$$

Add $(n+1)^2$ to both sides. Thus

$$1^2 + 2^2 + 3^2 + \cdots + n^2 + (n+1)^2$$
$$= \frac{1}{6}n(n+1)(2n+1) + (n+1)^2.$$

Now use some algebra to show that $\frac{1}{6}n(n+1)(2n+1)+(n+1)^2$ is equal to $\frac{1}{6}(n+1)(n+2)(2n+3)$. (That algebra *is* correct, though perhaps the beginning student should work out the details!) Thus we obtain

$$1^2 + 2^2 + 3^2 + \cdots + n^2 + (n+1)^2 = \frac{1}{6}(n+1)(n+2)(2n+3).$$

That completes the transition step, and hence the proof.

5.63. *Exercises.* Prove, for all positive integers n:
a. $1 + 3 + 5 + \cdots + (2n-1) = n^2$.
b. $1 + 2 + 3 + \cdots + n = \frac{1}{2}n(n+1)$.
c. $1^3 + 2^3 + 3^3 + \cdots + n^3 = \frac{1}{4}n^2(n+1)^2$.
d. $(1+x)^n \geq 1 + nx$, for any number $x > 0$.
e. $1! \cdot 1 + 2! \cdot 2 + 3! \cdot 3 + \cdots + n! \cdot n = (n+1)! - 1$, where $n!$ is defined as in 5.53.

5.64. *Example: prime factorization.* Every integer greater than 1 can be written as a product of prime numbers.

(Here we permit a single number to be considered a "product"; thus any prime number p is already a product of prime numbers.)

Proof. First let us review the relevant definitions. An integer $k > 1$ is *composite* if there exist two integers, both greater than 1, whose product is k. If no such pair of integers exists, then k is *prime*.

By induction, we shall show that for every positive integer n,

$P(n)$: $\begin{cases} \text{If } k \text{ is an integer and } 1 < k \leq n, \text{ then} \\ k \text{ can be written as a product of primes.} \end{cases}$

Induction examples (optional)

First, $P(1)$ is vacuously true, since there is no k satisfying $1 < k \leq 1$.

Now suppose $P(n)$ is true; we wish to prove $P(n+1)$. Let k be an integer, with $1 < k \leq n+1$; we wish to show k can be factored into primes. If $k < n+1$, then $k \leq n$, and so we know the result by $P(n)$. Assume, then, that $k = n+1$. If k is prime, then k is already factored, and we're done. Assume, then, that k is composite — i.e., $k = ab$ for some integers a, b with $1 < a < k$ and $1 < b < k$. Then $a, b \leq n$. By statement $P(n)$ we know that a and b can both be represented as products of primes. Combine those two representations; thus we write $k = ab$ as a product of primes.

5.65. Exercise. Prove that the prime factorization is unique, up to ordering. That is, if $p_1 p_2 \cdots p_m = q_1 q_2 \cdots q_n$, where the p_i's and q_j's are primes, then $m = n$, and the p_i's are the same as q_j's except possibly for the order in which they are listed.

5.66. Exercise/joke. We will prove that *all horses have the same color.* (See if you can find the error in this proof.) The proof will be by induction; for $P(n)$ we take the statement that "in any group of n horses, all the horses have the same color." Obviously $P(1)$ is true. Now, assume $P(n-1)$ and we'll prove $P(n)$. We have n horses to consider. If we remove any one of those horses, the remaining $n-1$ are of one color, C_1, by the induction hypothesis. If we remove a different one of those n, again the remaining $n-1$ are of one color, C_2, by the induction hypothesis. Those two sets of $n-1$ overlap, so those two colors C_1 and C_2 are the same. Thus all n horses have the same color.

5.67. The *binomial coefficient n over k* is the number denoted by $\binom{n}{k}$, defined recursively for all integers n and k with $0 \leq k \leq n$ as follows:

- $\binom{n}{0} = \binom{n}{n} = 1$ for $n \geq 0$, and
- $\binom{n}{k-1} + \binom{n}{k} = \binom{n+1}{k}$ for $1 \leq k \leq n$.

In particular, this yields $\binom{n}{1} = \binom{n}{n-1} = n$ and $\binom{n}{2} = \binom{n}{n-2} = \frac{1}{2}n(n-1)$ (using 5.63.b). The number $\binom{n}{k}$ is also known as *n choose k*, because it is the number of different ways that we can choose an unordered subset of k objects from a given collection of n different objects.

The recursive definition yields a triangular array of numbers known as *Pascal's Triangle*, shown below. It has 1's along its border, and each number inside the triangle is the sum of the two numbers above it. The row $1, n, \ldots, n, 1$ yields the numbers $\binom{n}{0}, \binom{n}{1}, \ldots, \binom{n}{n}$. For instance, $\binom{6}{2} = 15$.

Pascal's triangle (5.67)

```
                    1
                 1     1
              1     2     1
           1     3     3     1
        1     4     6     4     1
     1     5    10    10     5     1
  1     6    15    20    15     6     1
1    7    21    35    35    21     7     1
        ⋮           ⋮           ⋮
```

5.68. Exercises. By induction on n, show:

a. $\binom{n}{k} = \dfrac{n!}{k!(n-k)!}$, where $k!$ is defined as in 5.53. For instance,

$$\binom{5}{2} = \frac{5!}{2!\,3!} = \frac{1\cdot 2\cdot 3\cdot 4\cdot 5}{1\cdot 2\cdot 1\cdot 2\cdot 3} = \frac{120}{2\cdot 6} = 10.$$

b. $(a+b)^n = \sum_{k=0}^{n} \binom{n}{k} a^{(n-k)} b^k$. For instance,

$$(a+b)^6 = a^6 + 6a^5 b + 15 a^4 b^2 + 20 a^3 b^3 + 15 a^2 b^4 + 6 a b^5 + b^6.$$

c. $\dfrac{d^n}{dx^n}\bigl[p(x) q(x)\bigr] = \sum_{k=0}^{n} \binom{n}{k} \left[\dfrac{d^{n-k} p(x)}{dx^{n-k}}\right]\left[\dfrac{d^k q(x)}{dx^k}\right]$. For instance,

$$(pq)''' = p'''q + 3p''q' + 3p'q'' + pq'''.$$

5.69. Definition. The *Fibonacci sequence* is defined recursively by $f_0 = 0$, $f_1 = 1$, and $f_{n+1} = f_n + f_{n-1}$ for all positive integers n. Thus, the first few Fibonacci numbers are

n	0	1	2	3	4	5	6	7	8	9	10	11	12
f_n	0	1	1	2	3	5	8	13	21	34	55	89	144

Many interesting properties of this sequence can be proved by induction.

5.70. Exercises. Prove by induction that for all $n \in \mathbb{N}$:

a. $f_{n+1} f_{n-1} - f_n^2 = (-1)^n$.
b. $f_1^2 + f_2^2 + f_3^2 + \cdots + f_n^2 = f_n f_{n+1}$.
c. $(f_{n-1} + f_{n+1}) f_n = f_{2n}$.

Induction examples (optional)

d. Each of $f_{2n}, f_{3n}, f_{4n}, \ldots$ is divisible by f_n.

e. (Matrix multiplication.) $\begin{bmatrix} 0 & 1 \\ 1 & 1 \end{bmatrix}^n = \begin{bmatrix} f_{n-1} & f_n \\ f_n & f_{n+1} \end{bmatrix}$.

f. Summing along a *diagonal* of Pascal's triangle yields a Fibonacci number. For instance,

$$f_6 = \binom{5}{0} + \binom{4}{1} + \binom{3}{2},$$

$$f_7 = \binom{6}{0} + \binom{5}{1} + \binom{4}{2} + \binom{3}{3},$$

and in general $f_{n+1} = \binom{n}{0} + \binom{n-1}{1} + \binom{n-2}{2} + \binom{n-3}{3} + \cdots$ where the summation is continued through any remaining nonzero terms.

g. $f_n = (\alpha^n - \beta^n)/\sqrt{5}$, where $\alpha = (1+\sqrt{5})/2$ and $\beta = (1-\sqrt{5})/2$. Also, $\lim_{n\to\infty} f_{n+1}/f_n = \alpha$. *Hint:* First verify that α and β are the roots of the quadratic equation $x^2 = x + 1$. *Remark:* The number α is often called the *Golden Ratio*. It has many interesting geometric properties that can be found in the literature.

h. $1/f_{11} = 0.011235\cdots$. Write a few more digits of the decimal expansion, explain the pattern, and prove it.

Chapter 6

Definition of a formal language

6.1. We now begin our development of *formal* logic. In this chapter we will specify a formal language that will be used for all the logics studied throughout the remainder of this book. This language will be used first for semantics (beginning in Chapter 7), then for syntactics (beginning in Chapter 12), and finally for the ideas that combine semantics and syntactics (starting in Chapter 21).

We emphasize that this is *a* formal language, not *the* formal language. Different textbooks on logic may use slightly different formal languages. Presently there is no consensus on the best way to do this (unlike, say, the definition of "complete ordered field" or "topological space").

The Alphabet

6.2. Our formal language $\mathcal{L}$ will use some or all of the following primitive symbols:

$$\begin{array}{rl} \text{parentheses} & (\ ) \\ \text{propositional variable symbols} & \pi_1 \quad \pi_2 \quad \pi_3 \quad \ldots \\ \text{the implication (if\ldots then) symbol} & \rightarrow \\ \text{the conjunction (and) symbol} & \wedge \\ \text{the disjunction (or) symbol} & \vee \\ \text{the negation (not) symbol} & \neg \end{array}$$

The alphabet

(Informally, we will also use abbreviations discussed in 6.21.)

Initially, the symbols are meaningless, like the "slithy toves" mentioned in 2.20: We don't know what they are, but we will soon see how they fit into our grammar. The symbols will be endowed with meanings only as we adopt assumptions about those symbols; the assumptions and meanings will vary from one logic to another.

As a convenience for our discussions — so that we won't be reduced to grunting and pointing — we have attached names to our symbols. For instance, we might refer to "$\vee$" as "or" or "disjunction." But the student should put aside (i.e., temporarily forget) any usual meanings for these words or symbols.

More precisely, "$\vee$" is not disjunction. It is *the disjunction symbol*, and it will be used to represent different kinds of disjunctions in different logics. Those different disjunctions are different meanings for the symbol. Analogous remarks apply to the other symbols $\to, \wedge$, etc.

The symbol π_3 (for instance) is to be viewed as just one indivisible symbol; it is *not* to be viewed as somehow made up from components "π" and "3." Think of it as analogous to the letter "A," which is commonly drawn with three straight line segments but is considered to be just one symbol. You may also find it helpful to think of a computer whose keyboard has, not the usual 101 keys, but infinitely many keys; "π_3" is the label on one of the key caps.

In this book we'll study the "sentences" that can be typed on such a keyboard, but we won't disassemble any of the keys. That is, *within* our formal system, we won't separate the "π" from the "3." However, when we reason *about* our formal system, we can make use of the mathematical properties of the number 3. One other advantage of $\pi_1, \pi_2, \pi_3, \ldots$ over $a, b, c, \ldots$ is that we can't run out of symbols.

6.3. *Remarks (optional).* Note that "=" is *not* in our list of formal symbols. Equality is included in set theory and most other applied logics, and in some predicate logics, but it is generally not a part of propositional logic, the subject of this book. (See the distinctions in 2.19.)

The equality symbol continues to have its usual meaning in our metalanguage. Thus, in this book, the expression "$A = B$" means that the two formulas A and B are the same. It does not represent *one* formula that includes an equals sign as one of its symbols.

Also note that commas and periods are not among our formal symbols. When we refer to a string of symbols such as ")¬→π_1→π_1)(∨," the comma at the end is not actually part of the string of symbols we are discussing, despite its appearance inside the delimiting quotation marks. The comma is inside the quotation marks, rather than outside, in accordance with rules of English grammar; I apologize for any confusion that may be caused by my adherence to that rule.

6.4. *Remarks (optional).* For simplicity of notation, we have chosen an infinite alphabet, but a finite alphabet would also suffice for many of our needs. We could use the alphabet

$$ (\quad) \quad \neg \quad \vee \quad \wedge \quad \rightarrow \quad \pi \quad | $$

consisting of just *eight* symbols, and let the roles of $\pi_1, \pi_2, \pi_3, \pi_4, \ldots$ be taken by the finite sequences of symbols

$$ \pi| \quad \pi|| \quad \pi||| \quad \pi|||| \quad \cdots $$

respectively. (Thus, we could use a keyboard with only eight keys.) However, that notation has certain disadvantages:

- This book would then be longer, since expressions like $\pi|||||||||||$ take up more space.

- More importantly, the book would then be much harder to read, since $\pi||||||$ cannot be easily distinguished from $\pi|||||||$ by the unaided human eye.

- Most importantly, our use of $\pi_1, \pi_2, \pi_3, \ldots$ enables us to discuss π_j and π_k, where j and k are *variables* (in our metalanguage, not in our object language) — i.e., where j and k can represent any positive integers. Analogous concepts are less readily expressed with the notation $\pi||||||||||||$.

Also for simplicity of notation, we have chosen an alphabet that is only countably infinite (see 3.66). That alphabet is adequate for most applications of logic, but some logicians prefer to allow uncountable alphabets as well. (Imagine an even larger infinite computer keyboard, with *real numbers* written on the key caps!) Such a formal language facilitates certain kinds of advanced reasoning about uncountable sets (such as the set of real numbers). Switching to an uncountable alphabet does not significantly affect the reasoning carried out in this book, with one exception: the simple argument in 29.2 must be replaced with a more advanced argument involving Zorn's Lemma.

6.5. By a *string* (or *word*) we shall mean a finite sequence of (not necessarily distinct) symbols from a given alphabet. Its

The alphabet 209

length is the number of (not necessarily different) symbols in that sequence. For instance, $\to\pi_2\pi_{11}\land\neg\land$ is a string of length 6. (Length should not be confused with *rank*, defined in 6.9.)

For the moment, we view these symbols as meaningless; they are simply manipulable marks on paper. One operation that can be performed with meaningless strings is that of *concatenation* — i.e., following one string with another. For instance, if we concatenate the strings $\lor\pi_4)\pi_6$ and $\pi_3\neg\to\pi_{21}$, we obtain the string $\lor\pi_4)\pi_6\pi_3\neg\to\pi_{21}$. And we can even state a simple abstract principle:

> If A and B are strings of lengths m and n, respectively, then the result of concatenating A and B is a string of length $m+n$.

6.6. The symbols $(\,)\neg\lor\land\to\pi_j$ are the only symbols in our object language — i.e., the formal language that is part of the formal system we shall study. However, we shall also use some other symbols in our metalanguage — i.e., in the informal discussion *about* the object language. (See terminology in 2.17.) We will use the letters

$$A\quad B\quad C\quad D\quad E\quad\cdots$$

as *metavariables*; they are informal abbreviations for strings. (We will narrow the usage of metavariables slightly, starting in 6.27.) For instance, it is more convenient and more precise to say

> suppose C is a string containing ten symbols

than to say

> consider a string that is like $\lor\pi_4)\pi_6\pi_3\neg\to\pi_{21}\land\land$ or like $\pi_1\pi_2\lor\pi_3\lor\lor\to\pi_7\neg\pi_7$.

Consequently, we'll be writing metavariables more often than we write the formal proposition symbols π_j themselves. However, we emphasize that the expressions "$\lor\pi_4)\pi_6\pi_3\neg\to\pi_{21}\land\land$" and "$\pi_1\pi_2\lor\pi_3\lor\lor\to\pi_7\neg\pi_7$" are actual expressions in our object language, while the letter "C" is not.

In effect, we have two different alphabets of variable symbols:

$$\pi_1, \pi_2, \pi_3, \ldots \quad \text{and} \quad A, B, C, \ldots \, .$$

Some mathematicians manage with just one alphabet by using more complicated rules about substitutions, but in my opinion the two-alphabet approach is easier for beginners. (In fact, on rare occasions we will employ a *third* alphabet; see 6.29.)

When we mention "C," are we talking about the one letter "C," or are we talking about some string that the one letter "C" represents? That may sometimes be ambiguous, but the answer usually will be clear from the context. Answering questions like that becomes trickier in some topics of logic — e.g., in the proof of Gödel's Incompleteness Principles — but we won't run into those difficulties in this book.

We shall mix our formal symbols and metavariables together freely; generally the meaning will be clear from the context. For instance, if A and B represent strings, then the expression $A \vee B$ represents the string obtained by concatenating

- the string of formal symbols represented by A,
- "$\vee$" (a single formal symbol), and
- the string of formal symbols represented by B.

Though we will mix formal symbols and metavariables together in this fashion, the distinction between formal symbols and metavariables will remain important in our reasoning.

THE GRAMMAR

6.7. For simplicity we have begun by considering all strings, but we shall now narrow our attention to just some of the strings: the ones that are *formulas*. We define that term recursively, by these two rules:

- Each of the propositional variable symbols π_1, π_2, π_3, ... is a formula.

The grammar

- If A and B are any formulas, then

$$(A) \to (B), \qquad (A) \wedge (B), \qquad (A) \vee (B), \qquad \neg (A)$$

are also formulas. (In 6.15 and 14.18.g we shall discuss some conventions for reducing the number of parentheses.)

For emphasis, some mathematicians add a third rule (though others consider it as understood, and do not bother to mention it explicitly):

- There are no other formulas, except those obtained by the two rules given above.

Hereafter, the only strings that will concern us are the formulas, and the only strings that we shall represent by metavariables (introduced in 6.6) are formulas.

Remarks. A logic instructor may, on occasion, inadvertently apply the word "formula" to expressions other than those specified above. The following explanation is intended to reduce the students' bewilderment. The instructor is accustomed to using the word "formula" in a wider context. In each branch of mathematics, "formula" means a string of symbols from that branch's alphabet, assembled according to that branch's grammatical rules. Thus, "$x + x + 3 + 3 = 2x + 6$" is a formula in algebra, and "$S \subseteq S \cap T$" is a formula in set theory. It would be best to refer to these as "expressions," to emphasize that they do not fit this course's definition of "formula."

The expression "$A \to B \vdash (\neg B) \to (\neg A)$" is a bit trickier to deal with. It is a formula in some logics other than the ones studied in this book, but our logics do not include "$\vdash$" as a formal symbol. Students, please correct your instructor gently if he or she applies the term "formula" to an expression such as "$A \to B \vdash (\neg B) \to (\neg A)$."

6.8. The construction of a formula from smaller formulas can be illustrated in one and only one way by a *tree diagram*. For instance, the tree diagram for $(((\pi_6) \wedge (\pi_3)) \vee (\pi_{12})) \to (\neg (\pi_3))$ is shown here. Note that we read the original formula from left to right when we distinguish between the "first" π_3 and the "second" π_3. This distinction will be important in some of our uses of tree diagrams; in other uses we may omit the words "the first" and "the second."

The *subformulas* of a formula are the formulas that appear in its tree diagram. For instance, by counting formulas in the diagram above, we see that $(((\pi_6) \wedge (\pi_3)) \vee (\pi_{12})) \to (\neg (\pi_3))$

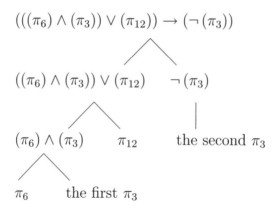

has eight subformulas, including itself and counting both π_3's. We emphasize that every formula is a subformula of itself, and that we count repeated subformulas repeatedly.

6.9. *Definitions.* We shall give two different definitions of the *rank* of a formula, and then show that they yield the same result.

- We may define the rank of a formula to be the number of subformulas that it has. For instance, we just noted in 6.8 that $(((\pi_6) \wedge (\pi_3)) \vee (\pi_{12})) \to (\neg (\pi_3))$ has eight subformulas.

- Alternatively, we may define the rank of a formula to be the number of (not necessarily different) symbols appearing in that formula *other than parentheses*. For instance, using the same example again,

$$(((\underset{1}{\pi_6}) \underset{2}{\wedge} (\underset{3}{\pi_3})) \underset{4}{\vee} (\underset{5}{\pi_{12}})) \underset{6}{\to} (\underset{7}{\neg} (\underset{8}{\pi_3}))$$

Theorem. In any formula G, the number of subformulas is equal to the number of non-parenthesis symbols.

Proof. We prove this by induction on the number of symbols. We shall use the customary principle of induction, given in 5.55. It suffices to show the following statement $S(n)$, for all positive integers n:

The grammar

$S(n)$: $\begin{cases} \text{If } G \text{ is a formula containing } n \text{ or fewer} \\ \text{non-parenthesis symbols, then the number} \\ \text{of non-parenthesis symbols in } G \text{ is equal} \\ \text{to the number of subformulas } G \text{ has.} \end{cases}$

Remark. The statement $S(n)$ has been worded very carefully. A more obvious formulation would begin "if G is a formula containing n non-parenthesis symbols," but saying "n or fewer" makes the proof much easier. See related remarks in 5.60.

For the initialization step, we must prove $S(1)$. If G is a formula with just one non-parenthesis symbol, then G is one of the propositional variable symbols $\pi_1, \pi_2, \pi_3, \ldots$. Then the only subformula of G is G itself, so the number of subformulas is 1.

For the transition step, let n be some positive integer, and assume $S(n)$; we must prove $S(n+1)$. Let G be a formula containing $n+1$ or fewer non-parenthesis symbols. We must prove the number of non-parenthesis symbols in G is the same as the number of subformulas of G.

If G actually has fewer than $n+1$ non-parenthesis we can apply $S(n)$, and we're done.

Thus, we may assume G has exactly $n+1$ non-parenthesis symbols. In particular, it has more than one non-parenthesis symbol. Hence G is a formula of the form $\neg(A)$ or $(A){\rightarrow}(B)$ or $(A){\wedge}(B)$ or $(A){\vee}(B)$. We now consider two cases:

(1) Suppose G is $\neg(A)$. Then A has exactly n non-parenthesis symbols. Hence we can apply $S(n)$ to A. Therefore A has exactly n subformulas. Then G has exactly $n+1$ subformulas — namely, G itself plus the subformulas of A.

(2) Suppose G is $(A){\rightarrow}(B)$ or $(A){\wedge}(B)$ or $(A){\vee}(B)$. Then each of A and B has n or fewer non-parenthesis symbols, and so we can apply $S(n)$ to each of those shorter formulas. Thus each of those shorter formulas has number of non-parenthesis symbols equal to number of subformulas. Moreover,

- the non-parenthesis symbols in G are those in A, together with those in B, together with one more — i.e., the one that appears between the two formulas A and B in the formula G; and

- the subformulas of G are those of A, together with those of B, together with one more: G itself.

Thus the number of non-parenthesis symbols in G equals the number of subformulas of G. This completes the demonstration of $S(n+1)$, and hence the proof.

6.10. *Further remarks.* The rank of a formula will be used in other induction proofs; see 6.14, 8.44, 10.10, 10.11, 10.13, 11.9, 14.16, 27.3, and 27.8.

Our two definitions of "rank" differ only slightly from our definition of "length" given in 6.5. Some mathematicians use just one of these definitions, or another definition slightly different from these. In each case, higher numbers are assigned to longer or more complicated formulas. Most such definitions can be used in induction proofs in essentially the same fashion.

6.11. *Exercise.* Use induction to prove that, in any formula, the number of left parentheses is equal to the number of right parentheses. (Hence the total number of parentheses is even.)

6.12. *Exercise.* For each of the following formulas, give a tree diagram showing all the subformulas, as in 6.8. Also give the ranks of the formula and each of its subformulas.
   a. $(\pi_4) \wedge ((\pi_3) \vee (\neg (\pi_1)))$.
   b. $((\pi_1) \wedge ((\pi_2) \to (\pi_1))) \wedge (\pi_2)$.
   c. $((\pi_4) \to ((\pi_1) \vee (\pi_3))) \to ((\pi_4) \wedge ((\pi_3) \vee (\neg (\pi_1))))$.

6.13. *There are only countably many formulas.* (That fact will be important in some later proofs.)

To show that the formulas are countable, let us temporarily relabel the key caps on our infinite keyboard (discussed in 6.2).

Replace with	(	)	$\to$	$\neg$	$\wedge$	$\vee$	π_1	π_2	π_3	π_4	$\cdots$
	s_1	s_2	s_3	s_4	s_5	s_6	s_7	s_8	s_9	s_{10}	$\cdots$

Then each formula is coded as a finite sequence of positive integers — just read off the subscripts on the s_j's. For example,

the formula is replaced with and coded with	(	π_3	)	∨	(	¬	(	π_1	)	)
	s_1	s_9	s_2	s_6	s_1	s_4	s_1	s_7	s_2	s_2
	1,	9,	2,	6,	1,	4,	1,	7,	2,	2.

Now apply the results of 3.67 and 3.68.

6.14. *Optional: another definition of "formula."* The following definition will not be used in this book, but it may be of interest to some readers. The set of all formulas can also be defined as the smallest set $\mathcal{F}$ (i.e., the intersection of all sets $\mathcal{F}$) having these two properties: (i) It contains all the one-character strings $\pi_1, \pi_2, \pi_3, \ldots$, and (ii) whenever A and B are members of $\mathcal{F}$, then

$$(A) \vee (B), \qquad (A) \wedge (B), \qquad (A) \to (B), \qquad \neg(A)$$

are also members of $\mathcal{F}$. An induction proof can be used to show that this definition yields the same result as 6.7.

REMOVING PARENTHESES

6.15. We use parentheses solely for grouping into tree diagrams. We now adopt this convention:

> *We will permit the omission of parentheses when this does not make the grouping ambiguous — i.e., whenever this does not make the choice of the tree diagram unclear.*

For instance, $(\pi_1) \vee ((\pi_3) \wedge (\pi_4))$ may be written more briefly as $\pi_1 \vee (\pi_3 \wedge \pi_4)$. On the other hand,

$$\pi_1 \vee (\pi_3 \wedge \pi_4) \qquad \text{and} \qquad (\pi_1 \vee \pi_3) \wedge \pi_4$$

have different tree diagrams, so they will be treated as different formulas; neither of them can be written with fewer parentheses.

There is a somewhat surprising asymmetry in the way that

parentheses surrounding a negation symbol can be removed. For instance, $\vee$ is commutative in most logics, and so

$$\pi_3 \vee (\neg \pi_1) \quad \text{and} \quad (\neg \pi_1) \vee \pi_3$$

will be interchangeable later in the book, after we introduce a few axioms governing the behavior of $\vee$. But with or without those axioms, the first of these formulas is easier to work with than the second one, because it can be shortened.

$\pi_3 \vee (\neg \pi_1)$ can be shortened to $\pi_3 \vee \neg \pi_1$, which is a perfectly acceptable, unambiguous formula.

$(\neg \pi_1) \vee \pi_3$ cannot be shortened. The expression $\neg \pi_1 \vee \pi_3$ is unacceptable, because it is ambiguous.

Exercise. What two acceptable formulas does $\neg \pi_1 \vee \pi_3$ ambiguously represent? Draw their tree diagrams, as well as the diagram of $\pi_3 \vee \neg \pi_1$.

6.16. In some branches of mathematics, it is customary to adopt certain conventions about the order of operations. For instance, in algebra, it is understood that

$$ax + b \quad \text{generally means} \quad (ax) + b, \quad \text{not} \quad a(x+b).$$

That is, multiplications are performed before additions except where explicitly indicated by parentheses. Such conventions reduce the need for parentheses.

Many logic books adopt order-of-operation conventions regarding $\vee, \wedge, \rightarrow, \neg$. That is helpful when one unvarying style of logic is used throughout a book. But it is less helpful in the present book, which follows different styles for different logics. Consequently

this book will not adopt any order-of-operation conventions regarding $\vee, \wedge, \rightarrow, \neg$.

Thus the expression $A \vee B \wedge C$ is ambiguous; we have not made it clear whether we intend $A \vee (B \wedge C)$ or $(A \vee B) \wedge C$. This ambiguity is unacceptable, so parentheses are required. Our lack of conventions means more parentheses but less memorizing.

Removing parentheses

6.17. In most logics of interest, the logical operators $\vee$ and $\wedge$ are *commutative* and *associative*. That is, in most logics the pairs

$A \vee B$ and $B \vee A$, $(A \vee B) \vee C$ and $A \vee (B \vee C)$,
$A \wedge B$ and $B \wedge A$, $(A \wedge B) \wedge C$ and $A \wedge (B \wedge C)$

are interchangeable in some sense, and consequently the parentheses usually can be omitted. We will establish a precise version of this interchangeability result in 14.18.f–14.18.g. However, we emphasize that this interchangeability depends on certain assumptions (made in 14.1), and does not apply to logics lacking those assumptions. For instance, $\vee$ and $\wedge$ are *not* commutative in some peculiar logics considered very briefly in 25.8. Consequently,

> in this book, logical connectives will not initially be assumed commutative or associative.

Indeed, we don't make *any* assumptions except where explicitly noted. That's the whole idea of the axiomatic method — see 2.16.

Unlike the other connectives. implication ($\to$) remains noncommutative and nonassociative in all of our major logics, even after we introduce customary axioms.

6.18. *Exercise.* The following formulas are wrong, because they are ambiguous (at least, if we follow this book's rules about parentheses, described in the preceding paragraphs). Rewrite each expression in two different unambiguous ways that have different meanings (i.e., different "tree diagrams"), by adding one pair of parentheses in two different ways. Thus, for each of the problems your answer should be two different, correctly written formulas.

a. $(\pi_1 \vee \pi_2) \to \pi_1 \to \pi_3$.
b. $(\neg \pi_3 \vee \pi_3) \to \neg \pi_1$.
c. $\pi_1 \to \pi_1 \to \pi_3$.
d. $\pi_1 \vee \pi_2 \vee \pi_3$.
e. $\neg \pi_1 \vee \neg \pi_2$.

6.19. *Exercise.* The following formulas are correct, but they can also be written correctly with fewer parentheses. Remove as many parentheses as you can (without making the formula ambiguous, that is). In other words, rewrite each of these formulas with as few parentheses as possible. For each of these problems, your answer should be just one correctly written formula.

a. $(\pi_1) \vee (\pi_2 \vee \pi_3)$.

b. $\pi_2 \vee (\pi_3 \wedge ((\pi_4) \to (\pi_2)))$.

c. $(\pi_1 \to \pi_2) \to ((\neg \pi_2) \to (\neg \pi_1))$.

d. $\pi_1 \vee (\neg \pi_2)$.

e. $\neg(\neg(\pi_1 \vee \pi_1))$.

6.20. As we noted in 6.11, grouping delimiters must occur in pairs — e.g., a left parenthesis must be paired with a right parenthesis. We can count from left to right, adding one whenever we get a left parenthesis, and subtracting one whenever we get a right parenthesis; in this fashion we can tell how deeply nested within the parentheses each symbol is:

$$(\; (\; (\pi_6) \wedge (\pi_3) \;) \vee (\pi_{12}) \;) \to (\neg (\pi_3) \;)$$
$$0 \; 1 \; 2 \; 3 \; 2 \; 3 \; 2 \; 1 \; 2 \; 1 \; 0 \; 1 \; 2 \; 1 \; 0.$$

In this example, for instance, the symbol π_{12} is two parentheses deep — it is contained in a pair of parentheses that is contained in another pair of parentheses.

However, that's more information than we need or want. For our work in logic, generally we don't need to know the actual number of the depth for each symbol; we just need to know which left parenthesis is paired with which right parenthesis, so that we can see which formulas are contained in which other formulas. In the following example, some pairs are visually emphasized with overbraces:

$$(\; (\; (\pi_6) \wedge (\pi_3) \;) \vee (\pi_{12}) \;) \to (\neg (\pi_4) \;).$$

But overbraces are typographically inconvenient. A more popular visual aid is the use of parentheses of different sizes, or other delimiters such as brackets [ ] or braces { } — for instance,

$$\Big([(\pi_6) \wedge (\pi_3)] \vee (\pi_{12}) \Big) \to [\neg (\pi_4)].$$

Defined symbols

Customarily, the larger delimiters are outside the smaller ones (as in the example above). Also, the order { [( )] } is slightly preferred over other orders by most mathematicians. However, that rule is easily overridden when we have some other uses for the delimiter symbols within the same paragraph — e.g., when we use ( ) for sequences or intervals, [ ] for intervals, { } for sets, or ⟦ ⟧ for valuations (in logic) or for the greatest integer function (in other parts of math).

In this book, we shall not specify any explicit rules about which delimiters should be used when; we'll just try to keep things readable. These delimiters should be understood as informal substitutions for ordinary parentheses, with no particular change in the information conveyed; all such expressions will be analyzed as though the delimiters were just parentheses ( ) of ordinary size.

DEFINED SYMBOLS

6.21. The symbols ( ) $\vee$ $\wedge$ $\to$ $\neg$ π_j introduced in 6.2 are *primitive* or *undefined symbols* in our language; they are our fundamental building blocks. We will also use some *defined symbols* as informal abbreviations for certain strings of defined symbols. In particular, whenever P and Q are any strings, we define

$\overline{P}$ as an abbreviation for $\neg(P)$,
$P \circ Q$ as an abbreviation for $\neg(P \to \neg Q)$,
$P \leftrightarrow Q$ as an abbreviation for $(P \to Q) \wedge (Q \to P)$.

Grammatically, $\circ$ and $\leftrightarrow$ behave like $\vee$, $\wedge$, and $\to$, in the sense that they combine two formulas into one longer formula.

The symbol & will also be used grammatically in that fashion, though we cannot describe so simply what it is an abbreviation for. In some logics & means $\wedge$; in other logics & means $\circ$. See 5.33, 11.2.h, and 18.2.

The symbols described above are not part of our formal language, and must be replaced by appropriate combinations of $\vee, \wedge, \to, \neg$ in any discussion concerning actual strings of our formal language — e.g., in determining the rank of a formula.

Prefix notation (optional)

6.22. Throughout this book, we are using $\to, \vee, \wedge$ in an *infix notation* — that is, the symbol goes *between* its two arguments. Some logicians prefer *prefix notation*, where every operation symbol goes *before* its arguments. For example, our infix expressions

$$(\pi_3 \vee \pi_2) \circ \pi_5, \qquad \pi_3 \vee (\pi_2 \circ \pi_5), \qquad (\neg \pi_3) \wedge \neg(\pi_1 \to \pi_5)$$

would be written in prefix notation as

$$\circ\bigl(\vee(\pi_3, \pi_2), \pi_5\bigr), \qquad \vee\bigl(\pi_3, \circ(\pi_2, \pi_5)\bigr), \qquad \wedge\bigl(\neg(\pi_3), \neg(\to(\pi_1, \pi_5))\bigr),$$

respectively. With prefix notation, there is no ambiguity about order of operations, and so parentheses actually are *altogether unnecessary*. The commas can also be omitted. Thus, the three expressions above would be written more briefly as

$$\circ \vee \pi_3 \pi_2 \pi_5, \qquad \vee \pi_3 \circ \pi_2 \pi_5, \qquad \wedge \neg \pi_3 \neg \to \pi_1 \pi_5.$$

That makes formulas shorter, so they may be a bit easier to handle after (if?) one becomes accustomed to the prefix notation. That may be why they are used in some of the research literature on logic. But I believe that infix grammar is easier for beginners, who are already accustomed to its use in arithmetic $(+, -, \times, \text{etc.})$ and in English (and, or).

6.23. *Exercise.* The following formulas are in infix notation. Rewrite them in prefix notation.

a. $\pi_1 \to (\pi_2 \vee (\pi_3 \wedge \pi_4))$.
b. $(\neg \pi_1) \circ (\pi_3 \wedge \pi_4)$.
c. $((\pi_1 \circ \pi_2) \vee \pi_3) \wedge \pi_4$.
d. $\pi_1 \vee \neg \pi_2$.
e. $(\neg \pi_2) \circ \pi_1$.

6.24. *Exercise.* The following formulas are in prefix notation. Rewrite them in infix notation. Insert parentheses as needed.

a. $\vee \pi_1 \to \pi_2 \neg \pi_3$.   **b.** $\vee \vee \pi_1 \pi_2 \neg \pi_3$.   **c.** $\vee \pi_1 \vee \pi_2 \pi_3$.   **d.** $\neg \vee \pi_1 \neg \pi_2$.

Variable sharing

6.25. Two formulas F_1 and F_2 are said to *share a propositional variable symbol* if there is at least one π_j that appears in both formulas. For instance, the formulas

$$(\pi_2 \to \pi_3) \vee \pi_7 \quad \text{and} \quad \pi_1 \to (\pi_7 \wedge \neg \pi_4)$$

share the symbol π_7, whereas the formulas

$$(\pi_2 \to \pi_3) \vee \pi_6 \quad \text{and} \quad \pi_1 \to (\pi_7 \wedge \neg \pi_4)$$

do not share any propositional symbols at all. If F_1 and F_2 are formulas that do not share any propositional symbols, then the formula $F_1 \to F_2$ commits a "fallacy of irrelevance," analogous to this statement:

> If it is now raining, then red is a color.

That statement is classically true, since red *is* a color. We shall see in later chapters that for $F_1 \to F_2$ to be a theorem of relevant logic, a necessary condition is that F_1 and F_2 must share at least one propositional variable symbol.

We remark that symbol sharing is only a necessary condition, not a sufficient condition, for relevance. For instance, the formulas π_2 and $\pi_1 \to \pi_2$ do share a symbol, but we shall see in later chapters that $\pi_2 \to (\pi_1 \to \pi_2)$ is not a theorem of relevant logic.

Preview. Different logics yield different results involving variable sharing. If $A \to B$ is a theorem, then:

- In classical or constructive logic, we can conclude that A and B share at least one variable or at least one of $\overline{A}$ or B is a theorem. See 8.15, 10.10, 29.12, and 29.30.

- In RM logic, we can conclude that A and B share at least one variable or both $\overline{A}$ and B are theorems. (That is not actually proved in this book, but references and part of the proof are given in 23.11.c.)

- In the six-valued crystal interpretation, or in relevant logic, we can conclude that A and B share at least one variable. See 9.12 and 23.8.

(See also related results in 8.43, 8.44, and 29.13.)

6.26. *Exercise.* For each of the following pairs of formulas, either write "nothing shared" or list all the propositional variable symbols shared by the two formulas.

a. $\pi_1 \to (\pi_2 \wedge \pi_3)$ and $\pi_2 \to (\pi_4 \vee \pi_5)$.
b. $\pi_1 \vee (\pi_2 \vee \neg \pi_3)$ and $\pi_4 \wedge (\pi_5 \wedge \neg \pi_6)$.
c. $\pi_1 \to \pi_1$ and $\pi_2 \to \pi_2$.
d. $\pi_1 \to \neg \pi_2$ and $\pi_2 \to \neg \pi_1$.

Formula schemes

6.27. A *formula scheme* is an expression involving some of the metavariables $A, B, C, \ldots$ and symbols $\vee, \wedge, \&, \to, \neg, \circ$ to represent *all* the formulas of a certain type. For instance, the expression
$$A \to (A \vee B)$$
is not a formula, but it is a formula scheme. It represents all the formulas that can be obtained from the expression $A \to (A \vee B)$ by replacing each metavariable with some particular formula. Both A's must be replaced by the same formula; the A's and B may be replaced by the same or different formulas. Some of the formulas represented by the scheme $A \to (A \vee B)$ are

$$\pi_6 \to (\pi_6 \vee \pi_6),$$

$$(\pi_1 \vee \pi_3) \to \Big((\pi_1 \vee \pi_3) \vee \pi_7\Big),$$

$$\Big((\neg \pi_2) \wedge \pi_6\Big) \to \Big(((\neg \pi_2) \wedge \pi_6) \vee (\pi_3 \to \pi_2)\Big).$$

We say each of those is an *instance* of the scheme $A \to (A \vee B)$. The formula $\pi_1 \vee (\pi_3 \to \pi_2)$ can *not* be an instance of the

Formula schemes 223

scheme $A \to (A \vee B)$. Indeed, $\to$ is the outermost operator of $A \to (A \vee B)$; it is the last operator used in the construction of $A \to (A \vee B)$, and also in the construction of any instance of $A \to (A \vee B)$. On the other hand, $\vee$ is the outermost operator of $\pi_1 \vee (\pi_3 \to \pi_2)$.

The formula $\pi_7 \to (\pi_7 \wedge \pi_3)$ is not an instance of the scheme $A \to (A \vee B)$, but determining that fact takes slightly longer. We first match the outermost operators; both are $\to$. Since those match, we must go to the next level: The expressions to the left of the main $\to$'s must match, and the expressions to the right of the main $\to$'s must match.

$$\underbrace{A}_{\pi_7} \to \underbrace{A \vee B}_{\pi_7 \wedge \pi_3}.$$

Thus A must be matched with π_7, and $A \vee B$ must be matched with $\pi_7 \wedge \pi_3$. The latter requirement is not met, since $A \vee B$ and $\pi_7 \wedge \pi_3$ have different outermost operators.

Usually a shorter formula scheme represents more formulas. The extreme case is a single metavariable, B; that formula scheme represents *all* formulas.

Out of sheer laziness, we may sometimes say "formula" when we really mean "formula scheme." Usually this will cause no confusion, because the correct meaning will be evident from the context.

6.28. *Exercise.* Following are three formula schemes:

(1) $A \to (B \to C)$,
(2) $(A \to B) \to (C \to D)$,
(3) $A \to \bigl(B \to (C \to D)\bigr)$,

and here are three formulas:

(i) $\bigl((\pi_1 \to \pi_6) \to \pi_2\bigr) \to \bigl((\pi_2 \to \pi_7) \to ((\pi_1 \to \pi_6) \to \pi_7)\bigr)$,

(ii) $\bigl(\pi_1 \to (\pi_6 \to \pi_2)\bigr) \to \bigl((\pi_2 \to \pi_7) \to ((\pi_1 \to \pi_6) \to \pi_7)\bigr)$,

(iii) $\bigl(((\pi_1 \to \pi_6) \to \pi_2) \to (\pi_2 \to \pi_7)\bigr) \to \bigl((\pi_1 \to \pi_6) \to \pi_7\bigr)$.

Determine which formulas can be instances of which formula schemes. (That makes *nine* yes-or-no questions.) Also, in each case where a formula *can* be an instance of a formula scheme, list what must be substituted for the metavariables to obtain the formula.

I'll do the first one as an example: Can formula (i) be an instance of formula scheme (1)? Yes, with

$$A = (\pi_1 \to \pi_6) \to \pi_2, \qquad B = \pi_2 \to \pi_7, \qquad C = (\pi_1 \to \pi_6) \to \pi_7.$$

Now you do the other eight problems. Do them in this order: the rest of (i); then all of (ii); then all of (iii). For each of those eight problems, either write "no" or give the formulas for A, B, C, and (where appropriate) D. (No proof or explanation is requested for the "no" results.)

6.29. *Remarks.* A formula scheme can be used to represent infinitely many formulas, as in this expression:

> For the axioms of our system we will use all formulas that fit the formula scheme $A \to (A \lor B)$.

On the other hand, a formula scheme can represent just one unspecified formula, as in this sort of reasoning:

> We will show that there does not exist a formula using only the connectives $\lor, \land, \neg$, and not the connective $\to$, that is intuitionistically equivalent to the formula $\pi_1 \to \pi_2$. Indeed, suppose that B were such a formula; we will arrive at a contradiction.

Sometimes we will need to talk *about* formula schemes. Occasionally we will find it useful to introduce yet a third alphabet; we shall use the *metametavariables* $\mathcal{A}, \mathcal{B}, \mathcal{C}, \ldots$ for this purpose. Thus we could say "Let $\mathcal{E}$ and $\mathcal{F}$ be formula schemes." Examples of this usage can be found in 6.30 and 29.15. However, in contexts where it is not likely to cause confusion, we shall use the metavariables $A, B, C, \ldots$ also for formula schemes.

Formula schemes

6.30. Let $\mathcal{E}$ and $\mathcal{F}$ be formula schemes. We will say that $\mathcal{F}$ is a *specialization* of $\mathcal{E}$, or that $\mathcal{E}$ is a *generalization* of $\mathcal{F}$, if

$$\{\text{instances of } \mathcal{F}\} \subseteq \{\text{instances of } \mathcal{E}\}.$$

If, moreover, those two sets of formulas are different, we say that the specialization or generalization is *proper* or *strict*. Here are some examples:

- $A \to (V \wedge W)$ is a strict specialization of $A \to B$. Indeed, every instance of the former scheme can be viewed as an instance of the latter scheme, via the substitution $B = V \wedge W$. But the converse does not hold — for instance, $\pi_1 \to \pi_1$ is an instance of the second scheme and not of the first scheme.

- $A \to A$ is a strict specialization of $A \to B$ — that is,

$$\{\text{instances of } A \to A\} \subsetneq \{\text{instances of } A \to B\}$$

  — because A and B might be the same or different.

- Each of $A \to (B \vee C)$ and $P \to (Q \vee R)$ is a specialization of the other, but not strict. Also, each is a generalization of the other, but not strict.

- The "specialized contraction" scheme

  (15.3.a) $\qquad [A \to (A \to \overline{A})] \to (A \to \overline{A})$

  is a strict specialization of the "weak contraction" scheme

  (15.2.b) $\qquad [A \to (A \to \overline{B})] \to (A \to \overline{B})$,

  which in turn is a strict specialization of "contraction,"

  (15.4.c) $\qquad [A \to (A \to B)] \to (A \to B)$.

  That, in turn, is a strict specialization of either of the schemes

  $(A \to (A \to B)) \to X \quad$ or $\quad (A \to W) \to (A \to B)$.

  Neither of those schemes is a specialization of the other, but each of them is a strict specialization of $(A \to W) \to X$, which is a strict specialization of $P \to X$, which is a strict specialization of Q.

6.31. Example. Consider the following formula schemes.

$$\mathcal{F}_1 = A \vee B, \quad \mathcal{F}_2 = A \wedge B, \quad \mathcal{F}_3 = (A \wedge B) \vee C,$$
$$\mathcal{F}_4 = A \vee (B \to C), \quad \mathcal{F}_5 = A \vee (B \to \overline{C}),$$
$$\mathcal{F}_6 = (A \wedge B) \vee (C \to D), \quad \mathcal{F}_7 = (A \wedge B) \vee (C \to \overline{D}).$$

Which of these schemes are specializations of which others?

Answer. The answer can be presented in several different forms. The most obvious of these is a list:

$\mathcal{F}_3$ is a specialization of $\mathcal{F}_1$, $\mathcal{F}_7$ is a specialization of $\mathcal{F}_3$,
$\mathcal{F}_4$ is a specialization of $\mathcal{F}_1$, $\mathcal{F}_5$ is a specialization of $\mathcal{F}_4$,
$\mathcal{F}_5$ is a specialization of $\mathcal{F}_1$, $\mathcal{F}_6$ is a specialization of $\mathcal{F}_4$,
$\mathcal{F}_6$ is a specialization of $\mathcal{F}_1$, $\mathcal{F}_7$ is a specialization of $\mathcal{F}_4$,
$\mathcal{F}_7$ is a specialization of $\mathcal{F}_1$, $\mathcal{F}_7$ is a specialization of $\mathcal{F}_5$,
$\mathcal{F}_6$ is a specialization of $\mathcal{F}_3$, $\mathcal{F}_7$ is a specialization of $\mathcal{F}_6$,

and each formula is a specialization of itself.

That same information can also be represented in a table: X is a specialization of Y, for the boxes where there is a checkmark in the table below.

$X =$ \ $Y =$	$\mathcal{F}_1$	$\mathcal{F}_2$	$\mathcal{F}_3$	$\mathcal{F}_4$	$\mathcal{F}_5$	$\mathcal{F}_6$	$\mathcal{F}_7$
$\mathcal{F}_1$	✓						
$\mathcal{F}_2$		✓					
$\mathcal{F}_3$	✓		✓				
$\mathcal{F}_4$	✓			✓			
$\mathcal{F}_5$	✓			✓	✓		
$\mathcal{F}_6$	✓		✓	✓		✓	
$\mathcal{F}_7$	✓		✓	✓	✓	✓	✓

A third way to present the information is as in the diagram below; perhaps it conveys the most insight. The formula at the bottom of any up/down path is a specialization of the formula at the top of the path. Paths may consist of more than one line segment. This simplifies our presentation, taking advantage

Formula schemes

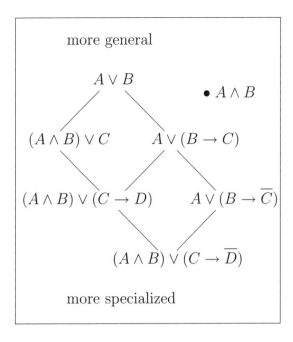

of the fact that "is a specialization of" is a transitive relationship — that is, if $\mathcal{D}$ is a specialization of $\mathcal{E}$, and $\mathcal{E}$ is a specialization of $\mathcal{F}$, then also $\mathcal{D}$ is a specialization of $\mathcal{F}$. For example,

$$A \vee (B \to C) \text{ is a specialization of } A \vee B,$$
$$\text{and } A \vee (B \to \overline{C}) \text{ is a specialization of } A \vee (B \to C);$$

each of these facts is represented by a line segment. Viewing those two line segments as a path gives us the consequence that

$$A \vee (B \to \overline{C}) \text{ is a specialization of } A \vee B.$$

Note that $A \wedge B$ is floating by itself in the diagram, not attached to the rest of the formulas. That is because $A \wedge B$ is neither a specialization nor a generalization of any of those other formulas.

Also note the phrases "more general" and "more specialized" at the top and bottom of the diagram; those are included to indicate the direction of the line segments. I consider those phrases to be essential, and I deduct a point when any of my students omit them (though other instructors may feel differently about this; we all have different opinions about what is "obvious").

6.32. *Exercise.* Consider the following formula schemes:

$$\mathcal{F}_1 = A \to (B \wedge C), \qquad \mathcal{F}_2 = (B \vee C) \to A,$$
$$\mathcal{F}_3 = A \to B, \qquad \mathcal{F}_4 = (P \vee Q) \to (R \wedge S).$$

Which of these schemes are specializations of which others? (You may omit mentioning that each scheme is a specialization of itself.)

6.33. *Exercise.* Consider the following formula schemes.

$$\mathcal{F}_1 = [P \to (P \to Q)] \to (P \to Q), \qquad \mathcal{F}_2 = F \to (G \to F),$$
$$\mathcal{F}_3 = [A \to (B \to C)] \to (D \to E), \qquad \mathcal{F}_4 = (T \to U) \to U,$$
$$\mathcal{F}_5 = H \to (I \to J), \qquad \mathcal{F}_6 = (K \to L) \to N,$$
$$\mathcal{F}_7 = (V \to W) \to (X \to Y), \qquad \mathcal{F}_8 = R \to S.$$

Which of these schemes are specializations of which others? (You may omit mentioning that each scheme is a specialization of itself.) *Caution.* Don't overlook the fact that some metavariables appear more than once in a single formula scheme.

ORDER PRESERVING OR REVERSING SUBFORMULAS (POSTPONABLE)

6.34. A formula may be viewed as a *formula-valued function* of one of its subformulas. For instance, define a function $\mathcal{R}$, with domain and codomain both equal to the set of all formulas, by

$$\mathcal{R}(X) = (\pi_3 \vee \pi_2) \to (X \wedge \pi_7).$$

Plugging in various formulas for X yields various values for $\mathcal{R}(X)$, such as

$$\mathcal{R}(\pi_7) = (\pi_3 \vee \pi_2) \to (\pi_7 \wedge \pi_7);$$
$$\mathcal{R}(\neg \pi_7) = (\pi_3 \vee \pi_2) \to \bigl((\neg \pi_7) \wedge \pi_7\bigr);$$
$$\mathcal{R}(\pi_1 \to \pi_3) = (\pi_3 \vee \pi_2) \to \bigl((\pi_1 \to \pi_3) \wedge \pi_7\bigr).$$

Similarly, we may define other formula-valued functions, such as

$$\mathcal{S}(X) = (\pi_6 \to X) \to \pi_6, \qquad \mathcal{T}(X) = X \vee (\pi_2 \wedge \neg \pi_1), \qquad \ldots \ .$$

Order preserving or reversing subformulas (postponable) 229

6.35. We defined the terms "order preserving" and "order reversing" a few chapters ago,

- for numerical functions, in 3.28, referring to the ordering of numbers (e.g., the fact that $2 \leq 3$), and
- for set-valued functions, in 3.45, referring to the ordering of sets (e.g., the fact that $\{2,5\} \subseteq \{2,5,7\}$).

For analogous use, we now wish to apply the terms "order preserving" and "order reversing" to formula-valued functions. However, we will have to postpone explaining just what is the "order" that is being preserved or reversed. (See 6.38.) Instead we define the terms this way:

- Each formula is an order preserving function of itself.
- $A \lor B$ and $A \land B$ are order preserving functions of A and B.
- $\neg A$ is an order reversing function of A.
- $A \to B$ is an order preserving function of B and an order reversing function of A.
- *Parity of reversals rule.* If $S(X)$ is obtained from subformula X by composing one or more applications of $\lor, \land, \to,$ and $\neg$, then (regardless of the number of preserving factors) the composition is
  * *order preserving* if the composition has an *even* number of reversing factors between X and $S(X)$, or
  * *order reversing* if it has an *odd* number of reversing factors.

Further terminology. If $S(X)$ is an order preserving or reversing function of X, we shall also say that X is (respectively) an order preserving or reversing *argument* of S.

6.36. *Example.* Let S be the formula $((\pi_6 \land \pi_3) \lor \neg \pi_{12}) \to \pi_3$. Which subformulas of S are order preserving or reversing arguments?

Note that π_3 occurs twice in the formula S. We refer to them as "first π_3" and "second π_3," reading from left to right in the given formula S. That is,

$$((\underset{\text{first}}{\pi_6} \wedge \pi_3) \vee \neg \pi_{12}) \to \underset{\text{second}}{\pi_3}$$

The first and second π_3's must be analyzed separately. They may yield different results, and in fact that will turn out to be the case in this example.

We begin by drawing a tree diagram of S, i.e., a diagram showing S and all of its subformulas. (One path in the diagram is drawn with thicker lines, to facilitate our discussion below.) We

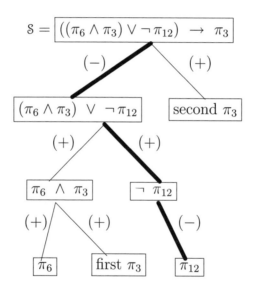

now label each connecting line segment with either (+) or (−), according to the type of connection it is. For instance, the two lines below $\pi_6 \wedge \pi_3$ are both labeled (+), because any formula of the form $A \wedge B$ is an order preserving function of both A and B. On the other hand, the two lines below $((\pi_6 \wedge \pi_3) \vee \neg \pi_{12}) \to \pi_3$ are labeled (−) and (+), respectively, because any formula of the form $A \to B$ is a order reversing function of A and an order preserving function of B.

Following are some conclusions that we can draw.

Order preserving or reversing subformulas (postponable)

- The path between S and π_{12} is shown with thicker lines, to indicate that path as a particular example. There are two $(-)$'s and a $(+)$ along that path. We can disregard the $(+)$; we are only interested in counting the $(-)$'s. We count two, which is an even number. Therefore S is an *order preserving* function of π_{12}.

- S is an *order preserving* function of itself and of the second π_3, because there are no $(-)$'s between S and those subformulas, and 0 is an even number.

- S is an *order reversing* function of each of the subformulas $(\pi_6 \wedge \pi_3) \vee \neg \pi_{12}$, $\pi_6 \wedge \pi_3$, $\neg \pi_{12}$, π_6, and the first π_3, because there is one $(-)$ between S and each of those subformulas, and 1 is an odd number.

6.37. *Exercise.* List all subformulas of the given formula. For each subformula, state whether is an order preserving or reversing argument of the given formula. Note that if a subformula appears more than once in a formula, the different occurrences must be treated separately (as with π_3 in the preceding example).

a. $[\pi_1 \wedge (\pi_2 \to \neg \pi_1)] \vee \pi_2$.
b. $\pi_5 \vee \neg (\pi_1 \to \neg \pi_2)$.

6.38. *Remarks (optional).* The notions of "order preserving" and "order reversing" subformulas will be applied in 11.2–11.6 (semantics) and 14.16 (syntactics). By the way, those subformulas are called *consequent* and *antecedent* (respectively) in the literature of relevant logic.

The last few pages have not explained just what "order" is being preserved or reversed. That explanation must wait until later chapters, but a preview can be given here. We have arranged our terminology so that

in any of this book's ...	... numerically valued semantics	... set-valued semantics	... main syntactics
"S is order preserving" means	$[\![A]\!] \leq [\![B]\!] \Rightarrow$ $[\![S(A)]\!] \leq [\![S(B)]\!]$	$[\![A]\!] \subseteq [\![B]\!] \Rightarrow$ $[\![S(A)]\!] \subseteq [\![S(B)]\!]$	$\vdash A \to B \Rightarrow$ $\vdash S(A) \to S(B)$
"S is order reversing" means	$[\![A]\!] \leq [\![B]\!] \Rightarrow$ $[\![S(B)]\!] \leq [\![S(A)]\!]$	$[\![A]\!] \subseteq [\![B]\!] \Rightarrow$ $[\![S(B)]\!] \subseteq [\![S(A)]\!]$	$\vdash A \to B \Rightarrow$ $\vdash S(B) \to S(A)$

We emphasize that the conclusions indicated in this table are valid for this book's logics, but not for *all* logics. Indeed, if we do not require our logic to have much resemblance to traditional logics, then we algebraically we can devise a logic that does *not* satisfy $\vdash A \to B \Rightarrow \vdash (\neg B) \to (\neg A)$. For such a logic it is not helpful to say that "negation is order reversing."

Part B

Semantics

Chapter 7

Definitions for semantics

7.1. *Remark.* This chapter *should be read simultaneously with the next two chapters* — i.e., the reader will have to flip pages back and forth. Apparently this arrangement is unavoidable. The present chapter contains abstract definitions; the next two chapters include many examples of those definitions.

7.2. *Preview.* Starting in this chapter, we distinguish between

- uninterpreted (meaningless) *symbols*, such as π_1 and $\vee$, and
- their interpretations or *meanings*, denoted by $[\![\pi_1]\!]$ and $\ovee$ in this book.

One symbol (such as the implication symbol, "$\to$") may have different meanings in different logics — e.g., $\ominus_f$ for fuzzy implication and $\ominus_S$ for Sugihara implication. Usually we will only be considering one logic at a time, and so subscripts will not be needed.

INTERPRETATIONS

7.3. Each interpretation will take values in some set, which we shall denote by Σ; this is the set of *semantic values*. It is partitioned into two subsets,

Σ_+ (the *true values*)   and Σ_- (the *false values*), as discussed in 5.8. Each of those must be nonempty; thus Σ must have at least two members.

Preview/examples. In principle the members of Σ_+ and Σ_- could be *anything*, but in practice we generally use mathematical objects for semantic values. Then we use the mathematical properties of those objects in our analysis of the resulting logics.

The simplest example is the classical two-valued interpretation. In that interpretation, Σ_+ contains just one member, called "true," and Σ_- contains just one member, called "false." Those might be denoted by "T" and "F," but any other labeling will work just as well, and some relabelings (particularly 1 and 0, or 1 and -1) will facilitate comparing this interpretation with other interpretations.

Other interpretations have more than two semantic values. Then the terms "true" and "false," though suggestive, may be misleadingly simple. It is best to think of Σ_+ as all the different degrees or kinds of truth, and Σ_- as all the different degrees or kinds of falsehood.

In Chapter 8 we will study examples in which the semantic values are real numbers; thus Σ_+, Σ_-, and Σ will be subsets of $\mathbb{R}$. For instance, in the fuzzy interpretation (8.16) we shall use $\Sigma = [0,1]$, and in the Sugihara interpretation (8.38) we shall use $\Sigma = \mathbb{Z}$. Interpretations of logical operator symbols $(\vee, \wedge, \rightarrow, \neg)$ will be defined in various ways in terms of arithmetic operations $(+, -, \min, \max, \text{etc.})$.

Starting with Chapter 9, we will study examples in which the semantic values are some (possibly all) subsets of a set Ω; thus $\Sigma \subseteq \mathcal{P}(\Omega)$. Moreover, starting in Chapter 10, the set Ω may be equipped with a topology. Interpretations of logical operator symbols $(\vee, \wedge, \rightarrow, \neg)$ will be defined in various ways, in terms of set operations $(\cup, \cap, \complement, \text{etc.})$ and possibly also a topological operation (the interior operator, "int"). For comparison purposes we relabel the classical two-valued interpretation, with "false" denoted by $\varnothing$ and "true" denoted by Ω. In the simplest examples, each member of Ω is a number (and so each semantic value of Σ is a set of numbers), but more complicated choices are sometimes

Functional interpretations 237

needed. For instance, in 28.11, each member of Ω is a finite set of formulas, and in 29.24, each member of Ω is an infinite set of formulas. In both those cases, a semantic value (i.e., a member of Σ) is then *a set of sets* of formulas, and Σ itself is a *set of sets of sets*.[1]

Though most of our interpretations use numbers or sets for members of Σ, the interpretation in 25.7 (needed for an independence proof) doesn't really fit either of those descriptions very well.

7.4. A Σ-*valued valuation* is a function $[\![\ ]\!]$, from the set of all formulas of our formal language, into some set Σ. A Σ-*valued interpretation* is simply any collection of Σ-valued valuations. Of course, we are mainly interested in collections that are related in some interesting ways, such as that in the next section.

FUNCTIONAL INTERPRETATIONS

7.5. Nearly all of the semantic logics presented in this book are described by functions. A Σ-valued *functional interpretation* consists of the following ingredients.

a. Some set Σ and subsets Σ_+, Σ_- are selected, as described in 7.3. (When these sets are finite, we say that the interpretation is a *finite* functional interpretation.)

b. Three binary operators $\ovee, \owedge, \ominusarrow$ and one unary operator $\ominus$ are specified on Σ. (See 3.30 for definition of unary and binary operators.)

We emphasize that the *same* function $\ovee$ is used for all the valuations in the interpretation. Likewise for $\owedge, \ominusarrow, \ominus$. This will be clear from the examples.

In many cases one or more of these functions is specified by a table or "matrix." Hence interpretations of this type are sometimes called *matrix interpretations*, especially when the matrix is finite in size.

---

[1] Sorry, I know that's hard to think about, but it can't be helped. But I promise to use deeply nested sets as seldom as possible.

c. A family of functions $[\![\ ]\!] : \{\text{formulas}\} \to \Sigma$, called *valuations*, is defined as follows:

(i) The values of $[\![\pi_1]\!], [\![\pi_2]\!], [\![\pi_3]\!], \ldots$ may be any values in Σ. The sequence

$$\bigl([\![\pi_1]\!], [\![\pi_2]\!], [\![\pi_3]\!], \ldots\bigr)$$

is a member of $\Sigma^{\mathbb{N}}$; it is called the *assignment* of the valuation. (To get all valuations of this interpretation, we must use all members of $\Sigma^{\mathbb{N}}$ as assignments.)

(ii) The values of longer formulas are determined from the values of shorter formulas, by these rules:

$$\boxed{\begin{array}{ll} [\![A \vee B]\!] = [\![A]\!] \, \ovee \, [\![B]\!], & [\![A \wedge B]\!] = [\![A]\!] \, \owedge \, [\![B]\!], \\ [\![A \to B]\!] = [\![A]\!] \, \ominus \, [\![B]\!], & [\![\neg A]\!] = \ominus [\![A]\!] \end{array}}$$

whenever A and B are any formulas of the formal language. I've added the box just for emphasis, because these rules are so important.[2]

In all of this book's examples, the functions $\ovee$ and $\owedge$ (and $\odot$, discussed further below) will be commutative — that is, each satisfies $x * y = y * x$. Consequently, their tables need not specify which side is x and which side is y. The table for $x \ominus y$ will specify which is x and which is y.

We may refer to an interpretation as an ordered sextuple

$$I \;=\; \bigl(\Sigma_+, \Sigma_-, \ovee, \owedge, \ominus, \ominus\bigr)$$

when we wish to display its ingredients; this notation will be helpful when we want to compare two or more interpretations.

By a *finite functional interpretation* we will mean one in which the set Σ of semantic values has only finitely many members.

---

[2] In the language of algebraists, these four rules say that $[\![\ ]\!]$ is a *homomorphism* (i.e., an operation-preserving function) from the algebra of uninterpreted, unevaluated formulas (defined as in 6.7) to the algebra Σ of values. Compare 11.9.

7.6. *Example: the rank valuation.* Let $\Sigma = \mathbb{N}$, and let

$$x \ominus y = x \owedge y = x \ovee y = x+y+1,$$

and $\ominus x = x+1$. When we use the assignment

$$[\![\pi_1]\!] = [\![\pi_2]\!] = [\![\pi_3]\!] = \cdots = 1,$$

we find that $[\![S]\!]$ is the *rank* of S (defined as in 6.9) for every formula S. For instance, it yields

$$[\![\neg((\pi_6 \wedge \pi_3) \to \pi_{12})]\!] = 6, \qquad [\![(\pi_6 \wedge \pi_3) \vee \pi_{12}]\!] = 5.$$

Here is that last computation, carried out in more detail:

$$
\begin{aligned}
& [\![(\pi_6 \wedge \pi_3) \vee \pi_{12}]\!] \\
= \; & [\![\pi_6 \wedge \pi_3]\!] \ovee [\![\pi_{12}]\!] \\
= \; & \big([\![\pi_6]\!] \owedge [\![\pi_3]\!]\big) \ovee [\![\pi_{12}]\!] \\
= \; & (1 \owedge 1) \ovee 1 \;=\; 3 \ovee 1 \;=\; 5.
\end{aligned}
$$

Or, if you prefer the steps in a slightly different order,

$$
\begin{aligned}
[\![\pi_6 \wedge \pi_3]\!] &= [\![\pi_6]\!] \owedge [\![\pi_3]\!] = 1 \owedge 1 = 3; \quad \text{hence} \\
[\![(\pi_6 \wedge \pi_3) \vee \pi_{12}]\!] &= [\![\pi_6 \wedge \pi_3]\!] \ovee [\![\pi_{12}]\!] = 3 \ovee 1 = 5.
\end{aligned}
$$

7.7. *Example: A nonfunctional valuation.* Let $\Sigma = \{0, 1\}$. Let

$$[\![S]\!] = \begin{cases} 1 & \text{if } S \text{ is the formula } \pi_3 \vee \neg \pi_7, \\ 0 & \text{if } S \text{ is any other formula.} \end{cases}$$

This yields, for instance,

$$[\![(\pi_6 \wedge \pi_3) \vee \pi_{12}]\!] = 0, \qquad [\![((\pi_6 \wedge \pi_3) \vee \pi_{12}) \to \pi_3]\!] = 0.$$

We shall show that this rather contrived valuation can *not* be part of a functional interpretation. Indeed, assume (for proof by contradiction) that it were. There would be some four functions $\ovee$, $\owedge$, $\ominus$, $\ominus$ satisfying the composition rules in 7.5.c(ii). Whatever those functions are, $0 \circledvee 0$ takes *some* particular value; it is either 0 or 1.

The formulas π_3, π_7, $\neg \pi_7$, $\pi_3 \vee \pi_7$ are all different from $\pi_3 \vee \neg \pi_7$; hence we have

$$\llbracket \pi_3 \rrbracket = \llbracket \pi_7 \rrbracket = \llbracket \neg \pi_7 \rrbracket = \llbracket \pi_3 \vee \pi_7 \rrbracket = 0.$$

Now we may compute simultaneously

$$1 = \llbracket \pi_3 \vee \neg \pi_7 \rrbracket = \llbracket \pi_3 \rrbracket \circledvee \llbracket \neg \pi_7 \rrbracket = 0 \circledvee 0, \text{ but also}$$
$$0 = \llbracket \pi_3 \vee \pi_7 \rrbracket = \llbracket \pi_3 \rrbracket \circledvee \llbracket \pi_7 \rrbracket = 0 \circledvee 0.$$

This contradiction proves our claim.

7.8. *Cotenability.* Recall from 6.21 that, for any formulas A and B, the expression $A \circ B$ is understood to be an abbreviation for the formula $\neg (A \to \neg B)$. Analogously, for any semantic values x and y, we now define

$$x \circledcirc y = \ominus (x \ominus \ominus y).$$

We have defined $\circ$ and $\circledcirc$ separately; it now follows — not by definition, but by a short and easy proof which is left as an exercise — that

$$\llbracket A \circ B \rrbracket = \llbracket A \rrbracket \circledcirc \llbracket B \rrbracket$$

for any formulas A and B. The specification of $\circledcirc$ follows from the specifications for $\ominus$ and $\ominus$, and therefore $\circledcirc$ need not be included in our description when we define an interpretation. Nevertheless, $\circledcirc$ is so important that we shall often list it along with that definition.

Tautology and truth preservation

7.9. *Definitions.* Let some interpretation be given. A formula C is said to be *tautologous* in that interpretation, or a *tautology* of that interpretation, if $\llbracket C \rrbracket \in \Sigma_+$ for every valuation $\llbracket \ \rrbracket$ of that interpretation — i.e., if C is *always true* in that interpretation. We write this as

$$\vDash C.$$

Tautology and truth preservation

For instance, we shall see later that $\models \pi_1 \to (\pi_2 \to \pi_1)$ in the classical two-valued interpretation but not in the Sugihara interpretation.

A *tautology scheme* is a formula scheme all of whose instances are tautologies. For instance, we shall see later that $\models A \to (B \to A)$ in the classical two-valued interpretation but not in the Sugihara interpretation. We remark that some, but not all, of the formulas represented by $A \to (B \to A)$ are tautological in the Sugihara interpretation.

If a formula C is *nontautologous* (i.e., not tautologous), we write $\not\models C$. This means that the interpretation has *at least one* valuation $[\![\ ]\!]$ for which $[\![C]\!] \in \Sigma_-$. That is, the formula is *false at least once*.

7.10. A semantic *inference rule* is an expression of the form

$$\mathcal{H} \models C$$

where C is a formula or formula scheme (the *conclusion* or *consequence*) and $\mathcal{H}$ is a nonempty set of formulas or formula schemes (the *hypotheses* or *assumptions*). This expression means that

> whenever $[\![\ ]\!]$ is a valuation that makes all the members of $\mathcal{H}$ true, then $[\![C]\!]$ is true also.

When this condition holds, we say that the inference rule $\mathcal{H} \models C$ is *truth-preserving* for this interpretation. We emphasize that we are not considering whether the members of $\mathcal{H}$ are tautological (i.e., always true); we are just considering those valuations in which the members of $\mathcal{H}$ *are* true.

When there is exactly one hypothesis, it is customary to omit the braces — e.g., we may write $\{A\} \models B \to A$ more briefly as $A \models B \to A$. (Beginners are cautioned that this is merely an abbreviation; a set $\{A\}$ and its member A are never equal. See 3.12.)

7.11. *Examples.* The two most fundamental inference rules in this book are

$$\begin{array}{rll} \text{detachment} & 13.2.\text{a} & \{A,\ A \to B\} \vDash B, \\ \text{adjunction} & 14.1.\text{a} & \{A,\ B\} \vDash A \wedge B. \end{array}$$

Detachment says that whenever A and $A \to B$ are true, then B is true. Adjunction says that whenever A and B are both true, then $A \wedge B$ is true. These rules both hold in most of the interpretations that we will consider in this book. An example where detachment does *not* hold is given in 7.14.

Some beginners initially will reject 7.14 as nonsense, thinking "Detachment *obviously* holds! How could it *not* hold??" But those readers are assuming that the "$\to$" symbol has some or all of its usual meaning. That is the wrong viewpoint to take in this book. Instead we start with "$\to$" as a completely meaningless symbol. We endow it with some of its usual or "obvious" meaning only by agreeing to formally accept some rules about it such as detachment. Our intuition, our sense of what is "obvious," may guide us in choosing such rules, but does not take the place of formalizing those rules.

7.12. *Definition.* By an *admissibility rule*, or a *tautology-preserving* rule, for an interpretation I, we shall mean an expression of this form:

$$\vDash H_1,\ \vDash H_2,\ \ldots,\ \vDash H_n \quad \Rightarrow \quad \vDash C.$$

That is,

> whenever all of $H_1, H_2, \ldots, H_n$ are tautologies (i.e., formulas that are always true in all valuations), then C is also a tautology.

7.13. *Observations and remarks.* From the definitions in 7.10 and 7.12 it follows easily that, in any given interpretation,

> any truth-preserving rule is also tautology-preserving.

In other words, *any valid inference rule is also a valid admissibility rule.*

The converse of that observation fails — i.e., there are some instances in which a tautology-preserving rule is not truth-preserving. Admittedly, such instances are rare among the examples of interest to us. One such pathological example is given below; it will be used in 25.6.

Tautology and truth preservation 243

7.14. *Example: a strange three-valued logic.* Following is an interpretation for which detachment, the rule $\{A, A \to B\} \vDash B$, is tautology-preserving but not truth-preserving.

Let $\Sigma = \{0, \frac{1}{2}, 1\}$ and $\Sigma_+ = \{1\}$. Define logic functions

$$x \malteseV y = \begin{cases} 1 & \text{if at least one of } x \text{ or } y \text{ is } 1, \\ 0 & \text{otherwise}; \end{cases}$$

$$x \malteseW y = x \malteseO y = \begin{cases} 1 & \text{if } x = y = 1, \\ 0 & \text{otherwise}; \end{cases}$$

$$x \malteseA y = \begin{cases} 0 & \text{if } x = 1 \text{ and } y = 0, \\ 0 & \text{if } x = y = \frac{1}{2}, \\ 1 & \text{otherwise}; \end{cases} \qquad \malteseN x = \begin{cases} 0 & \text{if } x = 1, \\ 1 & \text{if } x \neq 1. \end{cases}$$

Detachment is not truth-preserving: Immediate from $1 \malteseA \frac{1}{2} = 1$.

Detachment is tautology-preserving: Suppose that A and B are some particular formulas such that A and $A \to B$ are tautologies, but B is *not* a tautology; we shall arrive at a contradiction.

By assumption, $[\![A]\!] = 1$ and $[\![A \to B]\!] = 1$ for every valuation in the interpretation. The latter equation can be restated as: $[\![A]\!] \malteseA [\![B]\!] = 1$. Since $1 \malteseA 0 = 0$ and $[\![A]\!] = 1$, it follows that $[\![B]\!] \neq 0$. That is, $[\![B]\!] \in \{1, \frac{1}{2}\}$ in every valuation.

By assumption, B is not a tautology. Thus there is at least one particular valuation $[\![\ ]\!]_1$ such that $[\![B]\!]_1 \neq 1$. Since $[\![B]\!] \in \{1, \frac{1}{2}\}$ in every valuation, we can conclude that this particular valuation $[\![\ ]\!]_1$ yields $[\![B]\!]_1 = \frac{1}{2}$.

We observe that $\frac{1}{2}$ is not among the outputs of any of the logical operators. Since $[\![B]\!]_1 = \frac{1}{2}$, the particular formula represented by the metavariable B cannot be a formula involving any of the connectives. Thus the formula B must be one of the one-symbol formulas $\pi_1, \pi_2, \pi_3, \dots$.

Now let $[\![\ ]\!]_0$ be the assignment given by

$$0 = [\![\pi_1]\!]_0 = [\![\pi_2]\!]_0 = [\![\pi_3]\!]_0 = \cdots.$$

Then $[\![B]\!]_0 = 0$, contradicting our earlier assertion that $[\![B]\!] \in \{1, \frac{1}{2}\}$ in every valuation. This completes the demonstration.

7.15. *Exercise.* For the logic described in 7.14, evaluate

a. $0 \ \ovee\ (1 \ominus \frac{1}{2})$,

b. $\ominus (\frac{1}{2} \owedge \ominus 0)$,

c. $\frac{1}{2} \odot 0$, and

d. if $[\![Q]\!] = 1$, $[\![R]\!] = \frac{1}{2}$, and $[\![S]\!] = 0$, evaluate

$$\Big[\![[Q \to (R \to S)] \to [(Q \to R) \to (Q \to S)]]\!]\Big].$$

Chapter 8

Numerically valued interpretations

8.1. *Remark.* This chapter must be read simultaneously with the preceding chapter — i.e., the reader will have to flip pages back and forth. The preceding chapter contains several very abstract definitions; examples begin in this chapter.

In this chapter we will begin to use extensively the functions $\max\{x,y\}$ and $\min\{x,y\}$. Some students may find it helpful to review those functions' properties, listed in 3.29.

THE TWO-VALUED INTERPRETATION

8.2. The terms "two-valued" and "classical" are often used interchangeably in the literature, but we shall avoid that practice here. In this book, "two-valued" will refer to the semantic logic described in the next few paragraphs, and "classical" will refer to the syntactic logic described in Chapter 25. In 29.12 we shall show that those two logics are in some sense "the same"; but until we go through that proof, we should use terminology that refers to these two logics as though they might be different.

The *two-valued interpretation* was already described in 2.38, but hereafter we will replace the uninterpreted symbols $\land$, $\lor$, $\to$, $\neg$ with the symbols $\varominus$, $\varovee$, $\varoslash$, $\varominus$ which represent *particular interpretations*.

The two-valued interpretation can be represented in several

different ways, each of which will have its own advantages and uses:

a. It can be represented by a table like that in 2.38.

b. It can be represented by these 14 arithmetic facts:

$$0 \land 0 = 0, \quad 0 \lor 0 = 0, \quad 0 \to 0 = 1,$$
$$\neg 0 = 1, \quad 0 \land 1 = 0, \quad 0 \lor 1 = 1, \quad 0 \to 1 = 1,$$
$$\neg 1 = 0, \quad 1 \land 0 = 0, \quad 1 \lor 0 = 1, \quad 1 \to 0 = 0,$$
$$1 \land 1 = 1, \quad 1 \lor 1 = 1, \quad 1 \to 1 = 1.$$

c. It can be represented by these four algebraic rules:

$$[\![A \lor B]\!] = \max\left\{[\![A]\!], [\![B]\!]\right\}, \qquad [\![A \land B]\!] = \min\left\{[\![A]\!], [\![B]\!]\right\},$$
$$[\![\neg A]\!] = 1 - [\![A]\!], \qquad [\![A \to B]\!] = \max\left\{1 - [\![A]\!], [\![B]\!]\right\}.$$

8.3. *Numerical examples*

a. Evaluate $[\![0 \to (1 \land 0)]\!] \lor \neg 0$.

Answer: $[\![\underbrace{0 \to \underbrace{(1 \land 0)}_{0}}_{1}]\!] \lor \underbrace{\neg 0}_{1}$ so the answer is $\boxed{1}$.

b. Find $[\![(A \to B) \to (B \to A)]\!]$ when $[\![A]\!] = 0$ and $[\![B]\!] = 1$.

Answer: $\underbrace{\underbrace{(0 \to 1)}_{1} \to \underbrace{(1 \to 0)}_{0}}_{0}$ so the answer is 0.

8.4. *Exercises*

a. Evaluate $\Big[(0 \to 0) \land (1 \to \neg 0)\Big] \lor \Big[1 \lor \neg(0 \land 1)\Big]$.

b. Complete this table:

$[\![A]\!]$	$[\![B]\!]$	$[\![\overline{A \land B}]\!]$	$[\![\overline{A} \land \overline{B}]\!]$	$[\![\overline{A \to B}]\!]$	$[\![\overline{A} \to \overline{B}]\!]$
0	0				
0	1				
1	0				
1	1				

The two-valued interpretation 247

Remarks: Thus, negation does not distribute over conjunction or implication. Not shown here: Negation also does not distribute over disjunction.

8.5. *Nonassociativity of implication.* Make a truth table with eight input combinations, to show the values of $[\![A \to (B \to C)]\!]$ and $[\![(A \to B) \to C)]\!]$. In general, these are not the same; thus implication is not associative.

8.6. *Asymmetry of distribution.* Make a truth table with eight input combinations, to show the values of $[\![A \wedge (B \vee C)]\!]$ and $[\![(A \wedge B) \vee C]\!]$. If you do this correctly, you will find that

$$[\![A \wedge (B \vee C)]\!] \leq [\![(A \wedge B) \vee C]\!],$$

and that strict inequality ($<$) holds for some choices of $[\![A]\!]$, $[\![B]\!]$, $[\![C]\!]$. From this we can conclude that

$$\models [A \wedge (B \vee C)] \to [(A \wedge B) \vee C]$$
$$\text{but } \not\models [A \vee (B \wedge C)] \to [(A \vee B) \wedge C]$$

— that is, the first formula is tautologous but the second is not.

8.7. (*Postponable.*) All the assumptions of basic logic — listed in 13.2 and 14.1 — are valid in the two-valued interpretation. We could show that now using truth tables. But it is more efficient to postpone that demonstration. In Chapter 11 we shall employ more abstract methods, to simultaneously show the validity of those assumptions not only in the two-valued interpretation, but also in many other interpretations covered in the next few chapters.

Thus, we will now focus mainly on the ways in which two-valued logic *differs* from the other interpretations covered in the next few chapters.

8.8. *Some verifications by truth table.* The truth table below has been turned sideways, to better accommodate the names of the formulas. The nine bottom rows of the table are all 1's, showing that those formulas are tautologous in the two-valued interpretation. (Those formulas are *not* tautologous in some of the other interpretations that we shall study.) It is recommended that the student verify some of the computations for some of those bottom nine rows, but this is not suggested as a homework problem since the answer is already given.

Some of those recommended computations involve intermediate steps that are *not* all ones. As an exercise, fill in those intermediate steps — i.e., replace the twelve question marks with 0's and 1's.

	input A	0	0	1	1
	input B	0	1	0	1
intermediate (1):	$A \to B$	?	?	?	?
intermediate (2):	$A \to (A \to B)$	?	?	?	?
intermediate (3):	$(A \to B) \to A$	?	?	?	?
Contraction:	$(A \to (A \to B)) \to (A \to B)$	1	1	1	1
Expansion:	$(A \to B) \to (A \to (A \to B))$	1	1	1	1
Positive paradox:	$A \to (B \to A)$	1	1	1	1
Chain order:	$(A \to B) \lor (B \to A)$	1	1	1	1
Excluded middle:	$A \lor \overline{A}$	1	1	1	1
Noncontradiction:	$\overline{A \land \overline{A}}$	1	1	1	1
Explosion:	$(A \land \overline{A}) \to B$	1	1	1	1
Not-elimination:	$\overline{\overline{A}} \to A$	1	1	1	1
Peirce's law:	$((A \to B) \to A) \to A$	1	1	1	1

8.9. *Exercises.* We will see later that the following formulas are tautological in Abelian logic. Show that these formulas are *not* tautological in the two-valued interpretation, by writing out their truth tables.
   a. $\not\models (A \to B) \to \overline{B \to A}$.
   b. $\not\models ((A \to B) \to B) \to A$.
   c. $\not\models \overline{A \to A}$.
   d. $\not\models [(A \to B) \to C] \to [(C \to B) \to A]$.

8.10. *Explosion.* As we noted in 8.8, the explosion principle $(A \land \overline{A}) \to B$ is tautologous in the two-valued interpretation.

Another version of the explosion principle is $\{A, \overline{A}\} \models B$ — that is, whenever A and $\overline{A}$ are both true, then B is also true. But

The two-valued interpretation

in the two-valued semantics, A and $\overline{A}$ can *never* be simultaneously true. Thus the explosion principle $\{A, \overline{A}\} \vDash B$ is *vacuously* true (see 5.30).

8.11. *Lack of contradictions.* No formula that is an instance of the scheme $P \wedge \overline{P}$ can be a tautology in the two-valued interpretation. This means that any formula such as (for instance)
$$(\pi_1 \to (\pi_2 \vee \pi_4)) \wedge \overline{(\pi_1 \to (\pi_2 \vee \pi_4))}$$
cannot be an always-true formula in the two-valued interpretation.

This may look rather obvious. In fact, $P \wedge \overline{P}$ is always *false*, in all valuations of the two-valued interpretation. However, it has an important consequence that is less obvious. Later in this book we will study many nonclassical logics and multivalued interpretations, most of which have this property:

$$\left\{ \begin{array}{c} \text{tautologies of the} \\ \text{other interpretation} \end{array} \right\} \subseteq \left\{ \begin{array}{c} \text{tautologies of the two-} \\ \text{valued interpretation} \end{array} \right\}.$$

It follows that contradictions cannot be tautologous in those interpretations either.

On the other hand, for a logic that is *not* contained in the two-valued interpretation, a contradiction can be tautologous. See 8.37.g, for instance.

8.12. *Redundancy of symbols.* In some logics (see particularly 10.9) we need all four connective symbols $\vee, \wedge, \to, \neg$. But we don't really need all four for the two-valued interpretation. We could manage with just $\neg$ and any one of the other three symbols. For instance, we can express Ⓐ and Ⓥ in terms of ⊖ and ⊘:

$$[A] \text{Ⓐ} [B] = \ominus\big([A] \oslash \ominus[B]\big);$$
$$[A] \text{Ⓥ} [B] = \big(\ominus[A]\big) \oslash [B].$$

Exercise. Using the two-valued interpretation, and simplifying as much as possible, write
  a. Ⓐ, ⊘ in terms of Ⓥ, ⊖;
  b. Ⓥ, ⊘ in terms of Ⓐ, ⊖.

8.13. *Nonredundancy of negation.* Unlike the other three symbols, negation is irreplaceable: It is not possible to write $\ominus$ in terms of some two — or even all three — of $\wedge, \vee, \ominus$, in the two-valued interpretation. More precisely:

There is no formula A built from just $\wedge, \vee, \rightarrow, \pi_1, \pi_2, \pi_3, \ldots$ that satisfies $[\![A]\!] = [\![\neg \pi_1]\!]$ for all valuations $[\![\ ]\!]$ in the two-valued interpretation.

Proof. Consider the assignment $[\![\pi_1]\!] = [\![\pi_2]\!] = [\![\pi_3]\!] = \cdots = 1$. Then $[\![\neg \pi_1]\!] = 0$. However, $1 \vee 1 = 1 \wedge 1 = 1 \ominus 1 = 1$; hence this valuation yields $[\![A]\!] = 1$.

8.14. *Remarks (optional).* There are 16 different binary operations on $\{0, 1\}$ — that is, 16 different functions from $\{0, 1\} \times \{0, 1\}$ into $\{0, 1\}$. All 16 can be expressed in terms of $\vee, \ominus$ or in terms of $\wedge, \ominus$ or in terms of $\ominus, \ominus$. Some mathematicians refer to this fact by saying that each of the sets $\{\vee, \ominus\}$, $\{\wedge, \ominus\}$, $\{\ominus, \ominus\}$ is *functionally complete* in the two-valued interpretation; some mathematicians omit the word "functionally."

Of course, this notion of "completeness" is very different from, and should not be confused with, the main type of "completeness" studied in this book, i.e., that described in 2.26. See 2.27 for a general discussion of the word "completeness."

We could even get by with just *one* connective symbol. The most popular symbol for this purpose is the *Sheffer stroke*, defined by

$$A \mid B \;=\; \overline{A \wedge B}.$$

It is sometimes called *nand*, an abbreviation for "not and." We could "simplify" our treatment of two-valued logic — we could dispense with all of $\vee, \wedge, \neg, \rightarrow$, and just do everything in terms of $\mid$. (*Exercise.* Show how to do that.) However, our formulas would then be longer, harder to read, and less intuitive. Moreover, this "simplification" is not available for most nonclassical logics, and using it in the two-valued interpretation would create difficulties in our comparing that logic with others.

8.15. *Symbol sharing in two-valued logic.* If A and B are some particular formulas such that $A \rightarrow B$ is a tautology of the two-valued interpretation, then at least one of these three conditions holds:

- B is a tautology.
- $\overline{A}$ is a tautology.

- The formulas A and B share at least one symbol.

(Contrast this with other results listed in 6.25.)

Demonstration. Suppose that all three of those conditions fail. Since B is not a tautology, we can choose some assignment of 0's and 1's to $\pi_1, \pi_2, \ldots$ so that $[\![B]\!] = 0$. Moreover, we can arbitrarily assign values to the π_j's that do not appear in the formula B.

Likewise, since $\overline{A}$ is not a tautology, we can choose some assignment of 0's and 1's to $\pi_1, \pi_2, \ldots$ so that $[\![\neg A]\!] = 0$; and we are free to choose whatever values we like for the π_j's that do not appear in the formula $\overline{A}$.

Since A and B share no propositional symbols, we can meet both of those conditions simultaneously. That is, we can choose some assignment of 0's and 1's to $\pi_1, \pi_2, \ldots$ such that $[\![B]\!] = [\![\neg A]\!] = 0$. Then it follows easily that, at least for this assignment, $[\![A \to B]\!] = 0$. Thus $A \to B$ is not a tautology.

FUZZY INTERPRETATIONS

8.16. Some of the simplest and earliest nonclassical logics were those investigated by Łukasiewicz in the 1930s for philosophical reasons. These were extended by Zadeh in the 1960s for engineering purposes. See 2.39.

We shall apply the term *fuzzy logics* to two semantic systems that are very similar. In this book, we shall use the terms

- *Łukasiewicz logic* with $\Sigma = \{0, \frac{1}{2}, 1\}$, and
- *Zadeh logic* for the case of $\Sigma = [0, 1]$,

though both Łukasiewicz and Zadeh worked with other logics as well. In both cases, we take $[\![\bot]\!] = 0$ and $\Sigma_+ = \{1\}$; thus the only true value is 1. We use these functions:

$$x \varovee y = \max\{x, y\}, \qquad x \varowedge y = \min\{x, y\}, \qquad \ominus x = 1 - x,$$

$$x \ominus y = 1 - \max\{0, x - y\} = 1 + \min\{0, y - x\},$$

$$x \odot y = \ominus(x \ominus \ominus y) = \max\{0, x+y-1\}.$$

The rules for $\vee$, $\wedge$, $\ominus$, $\ominus$ agree with the ones in 8.2 when applied to just the values 0 and 1, but now we are applying those rules to numbers in the interval $(0,1)$ as well. Note that $1/2$ is equal to its own negation.

8.17. For Łukasiewicz logic, the logical operators can also be represented by finite tables. If we follow the format of 8.2.a, we would need a table with nine output rows. That is somewhat inconvenient; we find it easier to use a 3-by-1 table and several 3-by-3 tables:

$x \wedge y$	0	$\frac{1}{2}$	1
0	0	0	0
$\frac{1}{2}$	0	$\frac{1}{2}$	$\frac{1}{2}$
1	0	$\frac{1}{2}$	1

$x \vee y$	0	$\frac{1}{2}$	1
0	0	$\frac{1}{2}$	1
$\frac{1}{2}$	$\frac{1}{2}$	$\frac{1}{2}$	1
1	1	1	1

x	$\ominus x$
0	1
$\frac{1}{2}$	$\frac{1}{2}$
1	0

$x \ominus y$	0	$\frac{1}{2}$	$1 = y$
$x = 0$	1	1	1
$x = \frac{1}{2}$	$\frac{1}{2}$	1	1
$x = 1$	0	$\frac{1}{2}$	1

$x \odot y$	0	$\frac{1}{2}$	1
0	0	0	0
$\frac{1}{2}$	0	0	$\frac{1}{2}$
1	0	$\frac{1}{2}$	1

The tables for $\wedge$, $\vee$, and $\odot$ are symmetric in x and y, so we do not need to specify which input is x and which is y in those tables.

The tables for $\wedge, \vee, \ominus, \ominus$ can be completely described by this list of arithmetic properties, which will be more convenient for our purposes in 29.19. For all formulas A and B,

a. $[\![A]\!]$ is 0 or $\frac{1}{2}$ or 1;
b. $[\![\neg A]\!] = 1 - [\![A]\!]$;
c. $[\![A \wedge B]\!] = [\![B \wedge A]\!]$ and $[\![A \vee B]\!] = [\![B \vee A]\!]$;
d. if $[\![A]\!] = 0$ or $[\![B]\!] = 1$, then $[\![A \to B]\!] = 1$;
e. if $[\![A]\!] = 1$, then

$$[\![A \vee B]\!] = 1, \qquad [\![A \wedge B]\!] = [\![B]\!], \qquad [\![A \to B]\!] = [\![B]\!];$$

Fuzzy interpretations

f. if $[\![B]\!] = 0$, then

$$[\![A \wedge B]\!] = 0, \qquad [\![A \vee B]\!] = [\![A]\!], \qquad [\![A \to B]\!] = [\![\neg A]\!];$$

g. if $[\![A]\!] = [\![B]\!] = \frac{1}{2}$, then $[\![A \wedge B]\!] = [\![A \vee B]\!] = \frac{1}{2}$ and $[\![A \to B]\!] = 1$.

8.18. Of course, for Zadeh logic, with infinitely many semantic values, no such tables are possible. The algebraic formulas given in 8.16 for $\vee, \wedge, \ominus$ are simple enough, but $\ominus$ is complicated and warrants further consideration. Its graph may aid the intuition of some students; that graph is given below.

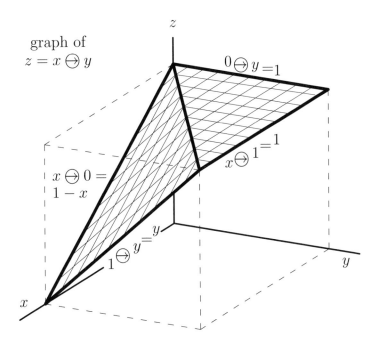

The graph is a surface in three dimensions. The surface consists of two triangles which meet along the line $x = y$, $z = 1$. The other edges of the triangles are along lines whose equations are shown in the illustration.

8.19. Why would anyone *want* to use such a peculiarly defined logic? One reason is to avoid the Sorites Paradox (from the Greek *soros*, "heap"):

- If you remove one grain from a heap of grains, you still have a heap — but if you do this sufficiently many times, what happens?
- Take "bald" to mean "having very little or no hair." If you remove one hair from a man who is not bald, you still have a man who is not bald — but if you repeat this procedure enough times, eventually you have a bald man.

We will analyze this statement in greater detail:

(∗) if person A is very tall, and person B is only $\frac{1}{4}$ inch shorter than person A, then person B is also very tall.

Suppose I have 101 students in one of my classes, and when I line them up in order of height, I find that the tallest is six and a half feet tall. Certainly that is "very tall." After that they go down by 1/4 inch per student. Thus the shortest student is four and a half feet tall. That is *not* "very tall." But induction would tell us that all the students in the class are "very tall."

To analyze this situation in fuzzy logic, let us number the students as $0, 1, 2, \ldots, 100$, in order of decreasing height. Define the statement

$$P_i \quad = \quad \text{"the } i\text{th person is very tall."}$$

This may be neither absolutely true nor absolutely false; it has some semantic value between 0 and 1. Similarly, the statement (∗) of the previous paragraph can be restated as the formula scheme

$$P_{i-1} \to P_i \qquad (i = 1, 2, 3, \ldots, 100)$$

which is neither absolutely true nor absolutely false. In fact, it is *mostly* true, so let us say[1] that $[\![P_{i-1} \to P_i]\!] = 0.99$. That equation can be restated as

$$1 + \min\{0, [\![P_i]\!] - [\![P_{i-1}]\!]\} = 0.99 \qquad (i = 1, 2, 3, \ldots, 100).$$

The tallest person in the class is indeed very tall, so $[\![P_0]\!] =$

---

[1] The numbers in this example are entirely contrived, of course.

Fuzzy interpretations

1. The system of equations that we have just described has a solution given by

$$[\![P_i]\!] = 1 - \frac{i}{100} \quad \text{for} \quad i = 0, 1, 2, 3, \ldots, 100.$$

But then $[\![P_{100}]\!] = 0$ — i.e., there is no truth at all to the statement that the last person is very tall.

8.20. Example. Evaluate $(\frac{1}{5} \wedge 1) \ominus (0 \vee \ominus \frac{1}{3})$.

Answer. We compute

$$\underbrace{(\underbrace{\tfrac{1}{5} \wedge 1}_{1/5}) \ominus (0 \vee \underbrace{\ominus \tfrac{1}{3}}_{2/3})}_{1}$$

with these steps:

- $\frac{1}{5} \wedge 1 = \min\{\frac{1}{5}, 1\} = \frac{1}{5}$,
- $\ominus \frac{1}{3} = 1 - \frac{1}{3} = \frac{2}{3}$,
- $0 \vee \frac{2}{3} = \max\{0, \frac{2}{3}\} = \frac{2}{3}$,
- $\frac{1}{5} \ominus \frac{2}{3} = 1 + \min\{0, \frac{2}{3} - \frac{1}{5}\} = 1 + \min\{0, \frac{7}{15}\} = 1 + 0 = 1$,

so the answer is 1.

8.21. Exercises. Evaluate

a. $\frac{2}{3} \vee (\frac{1}{2} \ominus \frac{1}{4})$.
b. $\ominus \left[\frac{1}{5} \wedge \left(\ominus \frac{2}{3} \right) \right]$.
c. (Excluded middle.) $[\![P \vee \overline{P}]\!]$ when $[\![P]\!] = \frac{1}{3}$.
d. (Noncontradiction.) $[\![\overline{P \wedge \overline{P}}]\!]$ when $[\![P]\!] = \frac{1}{3}$.
e. (Contraction.) $[\![(A \to (A \to B)) \to (A \to B)]\!]$,
   (i) when $[\![A]\!] = \frac{2}{3}$ and $[\![B]\!] = \frac{1}{3}$.
   (ii) when $[\![A]\!] = \frac{1}{2}$ and $[\![B]\!] = 0$.
f. (Weak reductio.) $[\![(A \to \overline{A}) \to \overline{A}]\!]$ when $[\![A]\!] = \frac{1}{3}$.

8.22. *Observations/exercises.* It is easy to verify that
a. $[\![P \vee P]\!] = [\![P \wedge P]\!] = [\![P]\!]$;
b. $[\![P \to P]\!] = 1$;
c. $[\![\overline{P \vee Q}]\!] = [\![\overline{P} \wedge \overline{Q}]\!]$ and $[\![\overline{P \wedge Q}]\!] = [\![\overline{P} \vee \overline{Q}]\!]$;
d. $[\![P \to Q]\!] = 1$ if and only if $[\![P]\!] \le [\![Q]\!]$.

Slightly longer computations also show that
e. $[\![P \to (Q \to P)]\!] = 1$;
f. $[\![(P \to Q) \to Q]\!] = [\![P \vee Q]\!]$.

8.23. *Cotenability and common tautologies (postponable).* From the formulas
$$\ominus x = 1 - x, \qquad x \odot y = \max\{0, x + y - 1\}$$
given in 8.16 we see that negation is involutive and cotenability is commutative. A brief computation yields
$$(x \odot y) \odot z = \max\{0, x + y + z - 2\}$$
from which it follows that $\odot$ is also associative. Hence 11.6 is applicable with $\&= \odot$. Therefore all the assumptions of basic logic — listed in 13.2 and 14.1 — are valid for the fuzzy logics.

Note also that $x \odot x = \max\{0, 2x-1\} \le x$, since x takes values in $[0,1]$. Therefore 11.7.a is applicable, and the formula scheme $(A \to B) \to (A \to (A \to B))$ is tautological in fuzzy logics.

8.24. *Example.* For which values of $[\![\pi_1]\!]$ and $[\![\pi_2]\!]$ does the formula
$$F = (\pi_1 \wedge \overline{\pi_1}) \to \overline{\pi_1 \to \pi_2}$$
take a false value in the Zadeh interpretation?

Solution. This problem cannot be solved by listing all the possible values in a table, for there are infinitely many possible semantic values to consider. Instead we shall solve some algebraic inequalities.

The following steps are all reversible, so we may work backwards from our goal. We seek values that will make
$$[\![A \to B]\!] < 1, \qquad \text{where} \quad A = \pi_1 \wedge \overline{\pi_1} \quad \text{and} \quad B = \overline{\pi_1 \to \pi_2}.$$

Fuzzy interpretations

In view of 8.22.d, that inequality is equivalent to $[\![A]\!] > [\![B]\!]$. That is,
$$[\![\pi_1 \wedge \overline{\pi_1}]\!] > [\![\overline{\pi_1 \to \pi_2}]\!]$$
which can be reformulated and simplified as
$$\min\{[\![\pi_1]\!], 1-[\![\pi_1]\!]\} > 1-\{1+\min\{0, [\![\pi_2]\!]-[\![\pi_1]\!]\}\},$$
$$\min\{[\![\pi_1]\!], 1-[\![\pi_1]\!]\} > -\min\{0, [\![\pi_2]\!]-[\![\pi_1]\!]\},$$
$$\min\{[\![\pi_1]\!], 1-[\![\pi_1]\!]\} > \max\{0, [\![\pi_1]\!]-[\![\pi_2]\!]\}.$$

That last inequality says each of the two numbers $[\![\pi_1]\!]$, $1-[\![\pi_1]\!]$ must be strictly greater than each of the two numbers 0, $[\![\pi_1]\!]-[\![\pi_2]\!]$. Thus all four of these conditions must be satisfied simultaneously:

$$1-[\![\pi_1]\!] > 0, \qquad [\![\pi_1]\!] > [\![\pi_1]\!]-[\![\pi_2]\!],$$
$$[\![\pi_1]\!] > 0, \qquad 1-[\![\pi_1]\!] > [\![\pi_1]\!]-[\![\pi_2]\!].$$

Those inequalities simplify; also toss in the fact that we must have $[\![\pi_1]\!], [\![\pi_2]\!]$ in $[0,1]$. Thus we arrive at

$$\boxed{0 < [\![\pi_2]\!] \leq 1 \text{ and } 0 < [\![\pi_1]\!] < \tfrac{1}{2}[\![\pi_2]\!] + \tfrac{1}{2}}.$$

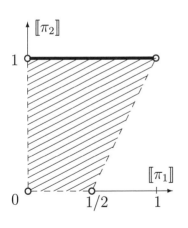

We can graph that region in the plane; see the diagram at left. The region consists of the inside of the trapezoid, plus the upper edge of the trapezoid, but omits the other three edges and all four vertices of the trapezoid.

Remarks. Incidentally, the fact that $(\pi_1 \wedge \overline{\pi_1}) \to \overline{\pi_1 \to \pi_2}$ is not tautologous can be used to show that fuzzy logic fails the classical deduction principle (introduced in 22.5). The formula $\pi_1 \wedge \overline{\pi_1}$ is never true, and therefore it is vacuously correct to say that whenever $\pi_1 \wedge \overline{\pi_1}$ is true, the formula $\overline{\pi_1 \to \pi_2}$ is also true. That is,

$$\pi_1 \wedge \overline{\pi_1} \vDash \overline{\pi_1 \to \pi_2}.$$

Nevertheless, $[\![\pi_1 \wedge \overline{\pi_1}]\!] \to [\![\overline{\pi_1 \to \pi_2}]\!]$ is not a tautology. This example, and further results about fuzzy logic, can be found in Epstein [1990].

8.25. *Exercise.* For which values of $[\![A]\!]$ does the formula
$$(A \vee \overline{A}) \vee [(A \to \overline{A}) \wedge (\overline{A} \to A)]$$
take on a true value?

8.26. *Further exercises on nontautologies.* The following formulas are tautological in the two-valued interpretation. Show that they are *not* tautologies of fuzzy logic, by giving values of $[\![A]\!]$ and $[\![B]\!]$ (and also $[\![C]\!]$, where applicable) that make the following formulas take values (which you should also give) less than 1.

  a. (A form of explosion.) $\not\models (A \wedge \overline{A}) \to B$.
  b. (Self-distribution.)
  $\not\models [A \to (B \to C)] \to [(A \to B) \to (A \to C)]$.
  c. (Proof by contradiction.) $\not\models (A \to B) \to [(A \to \overline{B}) \to \overline{A}]$.
  d. (Proof by dichotomy.) $\not\models (\overline{A} \to B) \to [(A \to B) \to B]$.
  e. (Contraction.) $\not\models [A \to (A \to B)] \to (A \to B)]$.
  f. (Conjunctive detachment.) $\not\models [A \wedge (A \to B)] \to B$.
  g. (Weak reductio, inferential form.) $A \to \overline{A} \not\models \overline{A}$. (To establish this, you need to give an example of a valuation in which $A \to \overline{A}$ is true but $\overline{A}$ is false, in the fuzzy interpretation.

8.27. *Fuzzy deduction principle (optional)*
  a. Suppose that $\mathcal{G}$ and $\mathcal{H}$ are two formula schemes related in such a way that, whenever some substitution of values for metavariables makes $[\![\mathcal{G}]\!] = 1$, then the same substitution also makes $[\![\mathcal{H}]\!] = 1$. Show that $(\mathcal{G} \circ \mathcal{G}) \to \mathcal{H}$ is tautological in the Łukasiewicz interpretation. In other words, $\mathcal{G} \models \mathcal{H} \Rightarrow \models (\mathcal{G} \circ \mathcal{G}) \to \mathcal{H}$.
  b. Show that the same conclusion does not hold in the Zadeh interpretation, by taking $\mathcal{H} = \mathcal{G} \circ \mathcal{G} \circ \mathcal{G}$.

TWO INTEGER-VALUED INTERPRETATIONS

8.28. The comparative and Sugihara interpretations have much in common, so we introduce them together in this subchapter.

Two integer-valued interpretations

The next two subchapters deal with further properties of comparative logic and further properties of Sugihara's logic, respectively.

Both interpretations have semantic values $\Sigma = \mathbb{Z} = \{\text{integers}\}$, subdivided as

$$\Sigma_+ = \{\text{nonnegative integers}\} = \{0, 1, 2, 3, 4, \ldots\},$$
$$\Sigma_- = \{\text{negative integers}\} = \{-1, -2, -3, \ldots\}.$$

(Thus both Σ_+ and Σ_- are infinite sets.) Both interpretations will have

$$x \wedge y = \min\{x, y\}, \qquad x \vee y = \max\{x, y\}, \qquad \ominus x = -x.$$

The two interpretations are defined differently only in their rules for implication:

comparative	Sugihara
$x \ominus_c y = y - x$	$x \ominus_s y = \begin{cases} \max\{-x, y\} & \text{if } x \leq y, \\ \min\{-x, y\} & \text{if } x > y. \end{cases}$

The different interpretations of implication lend themselves to different applications. Comparative logic is the type of reasoning used in our "beverages" example in 2.40, and Sugihara logic serves as semantics for a convenient intermediate step between relevant and classical logics; see 23.11.c.

Remarks. In each of these interpretations, note that 0 is its own negation. That is sometimes avoided in the research literature by modifying the interpretation slightly. Some variants omit 0 altogether, using $\Sigma = \mathbb{Z} \setminus \{0\}$. Other variants split 0 into two values, called $+0$ and -0. Still another approach would be to subdivide Σ into *three* sets — Σ_+, Σ_-, and Σ_0, which could be called "true," "false," and "neutral." But the definitions we have chosen — with 0 being its own negation — apparently yield the simplest comparisons of interpretations.

8.29. *Exercises.* Evaluate the following expressions in both integer-valued interpretations. (The expressions that do not involve implication will come out the same for both interpretations, of course.)

a. $0 \wedge (3 \vee \ominus 2)$.
b. $(1 \vee 2) \wedge 3$.

c. $[\![(A \vee B) \wedge (B \wedge (\neg C))]\!]$, if $[\![A]\!] = 5$, $[\![B]\!] = 4$, $[\![C]\!] = -7$.
d. $0 \ominus (3 \oslash \ominus 5)$.
e. $[\![\overline{A \wedge B} \to (\overline{A} \vee \overline{B})]\!]$, if $[\![A]\!] = 3$ and $[\![B]\!] = -4$.
f. $[\![A \to (B \to A)]\!]$, if $[\![A]\!] = 0$ and $[\![B]\!] = 1$.

8.30. *Further computations.* For either of our integer-valued interpretations:
a. *Nontautologies.* By finding suitable values of $[\![A]\!]$ and $[\![B]\!]$ or by giving a few equations and/or sentences, prove that
  (i) $\not\models A \vee (A \to B)$;
  (ii) $\not\models A \to (\overline{A} \to B)$;
  (iii) $\not\models A \to \left[A \wedge (B \vee \overline{B}) \right]$.
b. *Tautologies.*
  (i) (Identity.) $\models A \to A$.
  (ii) ($\neg$-elimination.) $[\![\overline{\overline{A}}]\!] = [\![A]\!]$, hence $\models \overline{\overline{A}} \to A$.
  (iii) (Excluded middle.) $[\![A \vee \overline{A}]\!] = |[\![A]\!]|$. (The vertical bars represent absolute value.) Hence $\models A \vee \overline{A}$.
  (iv) (Noncontradiction.) $[\![\overline{A \wedge \overline{A}}]\!] = |[\![A]\!]|$, hence $\models \overline{A \wedge \overline{A}}$.
  (v) (Chain order.) $\models (A \to B) \vee (B \to A)$.
  (vi) If $\overline{X}$ and Y are tautologies, then $X \to Y$ is a tautology. In particular,

  $\models (A \wedge \overline{A}) \to (B \vee \overline{B})$ and $\models \overline{A \to A} \to (B \to B)$.

  (vii) *Ackermann truth.* $\models A \leftrightarrow [\overline{B \to B} \to A]$.
  (viii) *Ackermann falsehood.* $\models \overline{A} \leftrightarrow [A \to (B \to B)]$.
c. *Semantic version of the relevant deduction principle.* Use a few sentences to show that $P \models Q \Rightarrow \models \overline{P} \vee Q$. That is, if P and Q are two formulas related so that

  in each valuation where P is true, we also have Q true,

then they are also related so that

$\overline{P \vee Q}$ is true in every valuation.

8.31. *Symbol sharing.* Suppose that X and Y are formulas that share no propositional variable symbols. Then $X \to Y$ is tautological if and only if both $\overline{X}$ and Y are tautological.

Remark. Contrast this with other results listed in 6.25.

Demonstration of the result. The "if" part is a consequence of 8.30.b(vi). For the "only if" part, we will prove the contrapositive: Assume that at least one of $\overline{X}$ or Y is not tautological; we shall show that $X \to Y$ is not tautological.

The proof is in two parts. For the first part, assume Y is not tautological. Then there is some valuation that yields $[\![Y]\!] < 0$. Since X and Y share no variables, we may choose the valuation to also satisfy $[\![\pi_j]\!] = 0$ for all the π_j's that appear in the formula X. It follows that $[\![X]\!] = 0$, and therefore $[\![X \to Y]\!]$ is false by 11.2.d.

For the second part, assume that $\overline{X}$ is not tautological. Then there is some valuation that yields $[\![\overline{X}]\!] < 0$; hence $[\![X]\!] > 0$. Since X and Y share no variables, we may choose the valuation to also satisfy $[\![\pi_j]\!] = 0$ for all the π_j's that appear in the formula Y. It follows that $[\![Y]\!] = 0$ by 11.2.f, and therefore $[\![X \to Y]\!]$ is false by 11.2.d.

8.32. *Corollaries.* The following results apply to both the comparative interpretation and the Sugihara interpretation.

a. The interpretation has no strongest formula in the sense of Church (see 5.34). That is, there does not exist a formula X with the property that $\vDash X \to P$ for all formulas P.

b. The interpretation has no weakest formula in the sense of Church (see 5.34). That is, there does not exist a formula Y with the property that $\vDash P \to Y$ for all formulas P.

c. *Positive paradox.* $\nvDash A \to (B \to A)$.

d. *Positive paradox (permuted).* $\nvDash B \to (A \to A)$.

e. *Explosion.* $\nvDash (A \wedge \overline{A}) \to B$.

More about comparative logic

8.33. This subchapter continues with the definitions in 8.28, but using only $x \ominus y = y - x$. (The following results will *not* apply to Sugihara logic.)

We begin with some computations.

a. $[\![A \to A]\!] = 0$.

b. We define $x \odot y = \ominus(x \ominus \ominus y)$, as usual. In the case of the comparative interpretation, that yields $x \odot y = x + y$.

8.34. *Common tautologies (postponable).* It is easy to verify the conditions of 11.6. Therefore all the assumptions of basic logic — listed in 13.2 and 14.1 — are tautological for this interpretation.

8.35. *Some nontautologies.* By choosing suitable values for $[\![A]\!]$ and $[\![B]\!]$, show that the following formulas are not tautological in comparative logic.

a. *Contraction.* $\not\vDash [A \to (A \to B)] \to (A \to B)$.
b. *Expansion.* $\not\vDash (A \to B) \to [A \to (A \to B)]$.
c. *Weak reductio.* $\not\vDash (A \to \overline{A}) \to \overline{A}$.

8.36. Show that the inference rule $A \to \overline{A} \vDash \overline{A}$ is valid in the comparative interpretation — i.e., show that whenever A is given a value such that $A \to \overline{A}$ is true, then $\overline{A}$ is also true.

Remarks. Contrasting this result with 8.35.c shows that the semantic version of the Herbrand-Tarski Deduction Principle is not applicable to the comparative interpretation. Also, it shows that weak reductio, though equivalent to other formulas in 15.2, has a detachmental corollary that is not equivalent to the detachmental corollaries of the other formulas in 15.2; see the remarks in that section for further discussion.

8.37. *Anticlassical tautologies.* Comparative logic is the only major logic studied in this book that is not a subset of classical logic. Show that the following formulas are tautological in the comparative interpretation, but are *not* tautological in the two-valued interpretation. These formulas are investigated syntactically in Chapters 20 and 26; see also 2.40.

a. *Coassertion.* $\vDash [(A \to B) \to B] \to A$.

b. Meredith's permutation. $\vDash [(Q \to R) \to S] \to [(S \to R) \to Q]$.

c. Suffix cancellation. $\vDash [(Y \to Z) \to (X \to Z)] \to (X \to Y)$.

d. Prefix cancellation. $\vDash [(C \to A) \to (C \to B)] \to (A \to B)$.

e. Anti-identity. $\vDash \overline{A \to A}$.

f. Centering. $\vDash (A \to A) \leftrightarrow \overline{A \to A}$.

g. Blatant contradiction. $\vDash (A \to A) \wedge \overline{(A \to A)}$.

h. Negated contradiction. $\vDash \overline{(A \to A) \wedge \overline{(A \to A)}}$.

More about Sugihara's interpretation

8.38. Implication for the *Sugihara interpretation* was defined in 8.28 by

$$x \ominus y = \begin{cases} \max\{-x, y\} & \text{if } x \leq y, \\ \min\{-x, y\} & \text{if } x > y. \end{cases}$$

An equivalent formula, more convenient in some contexts, is

$$x \ominus y = \begin{cases} -x & \text{if } |x| > |y|, \\ y & \text{if } |x| < |y|, \\ \max\{-x, y\} & \text{if } |x| = |y|. \end{cases}$$

In other contexts, it may be more convenient to dispense with $\ominus$ altogether; we can reformulate many questions involving $\ominus$ as questions involving $\odot$ (see 8.42), which is often easier to deal with.

Perhaps a little intuition can be gleaned from the graph of $\ominus$. Accompanying this section is a two-dimensional picture of the three-dimensional graph of the function $z = x \ominus y$. The graph consists of many dots (for integer values of x and y), lying in a surface that consists of four triangular dotted wedges — two in the slanted plane $z = y$ and two in the slanted plane $z = -x$.

There are also two other triangle wedges in the illustration, but those are dotless except along their edges. Those two wedges

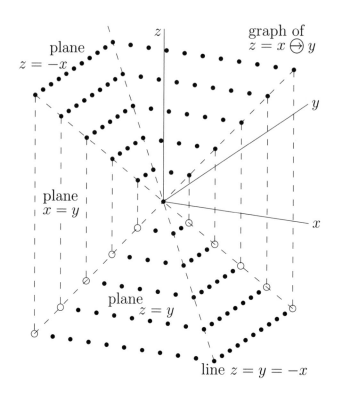

are in the *vertical* plane $x = y$. They indicate the discontinuity of the function $z = x \ominus y$ along the line $x = y$, where the graph jumps from one slanted (dotted) plane to the other. The dots along the lower edges of the vertical wedges don't count — they are hollow dots, used to indicate that they are *not* part of the graph of the function.

From the illustration (or from the algebraic formulas) it is not hard to see that $x \ominus y$ is an order reversing function of x and an order preserving function of y.

8.39. *Elementary observations about Sugihara's logic*

  a. *Computations.* (The vertical bars | | represent absolute value, as in trigonometry or calculus.)

  (i) $[\![A \to A]\!] = |[\![A]\!]|$.

  (ii) $[\![\overline{A} \to A]\!] = [\![A]\!]$ and $[\![A \to \overline{A}]\!] = [\![\overline{A}]\!]$.

  b. *Tautologies.* Using a few equations and/or sentences, show:

More about Sugihara's interpretation

(i) (Mingle.) $[\![A \to (A \to A)]\!] = |[\![A]\!]|$, hence $\vDash A \to (A \to A)$.

(ii) (Disjunctive consequence.) $\vDash (\overline{A} \to B) \to (A \vee B)$.

c. *Nontautologies.* The following formulas are tautological in the two-valued interpretation. By choosing suitable values for $[\![A]\!]$, $[\![B]\!]$, $[\![C]\!]$, show that these formula schemes are *not* tautological in Sugihara's interpretation.

(i) (Antilogism.) $\nvDash ((A \wedge B) \to C) \to ((A \wedge \overline{C}) \to \overline{B})$.

(ii) (Proof by contradiction.)
$\nvDash [(A \wedge \overline{B}) \to (C \wedge \overline{C})] \to (A \to B)$.

(iii) (Disjunctive syllogism — four versions.)
$\nvDash (A \vee B) \to (\overline{A} \to B)$, $A \vee B \nvDash \overline{A} \to B$,
$\nvDash [\overline{A} \wedge (A \vee B)] \to B$, $\overline{A} \wedge (A \vee B) \nvDash B$.

(iv) (Implicative disjunction.) $\nvDash ((A \to B) \to B) \to (A \vee B)$.

d. Show that $(A \to B) \to B \vDash A \vee B$. That is,

for any formulas A and B, in any valuation where $(A \to B) \to B$ is true, the same valuation also makes $A \vee B$ true.

Remark. Comparing this with 8.39.c(iv) shows that the semantic version of the Herbrand-Tarski deduction principle is not applicable to Sugihara's interpretation.

8.40. Suppose that A is a formula involving only $\pi_1, \pi_2, \ldots, \pi_n$ (and the symbols $\vee, \wedge, \neg, \to$). Let $[\![\ ]\!]$ be some valuation in the Sugihara interpretation. Then $[\![A]\!]$ is one of the numbers

$$\pm[\![\pi_1]\!], \ \pm[\![\pi_2]\!], \ \pm[\![\pi_3]\!], \ \ldots, \ \pm[\![\pi_n]\!].$$

(That's just $2n$ numbers — or fewer, if there are repetitions.)

Sketch of proof. This follows from the fact that $x \wedge y$ and $x \vee y$ are always equal to either x or y, and $x \to y$ is always equal to either $-x$ or y.

8.41. *Finite model property.* Suppose that A is a formula involving only $\pi_1, \pi_2, \ldots, \pi_n$ (and the symbols $\vee, \wedge, \neg, \rightarrow$). Then A is tautological for the Sugihara interpretation if and only if A is tautological for the restriction of the Sugihara interpretation to the smaller domain

$$\Sigma \;=\; \{-n,\ -n+1,\ -n+2,\ \ldots,\ n-2,\ n-1,\ n\},$$

an interpretation with only $2n + 1$ semantic values.

Sketch of proof. We don't really need to know the precise value of $[\![A]\!]$; we just need to know whether it is nonnegative. Consequently, we don't need to work with the precise values of the numbers $\pm[\![\pi_1]\!], \pm[\![\pi_2]\!], \ldots, \pm[\![\pi_n]\!]$; all that really matters is the relative order of those numbers and 0. We can replace those $2n + 1$ integers (or fewer, if there are repetitions) with integers in $[-n, n]$ that lie in the same order.

Remarks about the finite model property. Even though Σ is an infinite set, we can determine the tautologousness of A or lack thereof just by

n	$2n+1$	$(2n+1)^n$
2	5	25
3	7	343
4	9	6,561
5	11	161,051
6	13	4,826,809
7	15	170,859,375
8	17	6,975,757,441

testing finitely many values; we just need a "truth table" with n input columns each running through $2n + 1$ possible values. Admittedly, that's a total of $(2n + 1)^n$ rows in the table — a number that quickly grows too large for human calculation, and soon after that outgrows electronic computers too. Thus the procedure indicated in the preceding proof is not really practical for long formulas. Still, at least the procedure is finite.

8.42. *Cotenability and common tautologies (postponable).* As usual, we define $x \odot y = \ominus(x \ominus \ominus y)$. Our second formula for $\ominus$, given in 8.38, yields this formula for $\odot$:

$$x \odot y \;=\; \begin{cases} x & \text{if } |x| > |y|, \\ y & \text{if } |x| < |y|, \\ \min\{x, y\} & \text{if } |x| = |y|. \end{cases}$$

That formula can be restated in the following fashion: If x and y are any two

More about Sugihara's interpretation

integers, then $x \odot y$ is the one of those two integers that appears *later* in this list:
$$0, \ 1, \ -1, \ 2, \ -2, \ 3, \ -3, \ 4, \ -4, \ \ldots \ .$$
Likewise, $x \odot (y \odot z)$ is the one of x, y, z that appears latest on the list. This characterization makes it clear that cotenability is commutative and associative.

Hence 11.6 is applicable with $\& = \odot$ and $\boxed{\perp} = 0$. Therefore all the assumptions of basic logic — listed in 13.2 and 14.1 — are valid for the Sugihara interpretation.

Also, we have $x \odot x = x$; hence both the results in 11.7 are applicable. Thus both contraction and expansion are tautological in the Sugihara interpretation — i.e., we have

$$\models (A \to B) \leftrightarrow (A \to (A \to B)).$$

8.43. *Symbol sharing for positive tautologies (optional).* Suppose that A and B are formulas that do not involve negation — i.e., the formulas involve only proposition symbols and $\vee, \wedge, \to$. Suppose that $A \to B$ is a Sugihara tautology. Then A and B share at least one proposition symbol.

Proof. Suppose not. Choose an assignment that makes $[\![\pi_j]\!] = +2$ for every proposition symbol π_j appearing in formula A, and makes $[\![\pi_j]\!] = +1$ for every proposition symbol π_j appearing in formula B. It follows that $[\![A]\!] = +2$ and $[\![B]\!] = +1$. But then $[\![A \to B]\!] = -2$, so $A \to B$ is not a tautology.

8.44. *Symbol sharing for pure implication (optional).* Suppose that X is a purely implicational formula — i.e., it involves $\to$ but does not include any $\vee, \wedge, \neg$. Suppose that X is a tautology of the Sugihara interpretation.

Then each propositional variable symbol that appears in the formula must appear at least twice. In fact, it must appear both

- at least once as an order preserving argument of X, and
- at least once as an order reversing argument of X.

(The reader may wish to review the definitions of "order preserving" and "order reversing arguments" from 6.35.)

Proof. We will prove the contrapositive — i.e., assume that some variable symbol π_a appears at least once in X, but not as both order preserving and order reversing argument; we will prove that X is not a tautology. Use the following assignment:

$$[\![\pi_a]\!] = \begin{cases} -2 & \text{if } \pi_a \text{ is an order preserving argument of } X, \\ +2 & \text{if } \pi_a \text{ is an order reversing argument of } X, \end{cases}$$

$$[\![\pi_j]\!] = +1 \quad \text{for all } j \neq a.$$

We shall show that every subformula B in X has this valuation:

$$[\![B]\!] = \begin{cases} -2 & \text{if } \pi_a \text{ appears at least once in } B \text{ and} \\ & B \text{ is an order preserving subformula of } X, \\ +2 & \text{if } \pi_a \text{ appears at least once in } B \text{ and} \\ & B \text{ is an order reversing subformula of } X, \\ +1 & \text{if } \pi_a \text{ does not appear in } B. \end{cases}$$

The proof of that assertion is by induction on the rank of B. The initialization step follows immediately from our choice of the $[\![\pi_j]\!]$'s. The transition step is left as an exercise (fill in the details of the proof).

Since X is an order preserving subformula of itself, it follows that $[\![X]\!] = -2 \notin \Sigma_+$ so X is not a tautology.

8.45. *Exercise.* For each of the following formulas, find values of $[\![\pi_1]\!]$ and $[\![\pi_2]\!]$ to show that the formula is not a Sugihara tautology.

a. $(\pi_1 \to \pi_2) \to (\pi_1 \to \pi_1)$.
b. $(\pi_2 \to \pi_1) \to (\pi_1 \to \pi_1)$.

Chapter 9

Set-valued interpretations

9.1. In this chapter, we work with interpretations in which the semantic values are subsets of some set Ω. Thus Σ is a set whose members are *sets*, and we have $\Sigma \subseteq \mathcal{P}(\Omega)$.

POWERSET INTERPRETATIONS

9.2. *The singleton interpretation.* We begin with an extremely elementary example. Let $\Omega = \{1\}$ and $\Sigma = \mathcal{P}(\Omega) = \{\emptyset, \{1\}\}$. That is, there are two semantic values; one is the empty set; the other is a singleton. Take $\Sigma_+ = \{\{1\}\}$; that is, the only true value is the singleton. Interpret logical operators by the table below.

inputs		outputs			
x	y	$\ominus x$ (not x)	$x \otimes y$ (x and y)	$x \oslash y$ (x or y)	$x \ominus y$ (x implies y)
$\emptyset$	$\emptyset$	$\{1\}$	$\emptyset$	$\emptyset$	$\{1\}$
$\emptyset$	$\{1\}$	$\{1\}$	$\emptyset$	$\{1\}$	$\{1\}$
$\{1\}$	$\emptyset$	$\emptyset$	$\emptyset$	$\{1\}$	$\emptyset$
$\{1\}$	$\{1\}$	$\emptyset$	$\{1\}$	$\{1\}$	$\{1\}$

It is easy to see that this interpretation is essentially the same as the two-valued numerical interpretation presented in 8.2. The only differences are superficial. We have relabeled the semantic values, replacing 0 with $\varnothing$ and 1 with $\{1\}$; but all the computations are otherwise unchanged. This shows that *the classical two-valued numerical interpretation is a special case of the class of powerset interpretations described below.*

9.3. For a *powerset interpretation*, let Ω be any nonempty set, and let $\Sigma = \mathcal{P}(\Omega) = \{\text{subsets of } \Omega\}$. We take $\Sigma_+ = \{\Omega\}$; thus, the universal set is the only true value. Valuations will be built up from assignments by these formulas:

$$S \lor T = S \cup T, \qquad S \land T = S \cap T,$$

$$\ominus S = \complement S, \qquad S \ominus T = \complement(S \setminus T) = T \cup \complement S,$$

where $\complement S$ means $\Omega \setminus S$.

We emphasize that Ω may be *any* nonempty set. Thus the "powerset interpretation" is actually a whole class of interpretations; we get different interpretations with different choices of Ω. Observe that

- when Ω is an infinite set, then Σ is too;
- when Ω is a finite set, then Σ is too;
- when Ω contains just a single member, then the resulting powerset interpretation is the two-valued interpretation as described in 9.2. Thus the $\{0,1\}$-valued interpretation is *among* the powerset interpretations.

On the other hand, in 11.11 we shall show that all powerset interpretations have the same set of tautologies. Thus, in a certain sense, all the powerset interpretations are "the same."

One particular powerset interpretation, which is not precise enough to be legitimate mathematics, nevertheless will supply a great deal of intuition. We might call it the *fiction interpretation*. Let Ω represent the set of all conceivable worlds — i.e., the world that we live in, plus all the other worlds that did not happen historically but that can be imagined. There is a natural correspondence between $\lor$ and $\cup$. For instance,

Powerset interpretations 271

$$\left\{ \text{worlds where} \begin{bmatrix} \text{George Washington was the first president of the USA} \end{bmatrix} \vee \begin{bmatrix} \text{Elvis Presley was the second president of the USA} \end{bmatrix} \right\}$$

$$= \left\{ \begin{array}{c} \text{worlds where George} \\ \text{Washington was} \\ \text{the first president} \\ \text{of the USA} \end{array} \right\} \cup \left\{ \begin{array}{c} \text{worlds where Elvis} \\ \text{Presley was the} \\ \text{second president} \\ \text{of the USA} \end{array} \right\}.$$

Analogously, there is a natural correspondence between ∧ and ∩, and between ¬ and ∁. (There is also a correspondence for →, but it is more complicated and less helpful to our intuition.) A statement such as "0 = 0" is true in *every* conceivable world, so $[\![\text{``}0 = 0\text{''}]\!] = \Omega$.

For most purposes, the $\{0, 1\}$-valued interpretation is computationally simplest to work with. However, the more general class of powerset interpretations has this conceptual advantage: it reduces many questions about logic to more concrete questions about set theory; it makes Venn diagrams applicable to many questions about logic. For instance, from the set-theoretical distributive laws

$$(P \cap Q) \cup R = (P \cup R) \cap (Q \cup R),$$

$$(P \cup Q) \cap R = (P \cap R) \cup (Q \cap R)$$

introduced in 3.49, 3.54, and 3.57, we find that, at least in the powerset interpretation,

$$[\![(P \wedge Q) \vee R]\!] = [\![(P \vee R) \wedge (Q \vee R)]\!],$$

$$[\![(P \vee Q) \wedge R]\!] = [\![(P \wedge R) \vee (Q \wedge R)]\!]$$

in every valuation. We might describe this state of affairs by saying that *the powerset interpretation is distributive*. From those facts we find that, furthermore, the following logical distributive laws are tautologies — i.e., these formulas are a part of two-valued logic:

$$[(P \wedge Q) \vee R] \leftrightarrow [(P \vee R) \wedge (Q \vee R)],$$
$$[(P \vee Q) \wedge R] \leftrightarrow [(P \wedge R) \vee (Q \wedge R)].$$

(Recall that $A \leftrightarrow B$ is an abbreviation for $(A \to B) \wedge (B \to A)$.) We might describe this situation by saying that *the two-valued logic is distributive*.

It can be verified that the powerset interpretations satisfy the conditions of 11.6. Hence the assumptions of basic logic (13.2 and 14.1) are valid in the powerset interpretation. However, we might as well postpone that verification; it will follow as a consequence of a more general verification carried out in 10.2 and 10.8.

Hexagon interpretation (optional)

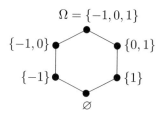

9.4. The set $\Omega = \{-1, 0, 1\}$ has eight subsets, but for the *hexagon interpretation* of Pavičić and Megill [1999] we will not use $\{-1, 1\}$ or $\{0\}$. The remaining six subsets make up Σ. The inclusion relations between those sets are summarized by the diagram at right; larger sets appear higher in the diagram.

Let $\Sigma_+ = \{\Omega\}$; the other five semantic values are false. Define logical operators as follows: For any sets $S, T \in \Sigma$,

$$S \varovee T = \begin{bmatrix} \text{the smallest member} \\ \text{of } \Sigma \text{ containing } S \cup T \end{bmatrix} = \begin{cases} S \cup T & \text{if } S \cup T \in \Sigma, \\ \Omega & \text{otherwise,} \end{cases}$$

$$S \varowedge T = \begin{bmatrix} \text{the largest member of} \\ \Sigma \text{ contained in } S \cap T \end{bmatrix} = \begin{cases} S \cap T & \text{if } S \cap T \in \Sigma, \\ \varnothing & \text{otherwise,} \end{cases}$$

$$\ominus T = \complement T = \Omega \setminus T, \qquad S \ominus T = T \varovee \ominus S.$$

These definitions are admittedly a bit more complicated than the ones in 9.3. Here are a few examples:

- $\{-1\} \varovee \{-1, 0\} = \{-1, 0\}$.
- $\{-1\} \varovee \{1\} = \Omega$, since $\{-1\} \cup \{1\} \notin \Sigma$.
- $\{-1, 0\} \varowedge \{0, 1\} = \varnothing$, since $\{-1, 0\} \cap \{0, 1\} = \{0\} \notin \Sigma$.

The crystal interpretation

9.5. *Optional remarks.* The hexagon interpretation does *not* satisfy conditions 11.2.c; hence Theorem 11.6 is not applicable. Nevertheless the assumptions of basic logic are valid in the hexagon interpretation. In fact, in 11.11 we shall show that

> the hexagon interpretation and the two-valued interpretation have the same tautologies and inference rules.

Nevertheless, the two interpretations do differ substantially in some ways. In particular:

- In the two-valued interpretation, if $[\![A \leftrightarrow B]\!]$ is true in some valuation, then $[\![A]\!] = [\![B]\!]$ in that valuation. But that result fails in the hexagon interpretation, as we can see from exercises 9.6.a and 9.6.b below.

- In the two-valued interpretation, if $A \leftrightarrow B$ is tautological (i.e., true in *every* valuation), then $[\![A]\!] = [\![B]\!]$ in every valuation. But that result fails in the hexagon interpretation. Indeed,

$$[(P \wedge Q) \vee R] \leftrightarrow [(P \vee R) \wedge (Q \vee R)]$$

is tautological in both interpretations, but

$$[\![(P \wedge Q) \vee R]\!] \neq [\![(P \vee R) \wedge (Q \vee R)]\!]$$

occurs in some hexagon valuations, as we can see from exercises 9.6.c and 9.6.d below. Thus, in a sense, the hexagon interpretation is not distributive. That fact came as a surprise to some logicians, since the two-valued logic itself is distributive, as noted in 9.3.

9.6. *Exercises.* Using the hexagon interpretation, evaluate each of these expressions:
   a. $\{0, 1\} \ominus \{1\}$.
   b. $\{1\} \ominus \{0, 1\}$.
   c. $\{1\} \ovee \big(\{-1, 0\} \owedge \{0, 1\}\big)$.
   d. $\big(\{1\} \ovee \{-1, 0\}\big) \owedge \big(\{1\} \ovee \{0, 1\}\big)$.

THE CRYSTAL INTERPRETATION

9.7. *The crystal interpretation.* Let $\Omega = \{-2, -1, +1, +2\}$. That set has 16 subsets, but for our semantic values (i.e., members of Σ) we shall use just the six subsets shown in the diagram below. As usual, the diagram is intended to display inclusions;

for instance, the line up from {+2} to {+1,+2} indicates that the former is a subset of the latter. For brevity we introduce the letters $\beta, \rho, \lambda, \tau$ (beta, rho, lambda, tau) as abbreviations for the sets at the Top, Left, Right, and Bottom vertices of the central square.

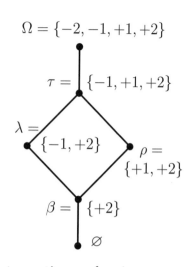

Let $\Sigma_{-} = \{\varnothing\}$; that is, the empty set is the only false value. The other five semantic values are all true. Conjunction and disjunction are interpreted to be intersection and union:

$$S \varowedge T = S \cap T, \qquad S \varovee T = S \cup T.$$

9.8. Negation in the crystal interpretation is defined as follows:

$$\ominus S = \{x \in \Omega : -x \notin S\}.$$

Thus, to obtain $\ominus S$ from the set S, perform these two steps (in either order, actually): multiply all the members of the set by -1, and take the complement (in Ω) of the set. That action is the same as turning the inclusion diagram upside down. Here are the results:

S	$\varnothing$	β	λ	ρ	τ	Ω
$\ominus S$	Ω	τ	λ	ρ	β	$\varnothing$

Note that $\ominus$ is an involution — that is, $\ominus\ominus x = x$. Hence $\overline{\overline{A \leftrightarrow A}}$ is a tautology of the crystal interpretation. Also note that λ and ρ are equal to their own negations.

We shall see below that $\ominus S$ is also equal to $S \ominus \tau$.

9.9. The definition of implication in the crystal interpretation is admittedly more complicated. For motivation we offer little more than the fact that it works — i.e., that it will enable us

The crystal interpretation

to prove some interesting results about relevant logic (23.11.b). Implication can be defined by this formula:

$$S \ominus T = \begin{cases} \Omega & \text{if } S = \emptyset \text{ or } T = \Omega, \\ \emptyset & \text{if } S \text{ is not a subset of } T, \\ T & \text{if } S = \beta, \\ \ominus S & \text{if } T = \tau, \\ S & \text{if } S = T \in \{\lambda, \rho\}, \end{cases}$$

or equivalently by the table below.

$S\ominus T$		$T=$					
		$\emptyset$	β	λ	ρ	τ	Ω
	$\emptyset$	Ω	Ω	Ω	Ω	Ω	Ω
	β	$\emptyset$	β	λ	ρ	τ	Ω
S	λ	$\emptyset$	$\emptyset$	λ	$\emptyset$	λ	Ω
$=$	ρ	$\emptyset$	$\emptyset$	$\emptyset$	ρ	ρ	Ω
	τ	$\emptyset$	$\emptyset$	$\emptyset$	$\emptyset$	β	Ω
	Ω	$\emptyset$	$\emptyset$	$\emptyset$	$\emptyset$	$\emptyset$	Ω

9.10. *Exercise.* Show that the following formulas are *not* tautological in the crystal interpretation:
 a. Mingle: $A \to (A \to A)$.
 b. Chain: $(A \to B) \vee (B \to A)$.
 c. Extremes: $(A \wedge \overline{A}) \to (B \vee \overline{B})$.

9.11. *Fusion and common tautologies (postponable).* As usual, we define $S \odot T = \ominus(S \ominus \ominus T)$. Here are its values, as formula and as table:

$$S \odot T = \begin{cases} \emptyset & \text{if } S = \emptyset \text{ or } T = \emptyset, \\ \Omega & \text{if some } s \in S, t \in T \\ & \text{satisfy } s + t = 0, \\ T & \text{if } S = \beta, \\ S & \text{if } T = \beta, \\ S & \text{if } S = T \neq \tau. \end{cases}$$

$\odot$	$\emptyset$	β	λ	ρ	τ	Ω
$\emptyset$	$\emptyset$	$\emptyset$	$\emptyset$	$\emptyset$	$\emptyset$	$\emptyset$
β	$\emptyset$	β	λ	ρ	τ	Ω
λ	$\emptyset$	λ	λ	Ω	Ω	Ω
ρ	$\emptyset$	ρ	Ω	ρ	Ω	Ω
τ	$\emptyset$	τ	Ω	Ω	Ω	Ω
Ω	$\emptyset$	Ω	Ω	Ω	Ω	Ω

A glance at the table makes it clear that ⊚ is commutative — i.e., $S \circledcirc T = T \circledcirc S$. We still need to verify associativity:

$$(S \circledcirc T) \circledcirc U = S \circledcirc (T \circledcirc U).$$

Rather than check all 216 possible combinations of S, T, U, we analyze the combinations in just a few cases, as follows:

- If any of S, T, U is ∅, then any fusion of them is also ∅, so the associativity equation holds. For the remaining cases we may assume that none of S, T, U is ∅.

- Since $S \circledcirc \beta = \beta \circledcirc S = S$ for any semantic value S, it is easy to verify the associativity equation when any of S, T, U is β. For the remaining cases we may assume that none of S, T, U is β.

It remains only to consider the values of $\lambda, \rho, \tau, \Omega$. That set is closed under the ⊚ operator, so we can consider a simpler table. That yields the following analysis:

⊚	λ	ρ	τ	Ω
λ	λ	Ω	Ω	Ω
ρ	Ω	ρ	Ω	Ω
τ	Ω	Ω	Ω	Ω
Ω	Ω	Ω	Ω	Ω

- If all of S, T, U are λ, then their cotenability (in any order) is also λ.
- If all of S, T, U are ρ, then their cotenability (in any order) is also ρ.
- With any other choice of S, T, U from $\lambda, \rho, \tau, \Omega$, the cotenability of S, T, U (in any order) is Ω.

Thus, we have established associativity of ⊚.

Hence 11.6 is applicable with $\& = \circledcirc$ and $\boxed{\bot} = \tau$. Therefore all the assumptions of basic logic — listed in 13.2 and 14.1 — are valid for the crystal interpretation.

We also observe that $S \circledcirc S \subseteq S$ for all six semantic values of S. Hence 11.7.b is applicable, and the contraction formula $(A \to (A \to B)) \to (A \to B)$ is tautologous in the crystal interpretation.

9.12. *Relevance property of the crystal interpretation.* If $A \to B$ is a tautology in the crystal interpretation, then the formulas A and B must share at least one propositional variable symbol π_j.

Proof. Suppose A and B share no propositional variable symbol. Define a valuation $[\![\ ]\!]$ by assigning

$$[\![\pi_j]\!] = \begin{cases} \lambda & \text{if } \pi_j \text{ appears in } A, \\ \rho & \text{otherwise,} \end{cases}$$

Church's diamond (optional)

for $j = 1, 2, 3, \ldots$. From the definitions of the logical functions it is easy to see that

$$x \lor x = x, \quad x \land x = x, \quad x \to x = x, \quad \lnot x = x$$

whenever x is either λ or ρ. It follows that $[\![A]\!] = \lambda$ and $[\![B]\!] = \rho$, hence $[\![A \to B]\!] = \emptyset \in \Sigma_-$. Thus the formula $A \to B$ is not tautological.

9.13. Church's chain (optional, postponable). If we restrict the operators $\lor, \land, \to, \lnot$ of the crystal interpretation (9.7–9.12) to the smaller set $\{\emptyset, \beta, \tau, \Omega\}$, we get a new functional interpretation, which we shall call *Church's chain* (for Alonzo Church). Again, we take $\emptyset$ to be the only false value.

Since Church's chain is just a restriction of the crystal interpretation, it follows from 11.8 that {crystal tautologies} $\subseteq$ {Church's chain tautologies}. Here are some further properties of Church's chain:

a. The chain formula, $(A \to B) \lor (B \to A)$, is tautological.
b. Unrelated extremes, $(A \land \overline{A}) \to (B \lor \overline{B})$, is tautological.
c. (*Exercise.*) Conjunctive explosion, $(A \land \overline{A}) \to B$, is not tautological.
d. (*Exercise.*) The mingle formula, $A \to (A \to A)$, is not tautological.

CHURCH'S DIAMOND (OPTIONAL)

9.14. This subchapter could easily be omitted in an abridged treatment. The two interpretations presented here are not to be viewed as "major" logics. In fact, they might well be described as "disposable" logics — we will use each of them just once.

Let $\Omega = \{-1, +1\}$. Let $\Sigma = \mathcal{P}(\Omega)$ and let the roles of $\lor, \land, \lnot$ be played by union, intersection, and complementation, respectively, just as in the power-set interpretation. We will depart from that interpretation in our remaining specifications.

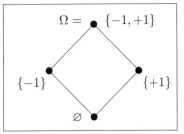

For both of the interpretations below, subdivide Σ into

$$\Sigma_- = \left\{\emptyset, \{-1\}\right\} \quad \text{and} \quad \Sigma_+ = \left\{\{+1\}, \{-1, +1\}\right\};$$

thus a set is true if and only if it contains $+1$. The definition of of $\ominus$ will be given by this formula:

$$S \ominus T = \begin{cases} \Omega & \text{if } S = \varnothing \text{ or } T = \Omega, \\ T & \text{if } S = \{+1\}, \\ \complement S & \text{if } T = \{-1\}, \\ \varnothing & \text{if } T = \varnothing \neq S \text{ or } S = \Omega \neq T, \\ ? & \text{if } S = \{-1\} \text{ and } T = \{+1\}, \end{cases}$$

(where the question mark indicates a value to be filled in below); or equivalently the following table:

$S \ominus T$	$T = \varnothing$	$T = \{-1\}$	$T = \{+1\}$	$T = \Omega$
$S = \varnothing$	Ω	Ω	Ω	Ω
$S = \{-1\}$	$\varnothing$	$\{+1\}$	?	Ω
$S = \{+1\}$	$\varnothing$	$\{-1\}$	$\{+1\}$	Ω
$S = \Omega$	$\varnothing$	$\varnothing$	$\varnothing$	Ω

The missing value (indicated by a question mark) is $\{-1\} \ominus \{+1\}$. We take that value to be false in either interpretation; more specifically, we take it to be

- $\varnothing$ for an interpretation known in some of the literature as *Church's diamond*, or
- $\{-1\}$ for an interpretation that we shall call *Church decontractioned*, for reasons explained in 9.18.b.

9.15. The resulting cotenability function is given by this formula:

$$S \odot T = \begin{cases} \varnothing & \text{if } S \text{ or } T \text{ is empty,} \\ S & \text{if } T = \{+1\}, \\ T & \text{if } S = \{+1\}, \\ \Omega & \text{if } S, T \text{ are both nonempty} \\ & \quad \text{and at least one of them is } \Omega, \\ \Omega & \text{if } S = T = \{-1\} \quad \text{(Church's diamond)}, \\ \{+1\} & \text{if } S = T = \{-1\} \quad \text{(decontractioned)}, \end{cases}$$

or equivalently by the table below.

$\odot$	$\varnothing$	$\{-1\}$	$\{+1\}$	Ω
$\varnothing$	$\varnothing$	$\varnothing$	$\varnothing$	$\varnothing$
$\{-1\}$	$\varnothing$	Ω or $\{+1\}$	$\{-1\}$	Ω
$\{+1\}$	$\varnothing$	$\{-1\}$	$\{+1\}$	Ω
Ω	$\varnothing$	Ω	Ω	Ω

It is easy to see that $\odot$ is commutative — that is, $S \odot T = T \odot S$.

Church's diamond (optional)

9.16. In both Church's diamond and Church decontractioned, the cotenability operation ⓞ is associative. That is,

$$(R \text{ ⓞ } S) \text{ ⓞ } T = R \text{ ⓞ } (S \text{ ⓞ } T).$$

Proof. Each of R, S, T can only be given the four values ∅, $\{-1\}$, $\{+1\}$, Ω, so there are only $4^3 = 64$ cases to consider; this can be done mechanically. For a much less tedious proof, however, we may eliminate cases as follows:

If any of R, S, T is empty, then so is any cotenability obtained from them. In the remaining cases below, we may assume that none of R, S, T is empty.

If any of R, S, T is $\{+1\}$, then the cotenability of that set with any set X is just X. In this case associativity follows easily. In the remaining cases below, we may assume that none of R, S, T is $\{+1\}$.

Thus we may assume that all of R, S, T take values $\{-1\}$ or Ω.

If at least one of them takes the value Ω, then the cotenability of all three in any order is Ω.

The only remaining case is where all three of R, S, T take the value $\{-1\}$. But ⓞ is commutative, so we have

$$\left(\{-1\} \text{ ⓞ } \{-1\}\right) \text{ ⓞ } \{-1\} \;=\; \{-1\} \text{ ⓞ } \left(\{-1\} \text{ ⓞ } \{-1\}\right).$$

This completes the proof.

9.17. *Common tautologies.* It follows from the preceding results that all of the hypotheses in 11.6 are satisfied by both Church's diamond and Church decontractioned; the assumptions of basic logic are satisfied in those two interpretations.

9.18. *More about Church's diamond and its decontractioned variant*

  a. Church's diamond satisfies $S \text{ ⓞ } S \supseteq S$. Hence 11.7.b is applicable, and the contraction axiom $[A \to (A \to B)] \to (A \to B)$ is tautological in Church's diamond.

  b. *Exercise.* By choosing particular values, show that the contraction axiom is *not* tautological in the "Church decontractioned" interpretation. (Hence its name.)

  c. Show that Church decontractioned makes these formulas tautological:

$$\vdash \overline{P \wedge \overline{P}}, \qquad \vdash \overline{P \vee \overline{P}}, \qquad \vdash [A \to (A \to \overline{A})] \to (A \to \overline{A})$$

and makes these inference rules valid:

$$\{A \to B,\; A \to \overline{B}\} \vDash \overline{A}; \qquad \{A \to B,\; \overline{A} \to B\} \vDash B.$$

(The discussion in 15.3 is related.)

d. Conjunctive explosion, $(A \wedge \overline{A}) \to B$, is tautological in both Church's diamond and decontractioned. Hence so is unrelated extremes: $(A \wedge \overline{A}) \to (B \vee \overline{B})$.

e. (Failure of chain order.) $(A \to B) \vee (B \to A)$ is not a tautology of either interpretation. *Exercise*: What values of $[\![A]\!]$ and $[\![B]\!]$ show this? What is the resulting false value of $[\![(A \to B) \vee (B \to A)]\!]$?

Chapter 10

Topological semantics (postponable)

This chapter is more technical, and may be postponed; it will not be needed until 22.14. This chapter requires definitions and results from Chapter 4; a review of that chapter might be recommended before proceeding further.

Topological interpretations

10.1. In the powerset interpretation 9.3, we used *all* subsets of a given set Ω. In a topological interpretation, we just use *some* of the subsets of Ω — specifically, those sets that are "open," as defined in 4.2.

Definition. A *topological interpretation* of the formal language $\mathcal{L}$ is a functional interpretation (defined as in 7.5) satisfying these further criteria:

a. The set Σ of semantic values is a topology on some nonempty set Ω; thus the semantic values are the *open sets*.
b. Ω itself is taken as the only true value — that is, $\Sigma_+ = \{\Omega\}$.
c. Valuations are built up from assignments by these formulas: For any open sets S and T,

$$S \varovee T = S \cup T, \qquad S \varoast T = S \cap T, \qquad \ominus S = \text{int}(\complement S),$$
$$S \ominus T = \text{int}\bigl(\complement(S \setminus T)\bigr) = \text{int}\bigl(T \cup \complement S\bigr).$$

Here "int" is the interior operator, which depends on the topology; see 4.10 and 4.16.

We have given two definitions for $S \ominus T$. The two definitions yield the same result, but one may be more convenient than the other in some contexts. Note also that $\ominus S = S \ominus \varnothing$; thus we may take $\boxed{\bot} = \varnothing$.

Further remarks. We are discussing a whole class of interpretations, not just one interpretation. Different sets Ω may be used, and even when we just look at one set Ω, it may be equipped with different topologies Σ.

The student is cautioned not to confuse $\ominus$ with $\complement$. (In my experience, that has been one of the most common errors made by beginners.) The two sets $\neg S$ and $\complement S$ coincide only if $\complement S$ happens to be an open set.

For motivation, it may be helpful to think of $\neg S = \text{int}(\complement S)$ as the open set that is closest to $\complement S$; thus the topological interpretation is a *modification* of the powerset interpretation — we just take interiors whenever we get a set that might not be open.

Unlike the two-valued interpretation, the topological interpretation generally has many false values. Most of those false values have false negations. The only value whose negation is true is the "totally false" value, $\varnothing$.

10.2. On the other hand, the powerset interpretation is a special case of the topological interpretation. Indeed, if we're given any set Ω, and we equip it with the discrete topology (discussed in 4.6.f), then every subset of Ω is open. Hence $\text{int}(S) = S$ for every set S, so the "int" can be omitted; the modification mentioned in the previous paragraph disappears.

EXAMPLES

10.3. *Chain is not tautological for the reals.* We will show that

$(\pi_1 \to \pi_2) \vee (\pi_2 \to \pi_1)$ is not a real line tautology — i.e., it is not tautological in the topological interpretation determined by the usual topology on $\mathbb{R}$.

Examples

Remark. Students should take care not to confuse the different orderings of $\mathbb{R}$. The real *numbers* themselves are chain-ordered by $\leq$; that is, if we compare any two real numbers, one is higher than or equal to the other. But that is irrelevant to our present purposes. What is relevant here is the fact that the open *sets* of reals are not chain-ordered by $\subseteq$. That is, if we compare any two sets of numbers, it is not necessarily the case that one of those sets is a subset of the other.

Proof. We will use a valuation $[\![\ ]\!]$ in which $[\![\pi_1]\!] = (0,1)$ and $[\![\pi_2]\!] = (1,2)$. (Those are intervals, not ordered pairs.) I chose those values after some trial-and-error experimentation which is not shown here — I am showing only the certification process, not the discovery process (see 2.13). Some students will understand the computation more easily by referring to the diagram below.

$$[\![\pi_1]\!] = (0,1)$$
$$[\![\pi_2]\!] = (1,2)$$
$$(\mathsf{C}[\![\pi_1]\!]) \cup [\![\pi_2]\!] = (-\infty, 0] \cup [1, \infty)$$
$$(\mathsf{C}[\![\pi_2]\!]) \cup [\![\pi_1]\!] = (-\infty, 1] \cup [2, \infty)$$
$$[\![\pi_1 \to \pi_2]\!] = (-\infty, 0) \cup (1, \infty)$$
$$[\![\pi_2 \to \pi_1]\!] = (-\infty, 1) \cup (2, \infty)$$
$$[\![(\pi_1 \to \pi_2) \vee (\pi_2 \to \pi_1)]\!]$$
$$= (-\infty, 1) \cup (1, \infty) = \mathbb{R} \setminus \{1\}$$

Compute as follows:

$$\left(\mathsf{C}[\![\pi_1]\!]\right) \cup [\![\pi_2]\!] = \left(\mathsf{C}(0,1)\right) \cup (1,2) = (-\infty, 0] \cup [1, \infty),$$
$$\left(\mathsf{C}[\![\pi_2]\!]\right) \cup [\![\pi_1]\!] = \left(\mathsf{C}(1,2)\right) \cup (0,1) = (-\infty, 1] \cup [2, \infty),$$

hence

$$[\![\pi_1 \to \pi_2]\!] = \mathrm{int}\Big((-\infty, 0] \cup [1, \infty)\Big) = (-\infty, 0) \cup (1, \infty),$$
$$[\![\pi_2 \to \pi_1]\!] = \mathrm{int}\Big((-\infty, 1] \cup [2, \infty)\Big) = (-\infty, 1) \cup (2, \infty).$$

Finally, compute

$$[\![(\pi_1 \to \pi_2) \vee (\pi_2 \to \pi_1)]\!] = [\![\pi_1 \to \pi_2]\!] \cup [\![\pi_2 \to \pi_1]\!]$$
$$= (-\infty, 1) \cup (1, \infty) = \mathbb{R} \setminus \{1\}.$$

That last value is not $\mathbb{R}$, so the formula $(\pi_1 \to \pi_2) \vee (\pi_2 \to \pi_1)$ is false in this valuation, and thus the formula is not a tautology.

10.4. *Exercise on nontautologies.* Evaluate each of the following in the topological interpretation with the usual topology of the reals, using $[\![P]\!] = (-1, 0) \cup (0, 1)$. In each case you should get something other than $\mathbb{R}$, thereby establishing that the formula in question is not a tautology.

a. Excluded middle. $[\![P \vee \overline{P}]\!]$.

b. Stone's formula. $[\![\overline{P} \vee \overline{\overline{P}}]\!]$.

c. Not-elimination. $[\![\overline{\overline{P}} \to P]\!]$.

d. Specialized De Morgan. $[\![\overline{P \wedge P} \to (\overline{P} \vee \overline{P})]\!]$.

10.5. *A three-set topology.* Equip $\Omega = \{0, 1\}$ with the topology $\Sigma = \{\varnothing, \{0\}, \Omega\}$. Note that this topology is *chain-ordered* — that is, $P \subseteq Q$ or $Q \subseteq P$ whenever $P, Q \in \Sigma$. Hence the resulting topological interpretation makes $(A \to B) \vee (B \to A)$ tautological, unlike 10.3.

On the other hand, (*exercise*) by finding a suitable valuation, show that each of the formulas listed in 22.17 is nontautological for this three-set topology.

10.6. *Exercise.* Use the set $\Omega = \{0, 1, 2, 3\}$. Then

$$\Sigma = \Big\{ S \subseteq \Omega \;:\; S = \Omega \text{ or } 0 \notin S \Big\}$$

is a topology on Ω.

a. List all its open sets; list the interiors of the nonopen sets.

b. Find the value of the formula

$$[\![\overline{C} \to (A \vee B)]\!] \quad \to \quad [\![(\overline{C} \to A) \vee (\overline{C} \to B)]\!]$$

using the topological interpretation with these assignments:

$$[\![A]\!] = \{1\}, \quad [\![B]\!] = \{2\}, \quad [\![C]\!] = \{3\}.$$

(You should get something other than Ω. We will use that fact in 27.15.)

COMMON TAUTOLOGIES

10.7. *Lemmas on valuations.* Assume that (Ω, Σ) is a topological space, and $[\![\ ]\!]$ is some valuation in the resulting topological interpretation. Then for any formulas A, B, C,

a. $[\![A]\!] \cap [\![\overline{A}]\!] = \varnothing$,
b. $[\![A \to B]\!]$ is true if and only if $[\![A]\!] \subseteq [\![B]\!]$,
c. $[\![A \to (B \to C)]\!] = [\![(A \wedge B) \to C]\!]$.

Hint for 10.7.a. Use 4.13.a.

Proof of 10.7.b. If we just unwind the definitions, the statement $[\![A \to B]\!] \in \Sigma_+$ turns into this statement: $\mathrm{int}\Big([\![B]\!] \cup \mathsf{C}[\![A]\!]\Big) = \Omega$. Now, the only subset of Ω that has interior equal to Ω is Ω itself, so that statement is equivalent to this one: $[\![B]\!] \cup \mathsf{C}[\![A]\!] = \Omega$. Now forget about topology, and just use set theory — i.e., draw some Venn diagrams. $[\![A]\!]$ and $[\![B]\!]$ are some subsets of a set Ω; the statement $[\![B]\!] \cup \mathsf{C}[\![A]\!] = \Omega$ is equivalent to $[\![A]\!] \subseteq [\![B]\!]$.

Proof of 10.7.c

$$[\![A \to (B \to C)]\!] \stackrel{(1)}{=\!=\!=} \mathrm{int}\Big\{\mathrm{int}\Big([\![C]\!] \cup \mathsf{C}[\![B]\!]\Big) \cup \mathsf{C}[\![A]\!]\Big\} \stackrel{(2)}{=\!=\!=}$$

$$\mathrm{int}\Big\{\Big([\![C]\!] \cup \mathsf{C}[\![B]\!]\Big) \cup \mathsf{C}[\![A]\!]\Big\} \stackrel{(3)}{=\!=\!=} \mathrm{int}\Big\{[\![C]\!] \cup \mathsf{C}\Big([\![A]\!] \cap [\![B]\!]\Big)\Big\}$$

$$\stackrel{(4)}{=\!=\!=} \mathrm{int}\Big\{[\![C]\!] \cup \mathsf{C}\Big([\![A \wedge B]\!]\Big)\Big\} \stackrel{(5)}{=\!=\!=} [\![(A \wedge B) \to C]\!]$$

where the equations are justified as follows:

(1) by two uses of the definition of $\ominus$;

(2) by 4.17 with $G = [\![A]\!]$ and $H = [\![C]\!] \cup \complement[\![B]\!]$;

(3) by elementary set theory (i.e., no topology involved);

(4) by definition of Ⓐ;

(5) by definition of ⊖.

10.8. *Verification of tautologies (postponable).* It is now easy to verify all the conditions of 11.6 (with Ⓑ = ∩ and ⊥ = ∅). Thus any topological interpretation satisfies all the axioms and assumed inference rules of basic logic, listed in 13.2 and 14.1.

In addition, it is easy to see that these formulas are tautological in every topological interpretation:

- $\overline{A \wedge \overline{A}}$, noncontradiction (by 10.7.a);
- $A \to (B \to A)$, positive paradox (by 10.7.c);
- $(A \to (A \to B)) \to (A \to B)$, contraction (by 11.7.b); and
- $(A \wedge \overline{A}) \to B$, conjunctive explosion (by 10.7.a and 10.7.b).

We will use these observations in 22.14.

Nonredundancy of symbols

10.9. *Nonredundancy of symbols in topological interpretations.* We saw in 8.12 that the four symbols $\vee, \wedge, \to, \neg$ are redundant in the classical two-valued interpretation: We don't need all four symbols; two will suffice. But we shall now show that *all four symbols are needed for topological interpretations*. More precisely:

a. There does not exist a single formula A, written just in terms of $\wedge, \to, \neg$, and the π_j's, that satisfies $[\![A]\!] = [\![\pi_1 \vee \pi_2]\!]$ for every valuation $[\![\ ]\!]$ in every topological interpretation.

b. There does not exist a single formula A, written just in terms of $\vee, \to, \neg$, and the π_j's, that satisfies $[\![A]\!] = [\![\pi_1 \wedge \pi_2]\!]$ for every valuation $[\![\ ]\!]$ in every topological interpretation.

c. There does not exist a single formula A, written just in terms of $\vee, \wedge, \neg$, and the π_j's, that satisfies $[\![A]\!] = [\![\pi_1 \to \pi_2]\!]$ for every valuation $[\![\ ]\!]$ in every topological interpretation.

d. There does not exist a single formula A, written just in terms of $\vee, \wedge, \rightarrow$, and the π_j's, that satisfies $[\![A]\!] = [\![\overline{\pi_1}]\!]$ for every valuation $[\![\ ]\!]$ in every topological interpretation.

Demonstration of a. For the topological space (Ω, Σ), use $\Omega = \{1, 2, 3\}$, with the topology $\Sigma = \{\varnothing, \{1\}, \{2\}, \{1, 2\}, \{1, 2, 3\}\}$. The operators $\mathbin{\ooalign{\hss\raise0.5ex\hbox{\scriptsize$\vee$}\hss\cr$\bigcirc$}}$ and $\mathbin{\ooalign{\hss\raise0.5ex\hbox{\scriptsize$\wedge$}\hss\cr$\bigcirc$}}$ are just union and intersection, but the operators $\ominus$ and $\ominus$ are less obvious and may require a table. We combine the $\ominus$ and $\ominus$ tables, since $\ominus x = x \ominus \varnothing$. The column for $\complement x$ is used in computing the table, since $x \ominus y = \text{int}(y \cup \complement x)$; that column is included as an aid for anyone who wishes to check the table.

$x \ominus y$	$\varnothing$	$\{1\}$	$\{2\}$	$\{1,2\}$	Ω	$\complement x$	$\ominus x$
$x = \varnothing$	Ω	Ω	Ω	Ω	Ω	Ω	Ω
$x = \{1\}$	$\{2\}$	Ω	$\{2\}$	Ω	Ω	$\{2,3\}$	$\{2\}$
$x = \{2\}$	$\{1\}$	$\{1\}$	Ω	Ω	Ω	$\{1,3\}$	$\{1\}$
$x = \{1,2\}$	$\varnothing$	$\{1\}$	$\{2\}$	Ω	Ω	$\{3\}$	$\varnothing$
$x = \Omega$	$\varnothing$	$\{1\}$	$\{2\}$	$\{1,2\}$	Ω	$\varnothing$	$\varnothing$

Observe that the subcollection

$$\Sigma_0 = \{\varnothing, \{1\}, \{2\}, \Omega\} = \{S \in \Sigma : S \neq \{1, 2\}\}$$

is *closed* under the three operators $\mathbin{\ooalign{\hss\raise0.5ex\hbox{\scriptsize$\wedge$}\hss\cr$\bigcirc$}}, \ominus, \ominus$, in the sense of 4.5. That is, we cannot get out of Σ_0 by applying those three operators to members of Σ_0.

We now consider an assignment in which

$$\{1\} = [\![\pi_1]\!], \qquad \{2\} = [\![\pi_2]\!] = [\![\pi_3]\!] = [\![\pi_4]\!] = \cdots.$$

Then $[\![\pi_1 \vee \pi_2]\!] = \{1, 2\}$. However, if A is any formula built from just $\wedge, \rightarrow, \neg$, and the proposition symbols, then $[\![A]\!] \in \Sigma_0$, by the remarks of the previous paragraph. Thus $[\![A]\!] \neq [\![\pi_1 \vee \pi_2]\!]$.

Demonstration of b. Again use $\Omega = \{1, 2, 3\}$, but this time use the topology

$$\Sigma = \{\varnothing, \{1\}, \{2\}, \{1,2\}, \{2,3\}, \Omega\}.$$

Again, $\lor$ and $\land$ are just union and intersection, but $\to$ and $\neg$ are less obvious and require a table.

$x \to y$	$\varnothing$	$\{1\}$	$\{2\}$	$\{1,2\}$	$\{2,3\}$	Ω	$\complement x$	$\neg x$
$x = \varnothing$	Ω	Ω	Ω	Ω	Ω	Ω	Ω	Ω
$x = \{1\}$	$\{2,3\}$	Ω	$\{2,3\}$	Ω	$\{2,3\}$	Ω	$\{2,3\}$	$\{2,3\}$
$x = \{2\}$	$\{1\}$	$\{1\}$	Ω	Ω	Ω	Ω	$\{1,3\}$	$\{1\}$
$x = \{1,2\}$	$\varnothing$	$\{1\}$	$\{2,3\}$	Ω	$\{2,3\}$	Ω	$\{3\}$	$\varnothing$
$x = \{2,3\}$	$\{1\}$	$\{1\}$	$\{1,2\}$	$\{1,2\}$	Ω	Ω	$\{1\}$	$\{1\}$
$x = \Omega$	$\varnothing$	$\{1\}$	$\{2\}$	$\{1,2\}$	$\{2,3\}$	Ω	$\varnothing$	$\varnothing$

Observe that the subcollection

$$\Sigma_0 = \{\varnothing, \{1\}, \{1,2\}, \{2,3\}, \Omega\} = \{S \in \Sigma : S \neq \{2\}\}$$

is closed under the operators $\lor, \to, \neg$ (in the sense of 4.5). Now consider the assignment with

$$\{1,2\} = [\![\pi_1]\!], \qquad \{2,3\} = [\![\pi_2]\!] = [\![\pi_3]\!] = [\![\pi_4]\!] = \cdots.$$

Then $[\![\pi_1 \land \pi_2]\!] = \{2\}$. However, any formula A built from just $\lor, \to, \neg$, and proposition symbols satisfies $[\![A]\!] \in \Sigma_0$. Thus $[\![\pi_1 \land \pi_2]\!] \neq [\![A]\!]$.

Demonstration of c. This time we use an underlying set $\Omega = \{1,2,3,4\}$ with four members. For the open sets, we use the nine subsets of Ω listed in the top row of the negations table below.

x	$\varnothing$	$\{1\}$	$\{2\}$	$\{1,2\}$	$\{1,3\}$	$\{2,4\}$	$\{1,2,3\}$	$\{1,2,4\}$	Ω
$\complement x$	Ω	$\{2,3,4\}$	$\{1,3,4\}$	$\{3,4\}$	$\{2,4\}$	$\{1,3\}$	$\{4\}$	$\{3\}$	$\varnothing$
$\neg x$	Ω	$\{2,4\}$	$\{1,3\}$	$\varnothing$	$\{2,4\}$	$\{1,3\}$	$\varnothing$	$\varnothing$	$\varnothing$

The implications table has nine rows and nine columns, so its construction is very tedious, but fortunately we don't need that whole table; we just need one value from that table. It is easy to compute $\{1,2,3\} \to \{1,2\} = \{1,2,4\}$.

Use the assignment

$$\{1,2,3\} = [\![\pi_1]\!], \qquad \{1,2\} = [\![\pi_2]\!] = [\![\pi_3]\!] = [\![\pi_4]\!] = \cdots.$$

Then $[\![\pi_1 \to \pi_2]\!] = \{1, 2, 4\}$. However, the set

$$\Sigma_0 = \Big\{ \varnothing, \{1, 2\}, \{1, 2, 3\}, \Omega \Big\}$$

is closed under $\ovee$, $\owedge$, $\ominus$ (in the sense of 4.5), and so those four semantic values are the only possible valuations for a formula built without the use of the $\to$ symbol.

Demonstration of d. Immediate from 8.13, since the two-valued interpretation is a topological interpretation (see 10.2).

VARIABLE SHARING

10.10. *Variable sharing in the reals.* Suppose A and B are formulas that share no variables, and $A \to B$ is a real line tautology — i.e., a tautology of the topological interpretation determined by the usual topology of the reals. Then at least one of $\overline{A}$ or B is also a tautology of that interpretation. (Contrast this with other results listed in 6.25.)

Proof. We shall prove the contrapositive — i.e., assume A, B share no variables, and neither $\overline{A}$ nor B is a tautology; we shall show that $A \to B$ is not a tautology.

Since $\overline{A}$ and B are not tautologies, there exist valuations $[\![\ ]\!]_1$ and $[\![\ ]\!]_2$ in the real line interpretation, satisfying $[\![\overline{A}]\!]_1 \neq \mathbb{R}$ and $[\![B]\!]_2 \neq \mathbb{R}$. The former assertion can be restated as $\mathrm{int}\Big(\mathsf{C}[\![A]\!]_1\Big) \neq \mathbb{R}$, which in turn implies $\mathsf{C}[\![A]\!]_1 \neq \mathbb{R}$ and $[\![A]\!]_1 \neq \varnothing$. The assertion about B can be restated as $\mathbb{R} \setminus [\![B]\!]_2 \neq \varnothing$. Hence we can choose real numbers $r_1 \in [\![A]\!]_1$ and $r_2 \in \mathbb{R} \setminus [\![B]\!]_2$.

The topology of $\mathbb{R}$ is translation-invariant — that is, S is an open set if and only if the set $S + x = \{s + x : s \in S\}$ is open. Also, all the functions $\ovee$, $\owedge$, etc., are translation-invariant — that is, $(S + x) \ovee (T + x) = (S \ovee T) + x$, etc. Thus we may form new valuations $[\![\ ]\!]'_1$, $[\![\ ]\!]'_2$ by taking

$$[\![F]\!]'_j \;=\; [\![F]\!]_j - r_j \;=\; \{x - r_j \,:\, x \in [\![F]\!]_x\} \qquad (j = 1, 2)$$

for every formula F. In particular, this yields $0 \in [\![A]\!]'_1$ and $0 \in \mathbb{R} \setminus [\![B]\!]'_2$.

Now define yet another valuation, by assigning propositional variable symbols according to this rule:

$$[\![\pi_j]\!]_3 = \begin{cases} [\![\pi_j]\!]'_1 & \text{if } \pi_j \text{ appears in formula } A, \\ [\![\pi_j]\!]'_2 & \text{if } \pi_j \text{ appears in formula } B, \\ \mathbb{R} & \text{otherwise.} \end{cases}$$

(This is possible since A and B share no variables.) By induction on the ranks of formulas F, it follows that

$$[\![F]\!]_3 = \begin{cases} [\![F]\!]'_1 & \text{if all of } F\text{'s variables appear in } A, \\ [\![F]\!]'_2 & \text{if all of } F\text{'s variables appear in } B, \\ \text{something more complicated, otherwise.} \end{cases}$$

In particular, $[\![A]\!]_3 = [\![A]\!]'_1 \ni 0$ and $[\![B]\!]_3 = [\![B]\!]'_2 \not\ni 0$. An easy computation now yields $0 \notin [\![A \to B]\!]_3$, hence $A \to B$ is not a tautology.

Adequacy of finite topologies (optional)

10.11. *Finite refutation principle.* Let X be some particular formula. Suppose that there exists a topological space (Ω, Σ_i) for which X is not tautological. Then there exists a *finite* topology Σ_g on the same set Ω, for which X is not tautological.

Remarks

- Different formulas X may require different finite topologies.
- The subscripts i and g stand for "initial" and "generated"; that will be relevant in the proof below.
- By "finite topology" we mean that there are only finitely many different open sets. The set Ω might still have infinitely many members, as in 4.6.i. But see also 4.26.

Demonstration. We shall present what mathematicians call a *proof by example*. This means that, instead of proving the result in an abstract and general

Adequacy of finite topologies (optional)

formulation that includes all formulas X and topological spaces (Ω, Σ_i) as special cases, we shall work out the details of one typical example, using methods that clearly would also work for all other examples. This style of presentation avoids the far more cumbersome notation that would be required in an abstract proof.

For our example, we will start from the fact — demonstrated in 10.3 — that the formula $(\pi_1 \to \pi_2) \vee (\pi_2 \to \pi_1)$ is not a tautology for the usual topology of the real numbers. Of course, that topology is infinite; we wish to find some *finite* topology for which $(\pi_1 \to \pi_2) \vee (\pi_2 \to \pi_1)$ is not a tautology. Thus, Σ_i is the usual topology on $\mathbb{R}$, and we must find some suitable finite topology Σ_g on the same set $\mathbb{R}$. We will use superscripts to distinguish between the interpretations determined by these two topologies — e.g., $[\![\ ]\!]^i$ and $[\![\ ]\!]^g$.

We begin by reviewing the demonstration that was given in 10.3. The demonstration consisted only of certification; the discovery process was not shown and will not be needed in the proof below. In 10.3, we were given values for $[\![\pi_1]\!]^i$ and $[\![\pi_2]\!]^i$ — not just any values, but values selected by the discovery process — and we used those to compute the value of $[\![(\pi_1 \to \pi_2) \vee (\pi_2 \to \pi_1)]\!]^i$, which turned out to be different from $\mathbb{R}$, thus proving that $(\pi_1 \to \pi_2) \vee (\pi_2 \to \pi_1)$ is not tautological. In the proof that we now wish to carry out, we are *given* the fact that $(\pi_1 \to \pi_2) \vee (\pi_2 \to \pi_1)$ is not tautological, and thus that suitable values of $[\![\pi_1]\!]^i$ and $[\![\pi_2]\!]^i$ exist.

The computation in 10.3 started from the values of $[\![\pi_1]\!]^i$ and $[\![\pi_2]\!]^i$, and used those to gradually evaluate more complicated formulas, arriving finally at $[\![(\pi_1 \to \pi_2) \vee (\pi_2 \to \pi_1)]\!]^i \neq \mathbb{R}$. This procedure used only finitely many steps, since the formula $(\pi_1 \to \pi_2) \vee (\pi_2 \to \pi_1)$ has finite rank; that fact will remain true not only in the example we are presenting but in the more general case.

Let us now list those finitely many steps. In our listing, we will omit any nonopen sets used in the intermediate computations, such as $\left(\mathcal{C}[\![\pi_1]\!]^i\right) \cup [\![\pi_2]\!]^i = (-\infty, 0] \cup [1, \infty)$. We will just list the five logical *formulas* that were used in our computation, and the open subsets of Ω that were obtained as values of those formulas.

> Formulas and open sets used:
> $[\![\pi_1]\!]^i = (0,1)$, $\quad [\![\pi_2]\!]^i = (1,2)$,
> $[\![\pi_1 \to \pi_2]\!]^i = (-\infty, 0) \cup (1, \infty)$,
> $[\![\pi_2 \to \pi_1]\!]^i = (-\infty, 1) \cup (2, \infty)$,
> $[\![(\pi_1 \to \pi_2) \vee (\pi_2 \to \pi_1)]\!]^i = \mathbb{R} \setminus \{1\}$.

Let $\mathcal{C}$ be the collection of open sets listed in the box; then $\mathcal{C}$ is a finite subset of Σ_i. (In this example, $\mathcal{C}$ consists of five open subsets of $\mathbb{R}$.) Now form the topology Σ_g on $\mathbb{R}$ that is generated by $\mathcal{C}$. Actually, we've already done that in 4.23, because I anticipated the example that we're going through now.

Since $\mathcal{C}$ is a finite set, Σ_g is a finite topology, as we remarked in 4.20. That is, Σ_g has only finitely many members (though it is a topology on an infinite set Ω). That topology can be used for a new topological interpretation. Let us denote its logical operators by $\ominus^g, \vee^g, \wedge^g, \ominus^g$. They are defined as in 10.1, but this time using the interior operation int_g of the topology Σ_g; it may be different from int_i. Of course, the set-theoretical complementation operator $\complement$ does not depend on the topology, so it is still the same: $\complement X$ means $\Omega \setminus X$.

We are to show that the formula $(\pi_1 \to \pi_2) \vee (\pi_2 \to \pi_1)$ is not a tautology for the Σ_g interpretation. To do that, we must exhibit at least one particular valuation $[\![\ ]\!]^g$ that satisfies $[\![(\pi_1 \to \pi_2) \vee (\pi_2 \to \pi_1)]\!]^g \neq \mathbb{R}$. The valuation will be determined by the four functions $\ominus^g, \vee^g, \wedge^g, \ominus^g$, together with this assignment of values:

$$[\![\pi_j]\!]^g = \begin{cases} [\![\pi_j]\!]^i, & \text{if this is one of the } \pi\text{'s in the box;} \\ \text{an arbitrary member of } \mathcal{C} \text{ otherwise.} \end{cases}$$

Since $[\![(\pi_1 \to \pi_2) \vee (\pi_2 \to \pi_1)]\!]^i \neq \mathbb{R}$, it suffices to prove that

$$[\![(\pi_1 \to \pi_2) \vee (\pi_2 \to \pi_1)]\!]^g = [\![(\pi_1 \to \pi_2) \vee (\pi_2 \to \pi_1)]\!]^i.$$

As is often the case in mathematics (particularly with induction proofs), it turns out to be *easier* to prove a *stronger* statement. Since $(\pi_1 \to \pi_2) \vee (\pi_2 \to \pi_1)$ is one of the formulas appearing in the box, it suffices to show this stronger statement:

(∗) for each of the formulas X listed in the box, we actually have $[\![X]\!]^g = [\![X]\!]^i$.

(We emphasize that this assertion is only for the formulas X that appear in the box, not for *all* formulas.)

The proof is by induction on the rank of the formula X. The assertion is clearly true for rank 1: If X has rank 1, then X is just a propositional variable symbol π_j, and so (∗) holds by our choice of $[\![\ ]\!]^g$.

What about formulas of higher rank? The formula X must be of one of these four types: $X_1 \vee X_2$, $X_1 \wedge X_2$, $X_1 \to X_2$, $\overline{X_1}$. In each case the subformulas X_1 and X_2 also appear in the box. They have lower rank, so we may apply to them the induction hypothesis — that is, $[\![X_1]\!]^g = [\![X_1]\!]^i$ and $[\![X_2]\!]^g = [\![X_2]\!]^i$.

Verification for $\vee$. This is quite easy because $\vee$ is just union, in both interpretations i and g; it does not depend on the topology. Thus

$$[\![X_1 \vee X_2]\!]^g = [\![X_1]\!]^g \cup [\![X_2]\!]^g = [\![X_1]\!]^i \cup [\![X_2]\!]^i = [\![X_1 \vee X_2]\!]^i.$$

Verification for $\wedge$. Similar to that for $\vee$.

Verification for →. This is considerably harder, since $\ominus^i$ and $\ominus^g$ do depend on the different topologies.

As noted above, we have $[\![X_1]\!]^g = [\![X_1]\!]^i$ and $[\![X_2]\!]^g = [\![X_2]\!]^i$. Consequently, the two sets $[\![X_2]\!]^g \cup \complement [\![X_1]\!]^g$ and $[\![X_2]\!]^i \cup \complement [\![X_1]\!]^i$ are the same; let us denote their common value by V. That set is not necessarily open in either topology, since the complement of an open set is not necessarily open. We are interested in the two sets

$$[\![X_1 \to X_2]\!]^g = \mathrm{int}_g(V), \qquad [\![X_1 \to X_2]\!]^i = \mathrm{int}_i(V)$$

which are open in the two topologies respectively; we wish to show that these two sets are equal. The interior operators int_g and int_i are not always the same, but they will give the same result *when applied to this particular set V*; we may show that as follows. The set $[\![X_1 \to X_2]\!]^i = \mathrm{int}_i(V)$ appeared in our original demonstration in the infinite topology; thus it appears in our boxed list of formulas and sets. Therefore the set $\mathrm{int}_i(V)$ is in the collection $\mathcal{C}$ of sets used to generate the topology Σ_g. Now 4.25 yields $\mathrm{int}_g(V) = \mathrm{int}_i(V)$.

Verification for ¬. Similar to that for $\ominus$, or use the fact that $\ominus S = S \ominus \varnothing$.

This completes the demonstration of the principle.

10.12. Corollary. Let X be some particular formula. Suppose that there exists a topological space for which X is not tautological. Then there exists a topological space (Ω, Σ) for which X is not tautological, and where the set Ω is finite.

Proof. Immediate from the preceding result and 4.26.

DISJUNCTION PROPERTY (OPTIONAL)

The disjunction property is important, but we will prove a stronger statement of it in 27.11 and 29.29. The version below is included only because it uses more elementary methods, and may be of interest to readers who won't have time to get to 27.11 and 29.29.

10.13. Let A_1 and A_2 be some formulas. Assume that A_1 is nontautological for some topology (Ω_1, Σ_1), and A_2 is nontautological for some topology (Ω_2, Σ_2). Then there exists a topology (Ω, Σ) for which $A_1 \vee A_2$ is nontautological.

Proof. By relabeling, we may assume that the given sets Ω_1 and Ω_2 are disjoint.[1] By assumption, there exist functions

$$[\![\ ]\!]_1 : \{\text{formulas}\} \to \Sigma_1, \qquad [\![\ ]\!]_2 : \{\text{formulas}\} \to \Sigma_2$$

that are valuations in the two topological interpretations, such that $[\![A_1]\!]_1 \neq \Omega_1$ and $[\![A_2]\!]_2 \neq \Omega_2$.

Let ξ be some object that is not a member of the set $\Omega_1 \cup \Omega_2$. We shall take $\Omega = \Omega_1 \cup \Omega_2 \cup \{\xi\}$. Now let Σ be the set whose members are Ω and all the sets of the form $S_1 \cup S_2$, where $S_1 \in \Sigma_1$ and $S_2 \in \Sigma_2$. Verify that Σ is a topology on Ω. Note that, in this topology, Ω itself is the only open set that contains ξ.

We will now specify a particular valuation $[\![\ ]\!]$ in the topological interpretation given by the topology Σ. Let

$$[\![\pi_j]\!] \;\; = \;\; [\![\pi_j]\!]_1 \cup [\![\pi_j]\!]_2.$$

That determines the values of $[\![B]\!]$ for all longer formulas B, since they are built up from the propositional variable symbols.

It is tempting to conjecture that $[\![B]\!]$ will be equal to $[\![B]\!]_1 \cup [\![B]\!]_2$ for all formulas B. That conjecture is almost correct. In fact, by a straightforward but tedious induction (not shown here) on the rank of formula B,

$$[\![B]\!] \text{ is equal to either } [\![B]\!]_1 \cup [\![B]\!]_2 \text{ or } [\![B]\!]_1 \cup [\![B]\!]_2 \cup \{\xi\}$$

for every formula B.

Since $[\![A_1]\!]_1 \subsetneq \Omega_1$, we have $[\![A_1]\!] \neq \Omega$, and therefore $\xi \notin [\![A_1]\!]$. Similarly $\xi \notin [\![A_2]\!]$. Therefore $\xi \notin [\![A_1 \vee A_2]\!]$. Therefore $A_1 \vee A_2$ is not tautological in (Ω, Σ).

---

[1] For instance, if Ω_1 and Ω_2 are both the real line, replace them (respectively) with the sets of ordered pairs $\{(r,1) : r \in \mathbb{R}\}$ and $\{(r,2) : r \in \mathbb{R}\}$.

Chapter 11

More advanced topics in semantics

COMMON TAUTOLOGIES

11.1. In the preceding chapters we introduced several different semantic logics. We demonstrated the validity of a few formulas and the invalidity of a few others, separately for each logic, by fairly concrete and ad hoc methods intended only for use within that logic. In the next few pages we will demonstrate the validity of dozen more formulas, not separately for each logic, but simultaneously for many logics, via more abstract methods.

The semantic logics developed in the last few chapters are fairly diverse — e.g., they have different definitions of $\ominus$. But they are not a representative sample of *all* logics. The logics developed in this book do share certain features — e.g., the fact that $x \ominus y$ is an order reversing function of x and an order preserving function of y, though we have two different notions of "ordering" — we use $\leq$ for some logics and $\subseteq$ for others.[1] In most cases, the direct verification of these shared properties of $\ominus, \ovee, \owedge, \ominus$ within each logic is fairly easy — much easier,

---

[1]Algebraists can explain both these cases in terms of an abstract partial ordering $\preccurlyeq$, but that may be too abstract for my intended audience of beginners. I recommend heavier use of algebra in more advanced studies of logic.

in fact, than the verification of the tautologousness of formulas such as $[(B\to A) \land (C\to A)] \to [(B \lor C)\to A]$.

Thus, our strategy is to

(i) list the shared properties of $\ominus, \lor, \land, \ominus$ (this will be done in 11.2 and subsequent sections),

(ii) verify those shared properties within each individual logic, and

(iii) use those shared properties of $\ominus, \lor, \land, \ominus$ in an abstract setting to prove the tautologousness of a number of formulas (this will be done in 11.3–11.7).

Actually, we have already carried out step (ii), in sections marked "postponable": 8.7, 8.23, 8.34, 8.42, 9.11, 9.17, and 10.8. Readers who have skipped over those sections may want to read them, either now or immediately after 11.7. (Omitted from that list is the hexagon interpretation 9.4, which does not satisfy some of those "shared properties"; we will deal with it by other methods in 11.12.)

The formulas selected for this investigation are the ones that will be axioms for "basic logic," the basis of all the syntactic logics covered in subsequent chapters. See 13.2 and 14.1.

11.2. *Easy verifications.* Following is a list of conditions that we will use in subsequent proofs. The first few conditions differ slightly in the numerically valued and set-valued cases.

	Numerical	Sets (including topological)
a.	$\Sigma \subseteq \mathbb{R}$. That is, the semantic values are real numbers.	$\Sigma \subseteq \mathcal{P}(\Omega)$. That is, the semantic values are some of the subsets of some given set Ω.
b.	If $x \in \Sigma_-$ and $y \in \Sigma_+$ then $x < y$.	If $x \in \Sigma_-$ and $y \in \Sigma_+$, then we cannot have $x \supseteq y$. Also, if $x, y \in \Sigma_+$, then $x \cap y \in \Sigma_+$.
c.	$x \lor y = \max\{x, y\}$, $x \land y = \min\{x, y\}$.	$x \lor y = x \cup y$, $x \land y = x \cap y$.

Common tautologies

d.	$x \ominus y$ is true (i.e., a member of Σ_+) if and only if
	$x \leq y$. $\quad\vert\quad$ $x \subseteq y$.

e. $x \ominus y$ is order reversing in x and order preserving in y, in the sense of 3.28 or 3.45.

f. $\ominus x$ is an order reversing function of x.

g. (Contrapositive) $\quad x \ominus \ominus y = y \ominus \ominus x$ for all $x, y \in \Sigma$.

h. There exists some binary operator $\&$ on Σ (hereafter called *fusion*), satisfying for all $x, y, z \in \Sigma$

   (i) (Commutative) $\quad x \& y = y \& x$,

   (ii) (Associative) $\quad (x \& y) \& z = x \& (y \& z)$,

   (iii) (Fusion property) $\quad (x \& y) \ominus z = x \ominus (y \ominus z)$,

   (iv) $x \& y$ is an order preserving function of x and of y.

Conditions g and h usually are the hardest if we try to verify them directly. But often they are quite easy to verify indirectly, using Lemma 11.3 or 11.4 below. Note that 11.3 generally is not applicable to topological interpretations, since their negation is not involutive, but 11.4 is still applicable to those interpretations.

Remark. It follows easily from 11.2.c, 11.2.e, and 11.2.f that a formula-valued function $\mathsf{s}(\cdot)$ is order preserving or order reversing (in the sense of 6.35) if and only if its valuation $[\![\mathsf{s}(\cdot)]\!]$ is order preserving or reversing (in the sense of 6.38).

11.3. *Lemma using cotenability.* Suppose that a functional interpretation is given, satisfying all of conditions 11.2.a–11.2.f. Also assume that

   a. Negation is involutive — that is, $\ominus \ominus q = q$ for all $q \in \Sigma$.

Also assume that the cotenability operator, defined as usual by

   b. $q \odot r = \ominus (q \ominus \ominus r)$,

satisfies both of these conditions:

   c. commutative: $\quad q \odot r = r \odot q$, and

d. associative: $(q \odot r) \odot s = q \odot (r \odot s)$

for all $q, r, s \in \Sigma$. Then 11.2.g is satisfied, and 11.2.h is satisfied by cotenability — i.e., by defining $x \otimes y = x \odot y$.

Proof. From a and b we obtain

$$(*) \qquad q \ominus r = \ominus(q \odot \ominus r)$$

Now the contrapositive law 11.2.g follows from the commutativity of cotenability (i.e., 11.3.c). Also,

$$x \ominus (y \ominus z) \stackrel{(*)}{=} \ominus\left[x \odot \ominus (y \ominus z)\right] \stackrel{(*)}{=}$$
$$\ominus\left[x \odot \ominus\ominus(y \odot \ominus z)\right] \stackrel{a}{=} \ominus\left[x \odot (y \odot \ominus z)\right]$$
$$\stackrel{d}{=} \ominus\left[(x \odot y) \odot \ominus z\right] \stackrel{(*)}{=} (x \odot y) \ominus z.$$

Thus $\odot$ satisfies the fusion property 11.2.h(iii). If we take fusion to be cotenability, then the commutativity and associativity of fusion follow from those of cotenability. Finally, the fact that cotenability (and hence fusion) is order preserving in both arguments follows from 11.2.e and 11.2.f.

11.4. *Lemma using falsehood.* Suppose that a functional interpretation is given, satisfying all of conditions 11.2.a–11.2.f and 11.2.h. Also assume that the interpretation has an Ackermann falsehood, $\bot$ — that is, a semantic value with the property that $\ominus x = x \ominus \bot$ for all $x \in \Sigma$.

Then 11.2.g is satisfied.

Proof. The second, third, and fourth equations use 11.2.h.
$$x \ominus \ominus y = x \ominus (y \ominus \bot) = (x \otimes y) \ominus \bot$$
$$= (y \otimes x) \ominus \bot = y \ominus (x \ominus \bot) = y \ominus \ominus x.$$

11.5. *Lemmas about implication.* Assume that some functional interpretation is given, satisfying all the conditions of 11.2. Then for all semantic values p and q:

a. *Fusion detachment.* $p \otimes (p \ominus q) \subseteq q$.

b. *Semantic assertion.* $p \subseteq (p \ominus q) \ominus q$.

(Similarly for $\leq$, in the case of numerical interpretations.)

Proof. Obviously $p \ominus q \subseteq p \ominus q$; hence $(p \ominus q) \to (p \ominus q) \in \Sigma_+$ by 11.2.d. By three different uses of 11.2.h, we have

$$(p \ominus q) \ominus (p \ominus q) \underset{}{\overset{11.2.\text{h(iii)}}{=\!=\!=}} [(p \ominus q) \,\&\, p] \ominus q$$
$$\underset{}{\overset{11.2.\text{h(i)}}{=\!=\!=}} [p \,\&\, (p \ominus q)] \ominus q \underset{}{\overset{11.2.\text{h(iii)}}{=\!=\!=}} p \ominus [(p \ominus q) \ominus q].$$

Thus both $[p \,\&\, (p \ominus q)] \ominus q$ and $p \ominus [(p \ominus q) \ominus q]$ lie in Σ_+. Apply 11.2.d to both of those.

11.6. *Validity of basic logic.* Assume that some functional interpretation is given, satisfying all the conditions of 11.2. Then all the assumptions of basic logic (listed in 13.2 and 14.1) are valid in that interpretation. More precisely,

- axiom schemes 13.2.b–13.2.d and 14.1.b–14.1.i are tautologous in the functional interpretation, and

- detachment and adjunction, the two assumed inference rules of basic logic, are truth-preserving (defined as in 7.10) in the functional interpretation.

Proofs. We will only carry out each proof for the set-valued case; the proof for the numerical case is similar.[2] The proof for detachment is carried out in great detail; the other proofs are sketched more briefly.

Verification of detachment. Let $[\![\ ]\!]$ be some valuation under which the formulas A and $A \to B$ both valuate to true values. That is, $[\![A]\!]$ and $[\![A \to B]\!]$ are both members of Σ_+. By 7.5.c(ii) we have $[\![A]\!] \ominus [\![B]\!] = [\![A \to B]\!]$; therefore $[\![A]\!] \ominus [\![B]\!] \in \Sigma_+$. By 11.2.d, it follows that $[\![A]\!] \subseteq [\![B]\!]$. Now 11.2.b tells us that $[\![B]\!] \in \Sigma_-$ cannot occur. Therefore $[\![B]\!] \in \Sigma_+$.

---

[2] Actually, the connection goes beyond mere similarity. It is possible to reformulate each numerical interpretation as a set-valued interpretation, e.g., by replacing each number r with the interval $(-\infty, r)$. But beginners may find that reduction cumbersome.

Verifications of identity, adjunction, ∧-elimination, ∨-introduction, and distribution. These are easy exercises; for distribution use 3.55.a.

Verification of permutation. Use the commutativity and fusion property in 11.2.h.

Verification of →-prefixing. We have $b \supseteq (a \ominus b) \,\&⃝\, a$ by fusion detachment 11.5.a. Now, $x \ominus c$ is an order reversing function of x (11.2.e); hence $b \ominus c \subseteq [(a \ominus b) \,\&⃝\, a] \ominus c$. We have also $[(a \ominus b) \,\&⃝\, a] \ominus c = (a \ominus b) \ominus (a \ominus c)$ as a special case of the fusion property 11.2.h(iii). Thus $b \ominus c \subseteq (a \ominus b) \ominus (a \ominus c)$. Now apply 11.2.d.

Verification of ∧-introduction. Begin with

$$a \,\&⃝\, [(a \ominus b) \cap (a \ominus c)] \underset{\text{11.2.h(iv)}}{\subseteq} a \,\&⃝\, (a \ominus b) \underset{\text{11.5.a}}{\subseteq} b.$$

Similarly $a \,\&⃝\, [(a \ominus b) \cap (a \ominus c)] \subseteq c$. Combining those two results, $a \,\&⃝\, [(a \ominus b) \cap (a \ominus c)] \subseteq b \cap c$. Therefore

$$\begin{aligned}
\Sigma_+ \ni \{a \,\&⃝\, [(a \ominus b) \cap (a \ominus c)]\} \ominus (b \cap c) &\quad \text{by 11.2.d} \\
= \{[(a \ominus b) \cap (a \ominus c)] \,\&⃝\, a\} \ominus (b \cap c) &\quad \text{by 11.2.h(i)} \\
= [(a \ominus b) \cap (a \ominus c)] \ominus [a \ominus (b \cap c)] &\quad \text{by 11.2.h(iii)} \\
= [(a \ominus b) \,⃝\!\wedge\, (a \ominus c)] \ominus [a \ominus (b \,⃝\!\wedge\, c)].
\end{aligned}$$

Verification of ∨-elimination. Begin with

$$b \underset{\text{11.5.b}}{\subseteq} (b \ominus a) \ominus a \underset{\text{11.2.e}}{\subseteq} [(b \ominus a) \cap (c \ominus a)] \ominus a$$

and $\quad c \subseteq (c \ominus a) \ominus a \subseteq [(b \ominus a) \cap (c \ominus a)] \ominus a$.

Combining those results,

$$(b \cup c) \subseteq [(b \ominus a) \cap (c \ominus a)] \ominus a.$$

Hence $(b \,⃝\!\vee\, c) \ominus \{[(b \ominus a) \,⃝\!\wedge\, (c \ominus a)] \ominus a\}$ is a member of Σ_+, by 11.2.d. Finally, apply permutation (proved earlier); thus

$$[(b \ominus a) \,⃝\!\wedge\, (c \ominus a)] \ominus [(b \,⃝\!\vee\, c) \ominus a] \quad \in \quad \Sigma_+.$$

Images of interpretations

11.7. *Slightly more specialized results.* (These will be useful in fewer logics.)

a. *Expansion.* If $x \,\&\, x \subseteq x$ then $x \ominus y \subseteq x \ominus (x \ominus y)$.

b. *Contraction.* If $x \,\&\, x \supseteq x$ then $x \ominus y \supseteq x \ominus (x \ominus y)$.

(Similarly for $\leq$ and $\geq$, in the case of numerical interpretations.)

Proof. We will prove only 11.7.a; the proof of 11.7.b is the same with the inclusions reversed. Let $u = x \,\&\, x$, and assume that $u \subseteq x$. Since $\ominus$ is order reversing in its first argument, it follows that $u \ominus y \supseteq x \ominus y$. That is, $x \ominus y \subseteq (x \,\&\, x) \ominus y$. This is just the conclusion of 11.7.a.

IMAGES OF INTERPRETATIONS

11.8. *Restriction principle.* Suppose I_1 and I_2 are two interpretations with sets of semantic values $\Sigma_1 \subseteq \Sigma_2$, and the functions $\vee_1, \wedge_1, \ominus_1, \ominus_1$ are restrictions of the functions of I_2 to the smaller domain Σ_1. Also suppose that $\Sigma_{2+} \cap \Sigma_1 \subseteq \Sigma_{1+}$; that is, every semantic value of interpretation I_1 that is true in I_2, is also true in I_1.

Then every I_2-tautology is also an I_1-tautology.

Proof. This is an application of 11.9, with $f(x) = x$ for all $x \in \Sigma_1$. (We have presented this result before 11.9 as a preview of an important special case, because 11.9 is complicated.)

11.9. *Image principle.* Suppose that

$$I_1 = (\ \Sigma_1,\ \Sigma_{1+},\ \vee_1,\ \wedge_1,\ \ominus_1,\ \ominus_1\ ),$$
$$I_2 = (\ \Sigma_2,\ \Sigma_{2+},\ \vee_2,\ \wedge_2,\ \ominus_2,\ \ominus_2\ )$$

are two functional interpretations of our formal language. Assume $f : \Sigma_1 \to \Sigma_2$ is some given function that preserves the logical operations³ in this sense:

$$\begin{aligned} f(x \vee_1 y) &= f(x) \vee_2 f(y), \\ f(x \wedge_1 y) &= f(x) \wedge_2 f(y), \\ f(x \ominus_1 y) &= f(x) \ominus_2 f(y), \\ f(\ominus_1 y) &= \ominus_2 f(y) \end{aligned}$$

for all $x, y \in \Sigma_1$. Define range and inverse image in their usual fashion (see 3.24). Then:

a. Suppose $f^{-1}(\Sigma_{2+}) \subseteq \Sigma_{1+}$. Then every I_2-tautology is also an I_1-tautology.

b. Suppose $f^{-1}(\Sigma_{2+}) \supseteq \Sigma_{1+}$ and $\mathrm{Range}(f) = \Sigma_2$. Then every I_1-tautology is also an I_2-tautology.

Combining those two results,

c. Suppose $f^{-1}(\Sigma_{2+}) = \Sigma_{1+}$ and $\mathrm{Range}(f) = \Sigma_2$. Then the two interpretations I_2 and I_1 have the same set of tautologies.

Demonstration. Some readers may find it helpful to review 3.25 before proceeding further.

The main ingredient in the proof is this preliminary result:

(†) $\begin{cases} \text{Suppose } [\![\ ]\!]_1 \text{ and } [\![\ ]\!]_2 \text{ are some particular valu-} \\ \text{ations in the two interpretations, satisfying} \\ (*) \quad f([\![\pi_j]\!]_1) = [\![\pi_j]\!]_2 \text{ for all } j = 1, 2, 3, \ldots. \\ \text{Then they also satisfy } f([\![S]\!]_1) = [\![S]\!]_2 \text{ for every} \\ \text{formula } S. \end{cases}$

The proof of (†) will follow by induction on the rank of the formula S. The initialization step is obvious: If S has rank 1, then S is the formula π_j for some j, and so the desired conclusion of (†) is one of the equations we assumed in (∗). For higher ranks, suppose that

$$f([\![Q]\!]_1) = [\![Q]\!]_2, \qquad f([\![R]\!]_1) = [\![R]\!]_2$$

for some formulas Q and R; we must establish the analogous result for each of $Q \vee R$, $Q \wedge R$, $Q \to R$, and $\neg Q$. We shall

---

³In the language of algebraists, f is a *homomorphism* from Σ_1 into Σ_2. Compare 7.5.c(ii).

prove this for $Q \vee R$; the proofs for the other three formulas are similar. We compute

$$f(\llbracket Q \vee R \rrbracket_1) = f(\llbracket Q \rrbracket_1 \heartsuit_1 \llbracket R \rrbracket_1)$$
$$= f(\llbracket Q \rrbracket_1) \heartsuit_2 f(\llbracket R \rrbracket_1) = \llbracket Q \rrbracket_2 \heartsuit_2 \llbracket R \rrbracket_2 = \llbracket Q \vee R \rrbracket_2$$

where the first and last equations make use of the fact that I_1 and I_2 are functional interpretations. This completes our sketch of the proof of (†).

To prove part a, suppose formula S is an I_2-tautology; we wish to show it is also a tautology in I_1. Fix any valuation $\llbracket \ \rrbracket_1$ in interpretation I_1; we wish to show that $\llbracket S \rrbracket_1 \in \Sigma_{1+}$. Define a corresponding valuation $\llbracket \ \rrbracket_2$ in interpretation I_2, by using equation (∗). Since S is an I_2-tautology, we have $\llbracket S \rrbracket_2 \in \Sigma_{2+}$. Applying (†), we have $f(\llbracket S \rrbracket_1) = \llbracket S \rrbracket_2 \in \Sigma_{2+}$, and therefore $\llbracket S \rrbracket_1 \in f^{-1}(\Sigma_{2+}) \subseteq \Sigma_{1+}$. This completes the proof of part a.

To prove part b, suppose formula S is tautological in the interpretation I_1; we wish to show that it is also tautological for I_2. Fix any valuation $\llbracket \ \rrbracket_2$ in the interpretation I_2; it suffices to show that $\llbracket S \rrbracket_2 \in \Sigma_{2+}$. That valuation specifies some particular values in Σ_2 for $\llbracket \pi_1 \rrbracket_2, \llbracket \pi_2 \rrbracket_2, \llbracket \pi_3 \rrbracket_2, \ldots$. Since $\text{Range}(f) = \Sigma_2$, that sequence of values lies in the range of f. That is, there exist choices $x_1, x_2, x_3, \ldots \in \Sigma_1$ (not necessarily unique) that satisfy

$$f(x_1) = \llbracket \pi_1 \rrbracket_2, \quad f(x_2) = \llbracket \pi_2 \rrbracket_2, \quad f(x_3) = \llbracket \pi_3 \rrbracket_2, \quad \cdots .$$

Choose any such sequence $x_1, x_2, x_3, \ldots$ in Σ_1. Define valuation $\llbracket \ \rrbracket_1$ in the interpretation I_1 by taking

$$\llbracket \pi_1 \rrbracket_1 = x_1, \quad \llbracket \pi_2 \rrbracket_1 = x_2, \quad \llbracket \pi_3 \rrbracket_1 = x_3, \quad \cdots .$$

Then (∗) is satisfied. Since S is tautological in I_1, we have $\llbracket S \rrbracket_1 \in \Sigma_{1+}$. By (†), then,

$$\llbracket S \rrbracket_2 = f(\llbracket S \rrbracket_1) \in f(\Sigma_{1+}) \subseteq \Sigma_{2+}$$

completing the proof of b.

Finally, c is immediate from a and b.

11.10. *Example: Subsets of classical logic.* We compare the two-valued interpretation (8.2) with the fuzzy, Sugihara, and topological interpretations (8.16, 8.38, 10.1). We shall show that

(f) $\quad$ {fuzzy tautologies} $\subsetneq$ {two-valued tautologies},

(s) $\quad$ {Sugihara tautologies} $\subsetneq$ {two-valued tautologies},

(t) $\quad$ {topological tautologies} $\subsetneq$ {two-valued tautologies}.

Proofs. All of the inclusions ($\subseteq$) follow easily from the restriction principle 11.8. We take I_1 to be the classical two-valued interpretation, but instead of using $\Sigma_1 = \{\text{"false," "true"}\}$, we relabel those semantic values. To prove (f), (s), or (t), respectively, we take Σ_1 equal to $\{0, 1\}$, $\{-1, +1\}$, or $\{\varnothing, \Omega\}$.

To show that the inclusions ($\subseteq$) are in fact strict ($\subsetneq$) it suffices to give examples of formulas that are tautological in the two-valued interpretation but not in the other interpretations. See 8.26, 8.39.c, 10.3, and 10.4.

11.11. *Powersets are classical.* Let Ω be any nonempty set. Then the tautologies of the $\mathcal{P}(\Omega)$-valued powerset interpretation are the same as the tautologies of the two-valued interpretation.

Proof. To show that every powerset tautology is a two-valued tautology, we shall relabel the two-valued interpretation so that its two semantic values are $\varnothing$ and Ω. Then its set of semantic values is a subset of $\mathcal{P}(\Omega)$, and we may apply the restriction principle 11.8.

Going the other way, suppose some formula K is not a tautology in the powerset interpretation. Thus there is at least one valuation $[\![\ ]\!]_\Omega$ in the powerset interpretation, with the property that $[\![K]\!]_\Omega \subsetneq \Omega$. Pick any point $z \in \Omega \setminus [\![K]\!]_\Omega$. Define a function $[\![\ ]\!]_z : \{\text{formulas}\} \to \{0, 1\}$ by taking

$$[\![A]\!]_z = \mathbb{1}_{[\![A]\!]_\Omega}(z) = \begin{cases} 1 & \text{if } z \in [\![A]\!]_\Omega, \\ 0 & \text{if } z \notin [\![A]\!]_\Omega \end{cases}$$

for each formula A; here $\mathbb{1}_{[\![A]\!]_\Omega}$ is the characteristic function of the set $[\![A]\!]_\Omega$. We easily verify that $[\![\ ]\!]_z$ is a valuation in the two-valued interpretation. Since $z \notin [\![K]\!]_\Omega$, we have $[\![K]\!]_z = 0$, and therefore K is not tautological in the two-valued interpretation.

11.12. *Hexagon is classical.* The hexagon interpretation, introduced in 9.4, has the same tautologies as the two-valued interpretation or any of the powerset interpretations.

Proof. Since we have already shown in 11.11 that all the powerset interpretations have the same set of tautologies, we may concentrate on just one powerset interpretation that is convenient. We shall take I_2 to be the the four-valued interpretation obtained as the powerset interpretation with $\Omega = \{-1, +1\}$ and

$$\Sigma_2 = \{\varnothing, \{-1\}, \{+1\}, \{-1, +1\}\}, \quad \Sigma_{2+} = \{\{-1, +1\}\}.$$

We take I_1 to be the hexagon interpretation. Define $f : \Sigma_1 \to \Sigma_2$ as follows:

$x \in \Sigma_1$	$\varnothing$	$\{-1\}$	$\{-1, 0\}$	$\{+1\}$	$\{0, +1\}$	$\{-1, 0, +1\}$
$f(x) \in \Sigma_2$	$\varnothing$	$\{-1\}$	$\{-1\}$	$\{+1\}$	$\{+1\}$	$\{-1, +1\}$

We will apply 11.9.c. It is easy to see that $f^{-1}(\Sigma_{2+}) = \Sigma_{1+}$ and $\text{Range}(f) = \Sigma_2$. To verify the remaining hypothesis, that f "preserves the logical operations" $\vee, \wedge, \to, \neg$, is much more tedious, but it is fairly elementary. We shall carry out some of the analysis for $\to$; the other operators can be analyzed similarly.

We must show that

$$f(x \ominus_1 y) = f(x) \ominus_2 f(y) \quad \text{for all} \quad x, y \in \Sigma_1.$$

This can be carried out on a case-by-case basis; there are just 36 combinations of possible values of x and y. Alternatively, this can be carried out using the formulas for $\ominus_2$ and $\ominus_1$, given in 9.3 and 9.4 respectively.

Both approaches are made easier, if we use this insight: The function $f(x) = x \setminus \{0\}$ has the effect of "forgetting" whether or not 0 is a member of a set. This modifies the 6×6 implication table of the hexagon interpretation, reducing it to the 4×4 table of what turns out to be the powerset interpretation. The procedure is made more transparent if we subdivide the 6×6 table to emphasize that $\{-1\}$ and $\{-1, 0\}$ are grouped together (since they are both mapped by f to the same member of Σ_2),

$S \ominus_1 T$	$T = \varnothing$	$\{-1\}$	$\{-1,0\}$	$\{+1\}$	$\{0,+1\}$	Ω
$S = \varnothing$	Ω	Ω	Ω	Ω	Ω	Ω
$\{-1\}$ $\{-1,0\}$	$\{0,+1\}$ $\{+1\}$	Ω Ω	Ω Ω	$\{0,+1\}$ $\{+1\}$	$\{0,+1\}$ $\{0,+1\}$	Ω Ω
$\{+1\}$ $\{0,+1\}$	$\{-1,0\}$ $\{-1\}$	$\{-1,0\}$ $\{-1\}$	$\{-1,0\}$ $\{-1,0\}$	Ω Ω	Ω Ω	Ω Ω
Ω	$\varnothing$	$\{-1\}$	$\{-1,0\}$	$\{+1\}$	$\{0,+1\}$	Ω

and likewise that $\{+1\}$ and $\{+1,0\}$ are grouped together. See the table.

By the "forgetting" process, $\{-1\}$ and $\{-1,0\}$ become indistinguishable from one another (but we may retain both in the table). Likewise $\{+1\}$ and $\{0,+1\}$ become indistinguishable from one another. Also, $\varnothing$ (which is a semantic value of the hexagon interpretation) becomes indistinguishable from $\{0\}$ (which was not a semantic value), and $\{-1,0,+1\}$ becomes indistinguishable from $\{-1,+1\}$.

A glance at the table now reveals that each of the smaller rectangles within the table now contains only sets that are indistinguishable from one another. This means that

    if S_1 and S_2 are indistinguishable,
    and T_1 and T_2 are indistinguishable,
    then $S_1 \ominus_1 T_1$ and $S_2 \ominus_1 T_2$ are indistinguishable.

That is, if $f(S_1) = f(S_2)$ and $f(T_1) = f(T_2)$, then $f(S_1 \ominus_1 T_1) = f(S_2 \ominus_1 T_2)$. In the terminology of algebraists, $\ominus_1$ "respects equivalence classes."

That observation greatly reduces our work: Instead of verifying $f(x \ominus_1 y) = f(x) \ominus_2 f(y)$ for 36 combinations of x and y, we need only to verify it for 16 combinations. We can consider just one of $\{-1\}$ or $\{-1,0\}$; we don't need to consider both. Likewise, we can consider just one of $\{+1\}$ or $\{+1,0\}$; we don't need to consider both. For simplicity, omit the 0. Our 6×6 table is reduced to the 4×4 table shown below.

Now 16 easy verifications show that that table is *identical* to the implication table for the $\mathcal{P}(\{-1,1\})$-valued powerset in-

Dugundji formulas

$S \ominus_2 T$?	$T = \varnothing$	$\{-1\}$	$\{+1\}$	Ω
$S = \varnothing$	Ω	Ω	Ω	Ω
$\{-1\}$	$\{+1\}$	Ω	$\{+1\}$	Ω
$\{+1\}$	$\{-1\}$	$\{-1\}$	Ω	Ω
Ω	$\varnothing$	$\{-1\}$	$\{+1\}$	Ω

terpretation, if we relabel Ω as $\{-1,+1\}$. Similar verifications can be carried out for $\vee, \wedge, \neg$.

Dugundji formulas

11.13. The *Dugundji formulas* are defined in this section and used in proofs in the next few sections. Let us abbreviate

$$\mathcal{A}_j^k \;=\; \pi_j \leftrightarrow \pi_k \;=\; (\pi_j \to \pi_k) \wedge (\pi_k \to \pi_j)$$

for positive integers j and k. For instance,

$$\mathcal{A}_1^2 = (\pi_1 \to \pi_2) \wedge (\pi_2 \to \pi_1), \qquad \mathcal{A}_1^3 = (\pi_1 \to \pi_3) \wedge (\pi_3 \to \pi_1).$$

Then define

$$\begin{aligned}
\mathcal{D}_2 &= \mathcal{A}_1^2, \\
\mathcal{D}_3 &= \mathcal{A}_1^2 \vee \mathcal{A}_1^3 \vee \mathcal{A}_2^3, \\
\mathcal{D}_4 &= \mathcal{A}_1^2 \vee \mathcal{A}_1^3 \vee \mathcal{A}_2^3 \vee \mathcal{A}_1^4 \vee \mathcal{A}_2^4 \vee \mathcal{A}_3^4, \\
\mathcal{D}_5 &= \mathcal{A}_1^2 \vee \mathcal{A}_1^3 \vee \mathcal{A}_2^3 \vee \mathcal{A}_1^4 \vee \mathcal{A}_2^4 \vee \mathcal{A}_3^4 \vee \mathcal{A}_1^5 \vee \mathcal{A}_2^5 \vee \mathcal{A}_3^5 \vee \mathcal{A}_4^5,
\end{aligned}$$

and so on. In general, formula $\mathcal{D}_n$ is the disjunction (the "or") of all the $\mathcal{A}_j^k$'s that satisfy $1 \leq j < k \leq n$. The formula $\mathcal{D}_n$ will be called the *Dugundji formula in n symbols*.

We have omitted the parentheses to improve readability. The placement of parentheses will not actually affect the proofs below; the disjunctions may be taken in any order. (See the remarks at the end of 14.18.) However, for definiteness we may specify a particular order. The simplest and most common specification is to read the disjunctions from left to right:

$$\begin{aligned} \mathcal{D}_2 &= \mathcal{A}_1^2, \\ \mathcal{D}_3 &= (\mathcal{A}_1^2 \vee \mathcal{A}_1^3) \vee \mathcal{A}_2^3, \\ \mathcal{D}_4 &= ((((\mathcal{A}_1^2 \vee \mathcal{A}_1^3) \vee \mathcal{A}_2^3) \vee \mathcal{A}_1^4) \vee \mathcal{A}_2^4) \vee \mathcal{A}_3^4, \end{aligned}$$

and so on.

The reader is cautioned not to confuse the Dugundji formula $\mathcal{D}_2 = (\pi_1 \to \pi_2) \wedge (\pi_2 \to \pi_1)$ with the "paradoxical tautology" $(\pi_1 \to \pi_2) \vee (\pi_2 \to \pi_1)$ discussed in 5.48.

Our next two results give sufficient conditions for a Dugundji formula $\mathcal{D}_m$ to be tautologous or nontautologous in a functional interpretation. Since the conditions on m are related to the number of members of Σ, this leads us to conclusions about that number.

11.14. Dugundji tautology lemma. Let I be a functional interpretation satisfying these conditions:

a. The formula scheme $(S \to S) \wedge (S \to S)$ is tautological in I.
b. If S and T are formulas, at least one of them tautological in I, then $S \vee T$ is tautological in I.

Let m be an integer greater than 1. Suppose that I has fewer than m semantic values — i.e., assume $|\Sigma| < m$.

Then the Dugundji formula $\mathcal{D}_m$ is tautological in I.

Proof. Temporarily fix any integers u, v with $1 \leq u < v \leq m$. Let $\mathcal{E}^{u,v}$ be the formula obtained from

$$\mathcal{D}_m = \mathcal{A}_1^2 \vee \mathcal{A}_1^3 \vee \cdots \vee \mathcal{A}_{m-1}^m$$

by replacing $\mathcal{A}_u^v$ with the formula

$$\mathcal{A}_u^u \quad = \quad \pi_u \leftrightarrow \pi_u \quad = \quad (\pi_u \to \pi_u) \wedge (\pi_u \to \pi_u).$$

We observe that $\mathcal{A}_u^u$ is tautological by hypothesis a; hence $\mathcal{E}^{u,v}$ is tautological by hypothesis b.

Now, let any valuation $[\![\ ]\!]$ in I be given; it suffices to show that

Dugundji formulas   309

$[\![\mathcal{D}_m]\!] \in \Sigma_+$. We may hold that valuation fixed for the remainder of the proof. The assigned values

$$[\![\pi_1]\!], \quad [\![\pi_2]\!], \quad [\![\pi_3]\!], \quad \ldots, \quad [\![\pi_m]\!]$$

are members of Σ. Since Σ has $m-1$ or fewer members, at least two of those values must be the same. That is, we have $[\![\pi_u]\!] = [\![\pi_v]\!]$ for some $u, v \in \{1, 2, \ldots, m\}$ with $u \neq v$. (Different valuations $[\![\ ]\!]$ might require different pairs u, v, but that will not cause any difficulties; we emphasize that our choice of $[\![\ ]\!]$ is fixed throughout the remainder of the proof.) The condition on u, v is preserved if we interchange u and v; hence without loss of generality we may assume $u < v$. Fix that pair u, v; it too will remain fixed throughout the rest of this proof.

The formulas $\mathcal{D}_m$ and $\mathcal{E}^{u,v}$ are nearly identical; we obtain $\mathcal{E}^{u,v}$ from $\mathcal{D}_m$ by replacing two π_v's with π_u's. Since $[\![\ ]\!]$ is a *functional* valuation and $[\![\pi_u]\!] = [\![\pi_v]\!]$, it follows that $[\![\mathcal{D}_m]\!] = [\![\mathcal{E}^{u,v}]\!]$.

The formula $\mathcal{E}^{u,v}$ is tautological in I, and therefore is true in each valuation of I. In particular, it takes a true value in *this* valuation — that is, $[\![\mathcal{E}^{u,v}]\!] \in \Sigma_+$. Hence $[\![\mathcal{D}_m]\!] \in \Sigma_+$, which completes the proof.

11.15. *Remark.* Hypotheses a and b of 11.14 and 11.16 are not very restrictive; they are satisfied by all the main logics of this book. But hypothesis 11.16.c is more restrictive; it rules out many set-valued interpretations.

11.16. *Dugundji nontautology lemma.* Suppose that J is a functional interpretation satisfying these conditions for all $x, y \in \Sigma$:

   a. If $x \wedge y$ is true (i.e., a member of Σ_+) then both x and y are true.
   b. If both $x \ominus y$ and $y \ominus x$ are true, then $x = y$.
   c. If $x \vee y$ is true then at least one of x, y is true.

Suppose, furthermore, m is an integer greater than 1, and the interpretation J has at least m different semantic values (perhaps infinitely many semantic values) — i.e., assume $|\Sigma| \geq m$.

Then the Dugundji formula $\mathcal{D}_m$ is not a tautology of J.

Proof. It suffices to produce one valuation $[\![\ ]\!]$ in the interpretation J, such that $[\![\mathcal{D}_m]\!]$ is false. Choose some assignment in which

$$[\![\pi_1]\!], [\![\pi_2]\!], [\![\pi_3]\!], \ldots, [\![\pi_m]\!]$$

are all different; this is possible since J has at least m different semantic values. Then for all $i, j \in \{1, 2, \ldots, m\}$ with $i \neq j$, at least one of $[\![\pi_i \to \pi_j]\!]$ or $[\![\pi_j \to \pi_i]\!]$ must be false, by hypothesis c. Hence

$$[\![(\pi_i \to \pi_j) \wedge (\pi_j \to \pi_i)]\!] \quad \text{is false}$$

by hypothesis a. Hence $[\![\mathcal{D}_m]\!]$ is false, by repeated uses of hypothesis b. Thus $\mathcal{D}_m$ is not tautological in J.

11.17. *Corollaries.* Most of the following results follow easily using one or both of 11.14, 11.16. (*Exercise.* Fill in the details of the proofs. What should we use for I, J, or m?)

a. None of the Dugundji formulas is tautological for the $[0, 1]$-valued fuzzy interpretation 8.16. Hence there is no finite functional interpretation I having the same tautologies as the $[0, 1]$-valued fuzzy interpretation.

b. None of the Dugundji formulas is tautological for the Sugihara interpretation 8.38. Hence there is no finite functional interpretation I having the same tautologies as the Sugihara interpretation.

c. Show that $\mathcal{D}_3, \mathcal{D}_4, \mathcal{D}_5, \ldots$ are tautological in the two-valued interpretation, but $\mathcal{D}_2$ is not.

d. Earlier in this chapter, we showed that the two-valued (8.2), powerset (9.3), and hexagon (9.4) interpretations all have the same set of tautologies. But those interpretations have different numbers of semantic values. Why does this not contradict 11.14 or 11.16? Discuss.

Part C

Basic syntactics

Chapter 12

Inference systems

12.1. *Remark.* This chapter must be read simultaneously with the next one or more chapters — i.e., the reader will have to flip pages back and forth. Apparently this arrangement is unavoidable. The present chapter contains several very abstract definitions; the next few chapters consist of many examples of those definitions.

12.2. *We now start over.* Though we shall occasionally use some semantic results (mainly for soundness arguments), our main focus in the next few chapters will not involve the semantic notions that we have laboriously developed over the last few chapters. We now (temporarily) discard all such notions.

We retain the terminology of Chapter 6 — our definitions of "formula," "generalization," "order preserving," etc. However, we take the symbols $\vee, \wedge, \rightarrow, \neg$ and formulas A as "meaningless," devoid of any semantic interpretations $\circledvee, \circledwedge, \ominus, \ominus$ or $[\![A]\!]$. We will be concerned with the manipulation of "meaningless" strings of symbols according to certain transformation rules.

12.3. An *inference system* (or *syntactic system* or *axiomatic system*) may be viewed as a process of *collecting*. Just as some people collect stamps, coins, butterflies, or bottlecaps, so we shall collect *theorems and inference rules*. We begin our collection with the inference system's *assumptions*:

- certain assumed formulas, called *axioms* (or *postulates*), and

- certain assumed inference rules, which (for lack of a better name) are called *assumed inference rules*.

We then enlarge our collection by using *derivations*, described in 12.5 below. This yields

- derived formulas, called *theorems*, and
- derived inference rules, which (for lack of a better name) are called *derived inference rules*.

Actually, every axiom is also considered to be a theorem (with a one-line derivation!), and every assumed inference rule is also considered to be a derived inference rule (with a very short derivation — one step for each hypothesis, and one step for the conclusion).

We will present only a few dozen theorems in the course of this book, but it must be understood that

> the term "theorem" applies to *every* formula that *can* be derived, regardless of whether we actually carry out the derivation explicitly in this book, or even whether we mention the formula explicitly in this book.

Thus, there are *infinitely* many theorems. This concept is important because we will investigate (among other things) some properties that are satisfied by *all* the theorems of an axiomatization.

Different inference systems start from different assumptions and yield different collections; one of our goals is to compare some of these collections. For instance, we shall see that the "positive paradox" formula $A \to (B \to A)$ is a theorem of classical, constructive, and fuzzy logics but not relevant logic.

12.4. We remind the reader of the separation between informal and formal logic.

In the next few chapters, we will introduce a variety of axioms. Some of these may seem to resemble the "natural" or "ordinary" logic of everyday reasoning. That resemblance may be useful in the discovery phase of our investigations. (Discovery and certification were contrasted in 2.13–2.15.)

Inference systems 315

But we will not attempt to make such a correspondence precise, nor to use it in the certification phase of our investigations. Our axioms may be informally justified by their resemblance to the "real world," but the axioms are formally justified only by the fact that we assume them.

The implication symbol, "$\to$," has no a priori meaning, so we will not "prove" or "verify" any of our assumptions about it. When we adopt those assumptions, we are merely agreeing to the convention that the symbol "$\to$" will be used in certain ways; thus we explicitly endow that symbol with some meaning.

12.5. Say $\mathcal{H}$ is a set of formulas (possibly the empty set), and C is some formula. A *derivation* of C from $\mathcal{H}$ is a finite sequence $A_1, A_2, \ldots, A_n$ of formulas, where the last formula A_n is just C, and where each formula A_j in the sequence is accompanied by a *justification* of one of these types:

- A_j is a *hypothesis* — i.e., a member of $\mathcal{H}$; or
- A_j is an instance or specialization of some already known theorem or theorem scheme (possibly an axiom or axiom scheme); or
- A_j is the consequence obtained by applying some previously established inference rule (i.e., one that was initially assumed or one that has already been derived) to some of the earlier formulas $A_1, A_2, \ldots, A_{j-1}$ in the sequence.

(Examples are given in 13.4, 13.5, etc.) We then say $\mathcal{H}$ is the set of *hypotheses*, and C is the *conclusion*. When such a derivation exists, we say that C can be *derived* from $\mathcal{H}$, or $\mathcal{H}$ *proves* C, or C is a *syntactic consequence* of $\mathcal{H}$. The existence of such a derivation can be stated more briefly as

- $\mathcal{H} \vdash C$, if the set $\mathcal{H}$ is nonempty. The expression "$\mathcal{H} \vdash C$" is called an *inference rule*.
- $\vdash C$, in cases where the set $\mathcal{H}$ is empty. The expression "$\vdash C$" can be read aloud as "C is a theorem."
- When there is exactly one hypothesis, it is customary to omit the braces — e.g., we may write $\{A\} \vdash B \to A$ more briefly as

$A \vdash B \to A$. (Beginners are cautioned that this does *not* mean $\{A\}$ is *equal* to A.)

We do not require that all the hypotheses (i.e., all the members of $\mathcal{H}$) get used in the derivation; we allow superfluous hypotheses. Indeed, we even allow $\mathcal{H}$ to be an infinite set, but a derivation has only finitely many steps so it can only *use* finitely many of the hypotheses. Because not all hypotheses need to be used, the notion of derivation used in this book is *monotone*, which means that it preserves inclusions:

if $\mathcal{H}_1 \subseteq \mathcal{H}_2$, then $\{C : \mathcal{H}_1 \vdash C\} \subseteq \{C : \mathcal{H}_2 \vdash C\}$.

The derivation is a list of steps; each step has several columns of information. The derivations in this book will usually follow a *three-column format*, though in a few early examples we shall also include a fourth column. The first column contains the step label, a reference number we can use when later steps refer back to the present step. The second column is the formula itself. The third column is the justification for the step. A fourth column, included in some proofs, gives additional information about the justification — e.g., what substitutions are being used.

We emphasize that each derivation in this book is a sequence of *formulas*, not a sequence of theorems or of inference rules. No turnstile symbols ($\vdash$ or $\vDash$) appear in the formulas column of our derivations.

12.6. The type of derivation described above is what one might call a *classical, propositional derivation*. That is the kind of derivation we shall use throughout this book. But other kinds of derivations are possible, and are used in some books and papers on logic.

- Some treatments of predicate logic use a more complicated format, permitting subproofs within proofs. Assumptions get adopted at the beginning of a subproof and discharged at the end of a subproof. However, we have no need for that complication, since this book only considers formal derivations in propositional logic.

- As we remarked above, our notion of derivations is monotone. However, *nonmonotonic logics* have been of great interest in the research literature in recent years. These are logics in which superfluous hypotheses are either prohibited entirely, or restricted in some way. Consequently, $\mathcal{H}_1 \subseteq \mathcal{H}_2$ does *not* imply $\{C : \mathcal{H}_1 \vdash C\} \subseteq \{C : \mathcal{H}_2 \vdash C\}$.

Inference systems

One of the main logics studied in this book is relevant logic. We present it with classical derivations, as described above. However, essentially the same logic can also be presented — perhaps more naturally — using a relevantist definition of "derivation." (It is sometimes called a "starring derivation," because it uses asterisks as markers.) With that definition, $\mathcal{H} \vdash C$ means that we can derive C by an argument that *uses* all the members of $\mathcal{H}$ in essential ways; no hypotheses can be superfluous. Evidently this type of derivation is not monotonic. See Anderson and Belnap [1975] and Kielkopf [1977] for some details. At least some traces of this non-monotone nature of relevant logic can be seen in 28.8.d, where constructive implication satisfies an isotonicity rule and relevant implication does not.

Chapter 13

Basic implication

13.1. *Remarks.* The logics of greatest philosophical interest in this book are classical, constructive, relevant, and fuzzy. However, there is a substantial amount of overlap between those logics. To reduce repetition in our syntactic development, we begin with that overlap — i.e., with assumptions satisfied by all of those logics. The resulting logic will be called *basic logic* in this book. It is weaker than any one of the logics mentioned above, in the sense that it is able to prove fewer theorems.

Even basic logic has a rather large number of axioms; we shall not attempt to present them all at one time. We shall begin with just the *implicational part* of basic logic — i.e., the assumptions that mention only implication. This is already enough to yield many interesting proofs, and moreover it is needed before we can do anything else.

It should be emphasized that, although the results considered in this chapter mention *explicitly* only implication, they may implicitly include other logical operators as well. For instance, in 13.2.b we adopt the axiom scheme $C \to C$, but this scheme includes as a particular axiom the formula

$$\left[\pi_2 \vee \left(\pi_1 \wedge \overline{\pi_3 \to \pi_2}\right)\right] \to \left[\pi_2 \vee \left(\pi_1 \wedge \overline{\pi_3 \to \pi_2}\right)\right]$$

which also involves $\vee, \wedge, \neg$. All the results in this chapter extend in that same fashion.

ASSUMPTIONS OF BASIC IMPLICATION

13.2. Our basic assumptions about implication consist of one inference rule and three axiom schemes.

a. *Detachment.* $\{A, A \to B\} \vdash B$.
b. *Identity.* $\vdash C \to C$.
c. *Permutation.* $\vdash [D \to (E \to F)] \to [E \to (D \to F)]$.
d. $\to$-*prefixing.* $\vdash (G \to H) \to [(I \to G) \to (I \to H)]$.

Discussion of the names. The names of these formulas vary somewhat from one book or article to another; usage is not uniform throughout the literature. I have tried to follow the consensus when it exists, and/or to use names that will aid the student's intuition.

"Detachment" refers to the fact that we detach A from the formula $A \to B$; the remaining term B is our conclusion. Detachment is also known in some of the literature as *modus ponens*.

"Identity" refers to the fact that the formulas to the left and right of the implication symbol are identical.

The idea behind "permutation" is that D and E are hypotheses which together imply F. In basic logic the order of those hypotheses doesn't matter, so we can permute (switch) them. This idea is more evident in the detachmental corollary, in 13.17.

Most "prefixing" or "suffixing" rules are of the form

$$(G \to H) \to \big[(\text{prefix } G) \to (\text{prefix } H)\big]$$
or $$(G \to H) \to \big[(G \text{ suffix}) \to (H \text{ suffix})\big],$$

respectively. The "prefix" or "suffix" is some expression that we attach to G and H; the same expression is attached to both metavariables. Many examples are given in this and later chapters. Two variants on this theme are

$\to$-suffixing $\quad (G \to H) \to [(H \to I) \to (G \to I)]$
and $\neg$-prefixing $\quad (G \to H) \to [(\neg H) \to (\neg G)]$,

which switch the order of G and H in the second clause of the implication.

A FEW EASY DERIVATIONS

13.3. *Specialized mingle.* $\vdash (A \to A) \to [(A \to A) \to (A \to A)]$.

Proof and remarks. The proof is just one line long: This is just a specialization of $\to$-prefixing (13.2.d).

Mingle is the formula scheme $X \to (X \to X)$. Though specialized mingle is a theorem of basic logic, mingle itself is not; we shall establish that in 23.8. Thus, *some, but not all*, of the formulas represented by $X \to (X \to X)$ are theorems of basic implication.

13.4. $\to$-*Suffixing.* $\vdash (I \to G) \to [(G \to H) \to (I \to H)]$.

Derivation

#	formula	justification
(1)	$\{(G \to H) \to [(I \to G) \to (I \to H)]\} \to$ $\{(I \to G) \to [(G \to H) \to (I \to H)]\}$	13.2.c (permut'n)
(2)	$(G \to H) \to [(I \to G) \to (I \to H)]$	13.2.d ($\to$-pref.)
(3)	$(I \to G) \to [(G \to H) \to (I \to H)]$	(2),(1),13.2.a (det.)

Remarks. We now discuss derivations in general, but referring to the derivation above as a particular example.

- **a.** Some students are accustomed to each step in a proof following from the preceding step. They may be confused by the derivation given above, because its step (2) does *not* follow in any way from step (1). Step (1) follows from 13.2.c, and step (2) follows from 13.2.d. Neither of those steps requires the other, so they could be presented in either order — i.e., we could switch the order of steps (1) and (2). Thus, the order of steps is not uniquely determined.

    On the other hand, step (3) must occur last, since its justification uses both (1) and (2).
- **b.** The presentation given above does not show the substitutions. Here are the substitutions that we've used.

Step (1) is a specialization of permutation, with the substitutions

$$D = G \to H, \qquad E = I \to G, \qquad F = I \to H.$$

The difference between the terms "specialization" and "generalization" (introduced in 6.30) confuses some beginners, but it becomes crucial at this point. We can justify a formula in a derivation by observing that it is a *specialization* (not a generalization) of any formula that we've already established earlier.

Step (2) is precisely $\to$-prefixing 13.2.d. We have chosen the letters so that *no* substitution is needed at this step (though we could have chosen other letters).

Step (3) is an application of detachment, with

$$\begin{aligned} A &= (2) = (G \to H) \to [(I \to G) \to (I \to H)], \\ B &= (3) = (I \to G) \to [(G \to H) \to (I \to H)], \end{aligned}$$

and hence $A \to B$ is the long formula of step (1).

My derivations generally do not explicitly include substitution information (such as that given above), in this book or in my lectures. Nevertheless, the substitutions are understood to be part of the derivation, and in fact they are an essential part. Figuring out what substitutions have been made is part of the reader's job.

c. Note that the justification column in step (3) has three ingredients: the labels (1) and (2), and the reference to detachment. I deduct a point if any of my students omits any of those three ingredients; all three ingredients are essential (at least, in the format that I use in this book and in my lectures).

d. Some instructors may prefer to include more information — e.g., include the substitutions as a fourth column. Other instructors might prefer to leave more information as "understood" — e.g., they might omit the "(1)" and "(2)" from the justification of step (3) above.

How much information is explicitly presented in a derivation is actually a matter of taste. My own inclination is toward a higher amount of explicitly presented information, as this leaves less room for confusion among beginners; but other instructors may feel differently about this. I also recommend uniformity of format: If the students' papers all look alike, this will reduce the workload of whoever has to grade the papers. At any rate, the instructor's taste must override the student's for the duration of the semester. Students are urged to ask their instructors for clarification if there is any doubt about what format will be used in their course.

e. In many of my justifications, I give a reference in *two* forms — both a name (e.g., "detachment") and a number (13.2.a). The numbers have less chance of being ambiguous or misinterpreted; but the names have more intuition or personality, and are easier to remember, thus making the proof easier to read. Generally I do not require my students to give both forms in their homework or tests; I feel that one is enough. And I caution them that they will not lose points in my class for a proof being hard to read, but they will lose points for ambiguity; so I urge them to use the numbers, which have much less chance of being ambiguous. Of course, other instructors may feel differently about this.

f. Throughout most of this book, I label the steps in the derivations as (1), (2), (3), etc. However, I have changed this numbering in a couple of the derivations below, to facilitate some later discussion in 13.12–13.15. Actually, it doesn't matter what we use for labels, provided they follow these rules: The labels in the "#" column must all be different, and a label can only appear in the "justification" column in a row *later* than where it appeared in the "#" column.

g. In the justification column in step (3), I have written "(2), (1)," rather than the more obvious "(1), (2)." This is neither accidental nor a misprint; it is intentional. It is a way of supplying the reader with a hint about what substitutions were used, without lengthening the derivation. Again, this is a matter of taste, and not all instructors will follow my lead on this.

A few easy derivations

When detachment was first introduced to the reader in 13.2.a, it was presented as $\{A, A \to B\} \vdash B$. Strictly speaking, $\{A, A \to B\}$ is a *set*, and thus unordered; but I like to temporarily treat it as a *list*. Then, when I apply detachment, I list the justifying hypotheses in the same order.

Thus, in the derivation of $\to$-suffixing, by listing "(2), (1)" in the justification of step (3), I am indicating this correspondence:

$$
\begin{array}{cccc}
(2) & (1) & \text{yields} & (3) \\
\updownarrow & \updownarrow & & \updownarrow \\
A & A \to B & \vdash & B.
\end{array}
$$

This tells us that the substitutions we want to use are $A = (2)$, $B = (3)$, and $A \to B = (1)$. Thus, if we haven't made any mistakes, the formula in step (1) of the derivation should be the concatenation $(2) \to (3)$. (Careful students will verify the "should be" in this and all other applications of inference rules in derivations.)

13.5. The last entry in the "formulas" column in any derivation is the formula that appears to the right of the turnstile in the expression we're trying to prove. Any formulas to the left of the turnstile are hypotheses, which can be used for free in the derivation. There were no hypotheses in $\to$-suffixing (13.4), but our next two results do have hypotheses.

a. $\to$-*suffixing-detached.* $A \to B \vdash (B \to C) \to (A \to C)$.
b. *Transitivity.* $\{A \to B, B \to C\} \vdash A \to C$.

Derivation of $\to$-suffixing-detached

(i)	$(A \to B) \to [(B \to C) \to (A \to C)]$	13.4 (suffixing)
(ii)	$A \to B$	hypothesis
(iii)	$(B \to C) \to (A \to C)$	(ii), (i), detachment

Derivation of transitivity

i	$A \to B$	hypothesis

ii	$B \to C$	hypothesis
iii	$(B \to C) \to (A \to C)$	i, 13.5.a ($\to$-suffixing-detached)
iv	$A \to C$	ii, iii, detachment

About the name. See 3.20 for usage of the term "transitive" in other mathematical contexts.

13.6. Exercises

a. Write out a derivation of *repeated transitivity*:

$$\{Q \to R,\ R \to S,\ S \to T\} \vdash Q \to T.$$

b. (Proof recommended only for advanced students; all others should just read this result.) Write a complete proof of *multiple transitivity*:

$$\{S_0 \to S_1,\ S_1 \to S_2,\ \ldots,\ S_{n-1} \to S_n\} \vdash S_0 \to S_n.$$

You will need to use induction and complete sentences.

13.7. *Meredith's lemma*
$$\vdash \{[(J \to L) \to (K \to L)] \to M\} \to [(K \to J) \to M].$$

Derivation

| (α) | $(K \to J) \to [(J \to L) \to (K \to L)]$ | 13.4 ($\to$-suffixing) |
| (β) | $\{[(J \to L) \to (K \to L)] \to M\}$ $\to [(K \to J) \to M]$ | (α), 13.5.a ($\to$-suffixing-det.) |

13.8. Exercise. Fill in the blanks in the derivations of the following two formulas.

a. *Assertion.* $\vdash N \to [(N \to O) \to O]$.
b. *Specialized assertion.* $\vdash [(P \to P) \to O] \to O$.

Derivation of assertion

| (1) | ? | ? |
| (2) | $(N \to O) \to (N \to O)$ | ? |

A few easy derivations 325

| (3) | $N \to [(N \to O) \to O]$ | (2), (1), 13.2.a (detachment) |

Derivation of specialized assertion

(1)	$P \to P$	?
(2)	?	13.8.a (assertion)
(3)	$[(P \to P) \to O] \to O$	(1), (2), detachment

13.9. *Some easy specializations*
  a. *Modalizer.* $\vdash [(A \to A) \to A] \to A$.
  b. *Demodalizer.* $\vdash A \to [(A \to A) \to A]$.

Proof and remarks. Those results are just specializations of 13.8.
  They are named for the fact that they can be used to convert between relevant logic and modal logic, but we shall not pursue that conversion in this book. However, we have another use for these two theorems below.

13.10. *Demodalization (postponable).* For any formula X, define

$$M(X) \;=\; (X \to X) \to X.$$

From 13.9.a and 13.9.b, it follows that X and $M(X)$ are syntactically equivalent — i.e., we have both $\vdash X \to M(X)$ and $\vdash M(X) \to X$. Repeating this procedure and applying transitivity, we find that any two of the formulas

$$X, \quad M(X), \quad M^2(X), \quad M^3(X), \quad \ldots$$

are syntactically equivalent, where

$$M^0(X) \;=\; X,$$
$$M^1(X) \;=\; M(X) = (X \to X) \to X,$$
$$\begin{aligned}M^2(X) \;&=\; M(M(X))\\ &= \{[(X \to X) \to X] \to [(X \to X) \to X]\} \to [(X \to X) \to X],\end{aligned}$$
$$\begin{aligned}M^3(X) \;&=\; M(M(M(X)))\\ &= (\langle\{[(X\to X)\to X]\to [(X\to X)\to X]\}\to [(X\to X)\to X]\rangle\\ &\quad \to \langle\{[(X\to X)\to X]\to [(X\to X)\to X]\}\to [(X\to X)\to X]\rangle)\\ &\quad \to \langle\{[(X\to X)\to X]\to [(X\to X)\to X]\}\to [(X\to X)\to X]\rangle,\end{aligned}$$
$$M^4(X) \;=\; M(M(M(M(X))))$$

and so on. This observation will be used in 28.12.

13.11. *Remarks (optional).* What if we used fewer axioms?

- Assume only $\to$-prefixing, $\to$-suffixing, and detachment. Then it is not possible to prove the identity formula scheme $A \to A$. In fact, it is not possible to prove even one of the formulas represented by that formula scheme.

- Assume only $\to$-prefixing, $\to$-suffixing, detachment, and the identity axiom scheme $A \to A$. Then, whenever P and Q are any two *different* formulas, it is not possible to prove both $P \to Q$ and $Q \to P$.

These interesting results were first proved by E. P. Martin [1978]. There are now several proofs of Martin's results, but unfortunately they are all too long and complicated to include here. The most recent of those proofs apparently is Hirokawa et al. [2000]; that paper also includes references to earlier proofs.

Lemmaless expansions

13.12. Earlier results can be used as justifications in later results. For instance, we used 13.4 ($\to$-suffixing) in our derivation of 13.5.a ($\to$-suffixing-detached), and we used both of those in our derivation of Meredith's lemma (13.7). In general, by a *lemma* we shall mean a derived result that is used as a justification in a later derivation.

A *lemmaless derivation* is one that uses no lemmas — i.e., where the only permitted justifications are hypotheses, axioms, and assumed inference rules. No derived formulas or derived inference rules are permitted as justifications. In particular, the special class of derived inference rules that we shall call "detachmental corollaries" (beginning in 13.16) are not permitted as justifications.

As we shall demonstrate below, any derivation can be expanded into a lemmaless derivation, by replacing each lemma with several steps analogous to the derivation which produced that lemma. Generally a lemmaless expansion is much longer — both in number of steps, and in length of each step — than a derivation with lemmas, and so we seldom would want to carry out such an expansion. However, just knowing that it *can* be carried out enables us to use certain kinds of reasoning about logic (as in 21.1, 22.5, 22.20, 23.6, and 24.26).

Lemmaless expansions

Rather than try to give a formal, abstract proof that such expansions can always be carried out, we will give a *proof by example*: We will carry out expansions for →-suffixing-detached and Meredith's lemma, but it should be evident that the methods used would work on any other derivations as well.

13.13. *Expansion of →-suffixing-detached.* We used →-suffixing (13.4) as a justification in step (i) of our earlier derivation of →-suffixing-detached (13.5.a). We wish to rewrite that derivation of →-suffixing-detached, replacing its step (i) with several steps that are analogous to the steps (1), (2), (3) of the derivation of →-suffixing.

It may be helpful to first rewrite the derivation of →-suffixing, changing its metavariables so that the letters in its concluding step (3) will match the letters in the expanding step (i). That involves replacing each I with A, each G with B, and each H with C. Here is the resulting restatement of →-suffixing:

$$\vdash (A \to B) \to [(B \to C) \to (A \to C)]$$

and its derivation:

(1)	$\{(B \to C) \to [(A \to B) \to (A \to C)]\} \to$ $\{(A \to B) \to [(B \to C) \to (A \to C)]\}$	13.2.c (permut'n)
(2)	$(B \to C) \to [(A \to B) \to (A \to C)]$	13.2.d (→-prefix.)
(3)	$(A \to B) \to [(B \to C) \to (A \to C)]$	(2),(1),13.2.a (det.)

Now we can rewrite the derivation of →-suffixing-detached, replacing its step (i) with those three steps. Note that we merge steps (3) and (i), and keep the justification of step (3) — not the justification of (i). Also note that the justification in step (iii) gets a label changed.

Lemmaless derivation of →-suffixing-detached

(1)	$\{(B \to C) \to [(A \to B) \to (A \to C)]\}$ $\to \{(A \to B) \to [(B \to C) \to (A \to C)]\}$	13.2.c (permut.)
(2)	$(B \to C) \to [(A \to B) \to (A \to C)]$	13.2.d (→-pref.)

(3/i)	$(A \to B) \to [(B \to C) \to (A \to C)]$	(2),(1), detach.
(ii)	$A \to B$	hypothesis
(iii)	$(B \to C) \to (A \to C)$	(ii), (3/i), det.

13.14. *Expansion of Meredith's lemma.* We used $\to$-suffixing and $\to$-suffixing-detached in steps (α) and (β) of our derivation of Meredith's lemma (13.7). We wish to rewrite that derivation of $\to$-suffixing-detached, replacing those two steps with steps analogous to the derivation of $\to$-suffixing (which was already lemmaless) and the lemmaless derivation of $\to$-suffixing-detached.

Let us first rewrite $\to$-suffixing, with metavariables changed to match the letters in Meredith's lemma:

$$\vdash (K \to J) \to [(J \to L) \to (K \to L)]$$

and also its derivation, with steps relabeled in anticipation of subsequent reasoning below:

(p)	$\{(J \to L) \to [(K \to J) \to (K \to L)]\} \to$ $\{(K \to J) \to [(J \to L) \to (K \to L)]\}$	13.2.c (permut.)
(q)	$(J \to L) \to [(K \to J) \to (K \to L)]$	13.2.d ($\to$-pref.)
(r)	$(K \to J) \to [(J \to L) \to (K \to L)]$	(q),(p),13.2.a (det.)

Next, we want to rewrite $\to$-suffixing-detached's lemmaless derivation (taken from 13.13). But to get (α) to match the hypothesis $A \to B$, and to get (β) to match the conclusion $(B \to C) \to (A \to C)$, we must do much more than just replace one metavariable with another metavariable. We replace C with M, but we also replace A with $K \to J$, and replace B with $(J \to L) \to (K \to L)$. Here is the formula:

$$\langle K \to J \rangle \to \langle (J \to L) \to (K \to L) \rangle \quad \vdash$$
$$(\langle (J \to L) \to (K \to L) \rangle \to M) \to (\langle K \to J \rangle \to M),$$

and its lemmaless derivation:

(1)	$\left\{ \left(\langle (J \to L) \to (K \to L) \rangle \to M \right) \to \right.$	

Lemmaless expansions

	$[(\langle K \to J\rangle \to \langle (J \to L) \to (K \to L)\rangle)$ $\to (\langle K \to J\rangle \to M)]\} \to \{(\langle K \to J\rangle \to$ $\langle (J \to L) \to (K \to L)\rangle) \to [(\langle (J \to L) \to$ $(K \to L)\rangle \to M) \to (\langle K \to J\rangle \to M)]\}$	13.2.c (perm.)
(2)	$(\langle (J \to L) \to (K \to L)\rangle \to M)$ $\to [(\langle K \to J\rangle \to \langle (J \to L) \to$ $(K \to L)\rangle) \to (\langle K \to J\rangle \to M)]$	13.2.d ($\to$-pref.)
(3/i)	$(\langle K \to J\rangle \to \langle (J \to L) \to$ $(K \to L)\rangle) \to [(\langle (J \to L) \to$ $(K \to L)\rangle \to M) \to (\langle K \to J\rangle \to M)]$	(2),(1), detach.
(ii)	$\langle K \to J\rangle \to \langle (J \to L) \to (K \to L)\rangle$	hypoth.
(iii)	$(\langle (J \to L) \to (K \to L)\rangle \to M)$ $\to (\langle K \to J\rangle \to M)$	(ii), (3/i), detach.

Finally, to get a lemmaless derivation of Meredith's lemma, we combine the two preceding derivations (three steps and five steps, respectively) with the derivation of Meredith's lemma given in 13.7 (two steps). That makes ten steps, but there are some reductions: Steps (r), (ii), and (α) get merged into a single step whose justification is the justification of (r). Steps (iii) and (β) are merged into a single step whose justification is the justification of (iii). The resulting derivation (not shown here, but easily written from the preceding description) is seven long steps.

By now it should be clear that any derivation *can* be expanded to a lemmaless derivation, but that we would prefer to do so as seldom as possible.

13.15. *Exercises.* The next few pages include several derivations that use lemmas. Expand one or more of them into lemmaless derivations. Particularly recommended for this purpose are 14.6,

14.7, and 14.12.a.

Detachmental corollaries

13.16. In mathematics, generally a *corollary* to a given result means an *easy consequence* of that result. Of course, this is a subjective term, and we shall not attempt to make it precise in general. However, we now describe precisely one particular type of corollary:

From any theorem or inference rule of the form

$$\mathcal{H} \vdash A \to B$$

(where $\mathcal{H}$ is some set of formulas, possibly empty), we can easily obtain the inference rule

$$\mathcal{H} \cup \{A\} \vdash B.$$

We shall say that the latter expression is the *detachmental corollary* of the former, because the proof uses detachment; that will be evident in examples below. If the former expression has theorem or inference number n in the labeling system of this book, we shall give its detachmental corollary the number $n.\delta$ — that is, n followed by the Greek lowercase letter delta.

13.17. Below are a few detachmental corollaries. Actually, we've already proved the first two in the list:

13.4.δ Detachmental corollary of $\to$-suffixing:
=13.5.a $I \to G \vdash (G \to H) \to (I \to H)$;

13.4.$\delta\delta$ Transitivity (2nd detach. corol. of $\to$-suffixing):
=13.5.b $\{I \to G, G \to H\} \vdash I \to H$;

13.2.c.δ Detachmental corollary of permutation:
$D \to (E \to F) \vdash E \to (D \to F)$;

Detachmental corollaries

13.2.d.δ Detachmental corollary of $\to$-prefixing:
$$G \to H \vdash (I \to G) \to (I \to H);$$

13.8.a.δ Detachmental corollary of assertion:
$$N \vdash (N \to O) \to O;$$

13.8.b.δ Detachmental corollary of specialized assertion:
$$(P \to P) \to O \vdash O.$$

Derivation of 13.2.c.δ

(1)	$D \to (E \to F)$	hypothesis
(2)	$[D \to (E \to F)] \to [E \to (D \to F)]$	13.2.c
(3)	$E \to (D \to F)$	(1), (2), detach.

Derivation of 13.2.d.δ

(1)	$G \to H$	hypothesis
(2)	$(G \to H) \to [(I \to G) \to (I \to H)]$	13.2.d
(3)	$(I \to G) \to (I \to H)$	(1), (2), detach.

Derivation of 13.8.a.δ, 13.8.b.δ. Omitted; left as exercises.

13.18. *Remarks.* Detachmental corollaries will often be useful for making derivations shorter — both in the number of steps, and in the lengths of the formulas appearing in those steps. But not all detachmental corollaries are equally useful. For instance, the detachmental corollary of the identity axiom is $C \vdash C$, which does not shorten any proofs. And the second detachmental corollary of assertion is just detachment itself, which was among our assumptions — and which is used to prove the detachmental corollaries.

It should now be evident that the *derivations* of detachmental corollaries are all essentially the same: state the hypothesis or hypotheses; state the theorem or apply the inference rule whose corollary we're proving; and finally, then apply detachment. Consequently,

> *hereafter we shall omit the derivations of detachmental corollaries.*

Moreover, the form that is taken by a detachmental corollary should be fairly evident: from $\mathcal{H} \vdash A \to B$ we obtain $\mathcal{H} \cup \{A\} \vdash B$. Consequently,

> *hereafter we shall consider detachmental corollaries of known results to also be known results, even when they are not stated explicitly.*

That is: Whenever any theorems or inference rules are stated, we shall view them as though their detachmental corollaries were also stated.

13.19. *Remarks on irreversibility.* Note that the process of forming a detachmental corollary, going

$$\text{from} \quad \mathcal{H} \vdash A \to B \quad \text{to} \quad \mathcal{H} \cup \{A\} \vdash B,$$

moves the formula A to the left, past the turnstile. This process is *not reversible* — i.e., in general, we have no method for moving the formula A to the *right* past the turnstile. We do have some partial methods for moving A to the right in some particular logics; that is the content of the deduction principles developed in 22.5, 23.6, and 24.26.

13.20. *Exercise.* $W \to X \vdash A \to [(A \to W) \to X)]$. (Postponable; this will be used in Chapter 24.)

ITERATED IMPLICATION (POSTPONABLE)

13.21. *Notation for right-nested formulas.* This subchapter is concerned with formulas that are *nested from the right*, such as $C \to (B \to (A \to Z))$, which will be abbreviated as $(A, B, C) \to Z$. We could say that Z has the three hypotheses A, B, C. More generally,

$$A_n \to (A_{n-1} \to \cdots \to (A_2 \to (A_1 \to Z))\cdots)$$

will be abbreviated as

$$(A_1, A_2, \ldots, A_n) \to Z;$$

Iterated implication (postponable)

here Z has n hypotheses. By "nested from the right" we mean a formula or formula scheme in which all the right parentheses appear after all the metavariables.

The results of this subchapter will also apply, at least indirectly and partially, to some formulas that are not strictly nested from the right. For instance, the formula scheme

$$P \to ((Q \to R) \to (S \to T))$$

has T preceded by four other metavariables, but they are not nested from the right. However, if we view $Q \to R$ as one inseparable unit, then the formula above can be abbreviated as

$$(S, \ Q \to R, \ P) \to T$$

following the conventions of the previous paragraph. Then we have T with the *three* hypotheses S, $Q \to R$, P, and we can apply to this formula any results of this subchapter.

13.22. *Repeated $\to$-prefixing.* Let Y, Z be any formulas, and let α be any finite *sequence* of formulas. Then, in the notation of 13.21,

$$\vdash (Y \to Z) \to \{[\alpha \to Y] \to [\alpha \to Z]\}.$$

Corollary 13.22.δ is also noteworthy: $Y \to Z \vdash [\alpha \to Y] \to [\alpha \to Z]$.

Proof of 13.22. The proof is by induction on n, the length of the sequence α. What we wish to prove is this statement, which we shall call $P(n)$:

$$\vdash (Y \to Z) \to \{[(A_1, \ldots, A_n) \to Y] \to [(A_1, \ldots, A_n) \to Z]\}$$

for any formulas $A_1, A_2, \ldots, A_n$. For the initialization step, note that statement $P(1)$ is just $\to$-prefixing (13.2.d). For the transition step, begin by noting that

$$\vdash \Big\{[(A_1, \ldots, A_n) \to Y] \to [(A_1, \ldots, A_n) \to Z]\Big\}$$
$$\to \Big(\{A_{n+1} \to [(A_1, \ldots, A_n) \to Y]\} \to \{A_{n+1} \to [(A_1, \ldots, A_n) \to Z]\}\Big)$$

is an instance of $\to$-prefixing (13.2.d); but that line can be restated as

$$\vdash \Big\{[(A_1, \ldots, A_n) \to Y] \to [(A_1, \ldots, A_n) \to Z]\Big\}$$
$$\to \Big\{[(A_1, \ldots, A_{n+1}) \to Y] \to [(A_1, \ldots, A_{n+1}) \to Z]\Big\}.$$

On the other hand, we are given statement $P(n)$, which is

$$\vdash (Y \to Z) \to \Big\{[(A_1, \ldots, A_n) \to Y] \to [(A_1, \ldots, A_n) \to Z]\Big\}.$$

Combine those last two lines, using transitivity 13.5.b. Thus we obtain

$$\vdash (Y \to Z) \to \Big\{[(A_1, \ldots, A_{n+1}) \to Y] \to [(A_1, \ldots, A_{n+1}) \to Z]\Big\},$$

which is $P(n+1)$. That completes the induction.

13.23. *Generalized permutation.* If α and β are the same finite sequence of formulas in different orders,[1] and Z is any formula, then (in the notation of 13.21)
$$\vdash [\alpha \to Z] \to [\beta \to Z].$$

Corollary 13.23.δ. With the same notation, $\alpha \to Z \vdash \beta \to Z$.

Proof of 13.23. We shall prove, by induction on n, that

$$P(n): \qquad \vdash [(A_1, A_2, \ldots, A_n) \to Z] \to [(B_1, B_2, \ldots, B_n) \to Z]$$

holds for $n = 1, 2, 3, \ldots$. For initialization, the statement $P(1)$ is simply
$$\vdash (A_1 \to Z) \to (A_1 \to Z),$$

which is an instance of the identity axiom (13.2.b). For the transition step, assume $P(n)$ for some positive integer n; we now wish to prove $P(n+1)$.

We obtain the sequence $(B_1, B_2, \ldots, B_{n+1})$ from $(A_1, A_2, \ldots, A_{n+1})$ by switching the positions of some of the A_i's. We first consider the special case where the switch is merely a swapping of two adjacent terms A_i, A_{i+1} in the list — i.e., where those two terms interchange positions, and no other terms are moved. We consider two subcases.

- *Neither of the terms being swapped is A_{n+1}.* In this case, $A_{n+1} = B_{n+1}$, and the list $(B_1, \ldots, B_n)$ is a rearrangement of the list $(A_1, \ldots, A_n)$. We know $P(n)$ by the induction hypothesis, and now $P(n+1)$ follows from $P(n)$ by $\to$-prefixing 13.2.d.δ.

- *One of the terms being swapped is A_{n+1}.* In this case the other term being swapped must be A_n, since that is the only term adjacent to A_{n+1}. Hence none of the terms with lower subscripts are changing their places — i.e., we have
$$A_1 = B_1, \qquad A_2 = B_2, \qquad \ldots, \qquad A_{n-1} = B_{n-1}.$$
Then the two formulas
$$(A_1, A_2, \ldots, A_{n-1}) \to Z, \qquad (B_1, B_2, \ldots, B_{n-1}) \to Z$$
are actually the same; let us denote them both by S. Then $P(n+1)$ is just the statement
$$\vdash [A_{n+1} \to (A_n \to S)] \to [A_n \to (A_{n+1} \to S)]$$
which is an instance of permutation 13.2.c.

---

[1] Or, in the terminology of 3.1.c, α and β are two different sequential representations of the same multiset of formulas.

Iterated implication (postponable)

Having covered the case of swaps, we now turn to the general case, where the sequence $B_1, B_2, \ldots, B_{n+1}$ is any reordering of $A_1, A_2, \ldots, A_{n+1}$. That reordering can be accomplished by composing many swaps, and the resulting implications can be composed using multiple transitivity (13.6.b). This proves $P(n+1)$, completing the induction and completing our proof of 13.23.

13.24. Prefixed detachment. Let Y, Z, $A_1, A_2, \ldots, A_m$, and $B_1, B_2, \ldots, B_n$ be any formulas. Assume both

(1) $\quad\quad \vdash (A_1, A_2, \ldots, A_{m-1}, A_m) \to Y \quad\quad\quad$ and
(2) $\quad\quad \vdash (B_1, B_2, \ldots, B_{n-1}, B_n) \to (Y \to Z)$.

Then

$$\vdash (B_1, B_2, \ldots, B_{n-1}, B_n, A_1, A_2, \ldots, A_{m-1}, A_m) \to Z.$$

Proof. Rewrite (2) as $\vdash Y \to [(B_1, B_2, \ldots, B_{n-1}, B_n) \to Z]$. Applying 13.22.δ to that yields

$$\vdash [(A_1, A_2, \ldots, A_m) \to Y] \to \\ \{(A_1, A_2, \ldots, A_m) \to [(B_1, B_2, \ldots, B_{n-1}, B_n) \to Z]\}.$$

The left half of that is just hypothesis (1). The right half therefore follows by detachment:

$$\vdash (A_1, A_2, \ldots, A_m) \to [(B_1, B_2, \ldots, B_{n-1}, B_n) \to Z]$$

Unwind the notation; this is just the conclusion of 13.24.

Chapter 14

Basic logic

Further assumptions

14.1. By *basic logic* we shall mean basic implication, already studied in the previous chapter, plus one more inference rule and several more axiom schemes, listed below:

a. *Adjunction.* $\{A, B\} \vdash A \wedge B$.
b. *Left $\wedge$-elimination.* $\vdash (A \wedge B) \to A$.
c. *Right $\wedge$-elimination.* $\vdash (A \wedge B) \to B$.
d. *Left $\vee$-introduction.* $\vdash A \to (A \vee B)$.
e. *Right $\vee$-introduction.* $\vdash B \to (A \vee B)$.
f. *$\wedge$-introduction.* $\vdash [(A \to B) \wedge (A \to C)] \to [A \to (B \wedge C)]$.
g. *$\vee$-elimination.* $\vdash [(B \to A) \wedge (C \to A)] \to [(B \vee C) \to A]$.
h. *Distributive law.* $\vdash [A \wedge (B \vee C)] \to [(A \wedge B) \vee C]$.
i. *First contrapositive law.* $\vdash (A \to \overline{B}) \to (B \to \overline{A})$.

Discussion of the names of the formulas. In English, "conjoin" and "adjoin" mean roughly the same thing. The meanings of "conjunction" and "adjunction" in logic are related but different. Adjunction should be also compared with *strong adjunction*, 16.3.c, which is *not* a part of basic logic — i.e., it is neither one of the assumptions of basic logic, nor something that can be proved from those assumptions; to use it we will need to make further assumptions.

Further assumptions 337

The distributive law given in 14.1.h yields several other distributive laws, listed in 14.20, that may be closer to the more customary meaning of "distributive" given in 3.32.

The meanings of the terms "elimination" and "introduction" may be more evident when we turn to inference rules.

- The rule $A \vdash A \vee B$ goes from a formula not involving $\vee$ to a formula involving $\vee$, so we call it $\vee$-*introduction* (which I pronounce as "or-introduction").

- Similarly, the rule $A \wedge B \vdash A$ goes from a formula involving $\wedge$ to one not involving $\wedge$, so we call it a $\wedge$-*elimination* (which I pronounce as "and-elimination").

Most of our introduction and elimination rules in at least two forms: an inferential form and a theorem form. The forms are not equivalent, but they are closely related; they have different numbers but may sometimes be referred to by the same name. The relation between inferential and theorem forms will be discussed further in 13.16–13.19.

Following are the main introductions and eliminations. The four items marked with asterisks (∗) are *not part of basic logic*; they require additional assumptions and are included in the following table merely as a preview of later results.

	inference rules	theorems
$\neg$-intro.	$A \vdash \overline{\overline{A}}$	$\vdash A \to \overline{\overline{A}}$
$\neg$-elim.	$\overline{\overline{A}} \vdash A$ (∗)	$\vdash \overline{\overline{A}} \to A$ (∗)
$\vee$-intro.	$A \vdash A \vee B$, $B \vdash A \vee B$	$\vdash A \to (A \vee B)$, $\vdash B \to (A \vee B)$
$\wedge$-elim.	$A \wedge B \vdash A$, $A \wedge B \vdash B$	$\vdash (A \wedge B) \to A$, $\vdash (A \wedge B) \to B$
$\vee$-elim.	$\{B \to A,\ C \to A,\ B \vee C\} \vdash A$	$\vdash [(B \to A) \wedge (C \to A)]$ $\to [(B \vee C) \to A]$

$\wedge$-intro.	$\{A \to B, A \to C, A\} \vdash B \wedge C$	$\vdash [(A \to B) \wedge (A \to C)]$ $\to [A \to (B \wedge C)]$
another $\wedge$-intro.	$\{A, B\} \vdash A \wedge B$ (adjunction)	$\vdash A \to (B \to (A \wedge B))$ (strong adjunction) $\quad (*)$
$\to$-elim.	$\{A, A \to B\} \vdash B$ (detachment)	$\vdash [A \wedge (A \to B)] \to B$ (conjunctive detach.) $\quad (*)$

14.2. Remarks on duality (optional). To some extent, the assumptions of basic logic occur in dual pairs: If we swap the roles of $\vee$ and $\wedge$ and reverse the implications, it turns some of the axioms into other axioms. For instance, left $\wedge$-elimination and left $\vee$-introduction are dual to each other.

The dual of the distributive law (as stated in 14.1.h) is $[(A \vee B) \wedge C] \to [A \vee (B \wedge C)]$. If we also permit commutativity and substitution (developed later in this chapter), we obtain $[C \wedge (B \vee A)] \to [(C \wedge B) \vee A]$. But that's just the distributive law in different letters. Thus, in some sense the distributive law is self-dual.

The two most complicated axiom schemes, $\wedge$-introduction and $\vee$-elimination, are not dual to each other. But their detachmental corollaries are — or more precisely, the parts on either side of the turnstile are dual to each other.

$$(A \to B) \wedge (A \to C) \quad \vdash \quad A \to (B \wedge C),$$
$$(B \to A) \wedge (C \to A) \quad \vdash \quad (B \vee C) \to A.$$

The pairing of formulas persists through many of the proofs in the following chapters; hence modifying a proof of one formula often yields the proof of its dual.

However, there are some asymmetries inherent in our assumptions. In particular, adjunction is an inference rule about $\wedge$; there is no analogous assumption about $\vee$. This is unavoidable, because any commas appearing to the left of the turnstile in any inference rule are playing a role much like "and." See 14.19.a.

There is also some asymmetry in our assumptions about negation. That is made evident in 14.12, where we can prove three of De Morgan's laws but not the fourth. The asymmetry about negation is only built into our basic logic because we are preparing for constructive logic (Chapter 22). All our other main logics include not-elimination (Chapter 19), which removes the asymmetry — e.g., it makes the fourth De Morgan's law provable.

BASIC POSITIVE LOGIC

14.3. To simplify our presentation, we shall first study the part of basic logic that does not involve negations. Any negation-free logic (whether basic or not) is called *positive logic*.

14.4. *Some corollaries.* The following results are *not* "detachmental corollaries" (defined in 13.16), since the proofs require not only Detachment but also Adjunction.

  a. *Multiple consequences.* $\{A \to B,\ A \to C\} \vdash A \to (B \wedge C)$.
  This result follows easily from adjunction, detachment, and $\wedge$-introduction. Its corollary (14.4.a.δ) is noteworthy: $\{A \to B,\ A \to C,\ A\} \vdash B \wedge C$.

  b. *Proof by cases.* $\{B \to A,\ C \to A\} \vdash (B \vee C) \to A$.
  This result follows easily from adjunction, detachment, and $\vee$-elimination. Its corollary (14.4.b.δ) is noteworthy: $\{B \to A,\ C \to A,\ B \vee C\} \vdash A$.

14.5. *Some converses.* We can turn 14.1.f and 14.1.g around:
  a. $\vdash [A \to (B \wedge C)] \to [(A \to B) \wedge (A \to C)]$.
  b. $\vdash [(B \vee C) \to A] \to [(B \to A) \wedge (C \to A)]$.

Outline of proof. Show $[A \to (B \wedge C)] \to (A \to B)$ using $\wedge$-elimination and $\to$-prefixing, Similarly, $[A \to (B \wedge C)] \to (A \to C)$. Now proving multiple consequences yields 14.5.a. The proof of 14.5.b is analogous. *Exercise:* Fill in the details.

14.6. *Exercises on conjunction ($\wedge$) and disjunction ($\vee$)*
  a. $\wedge$-*idempotency.* $\vdash R \leftrightarrow (R \wedge R)$.
  b. $\vee$-*idempotency.* $\vdash R \leftrightarrow (R \vee R)$.
  c. $\wedge$-*commutativity.* $\vdash (R \wedge S) \leftrightarrow (S \wedge R)$.
  d. $\vee$-*commutativity.* $\vdash (R \vee S) \leftrightarrow (S \vee R)$.
  e. $\wedge$-*associativity.* $\vdash [(R \wedge S) \wedge T] \leftrightarrow [R \wedge (S \wedge T)]$.
  f. $\vee$-*associativity.* $\vdash [(R \vee S) \vee T] \leftrightarrow [R \vee (S \vee T)]$.

14.7. (Compare to 16.10, 16.11, 23.18.h, and 23.18.i.)
  a. *Weak $\wedge$-prefixing.* $S \to T \vdash (R \wedge S) \to (R \wedge T)$.
  b. *Weak $\vee$-prefixing.* $T \to S \vdash (R \vee T) \to (R \vee S)$.
  c. *Weak $\wedge$-suffixing.* $S \to T \vdash (S \wedge R) \to (T \wedge R)$.
  d. *Weak $\vee$-suffixing.* $S \to T \vdash (S \vee R) \to (T \vee R)$.

We will prove weak $\wedge$-prefixing as an example; the others are left as exercises.

(1)	$S \to T$	hypothesis
(2)	$(R \wedge S) \to S$	right $\wedge$-elimination (14.1.c)
(3)	$(R \wedge S) \to R$	left $\wedge$-elimination (14.1.b)
(4)	$(R \wedge S) \to T$	(2), (1), transitivity (13.5.b)
(5)	$(R \wedge S) \to (R \wedge T)$	(3), (4), mult. conseq. (14.4.a)

14.8. *Miscellaneous exercises (optional)*
  a. $X \wedge (X \to Y) \vdash Y$. (See related remarks in 15.2.)
  b. $\{A \vee B,\ A \to X,\ B \to Y\} \vdash X \vee Y$.
  c. *$\wedge$-contraction.* $\vdash [A \to (A \wedge B)] \to (A \to B)$.
  d. *$\vee$-contraction.* $\vdash [(A \vee B) \to B] \to (A \to B)$.
  e. *Weak suffix cancellation.* Suppose that A and B are some formulas with the property that $\vdash (A \to X) \leftrightarrow (B \to X)$ for all formulas X. Then in fact $\vdash A \leftrightarrow B$. (*Hint.* Apply the assumption twice, with what choices of X?)
  f. $\vdash [A \to (B \to C)] \to \{(A \wedge B) \to [(A \wedge B) \to C]\}$. *Hints.* My derivation used $\wedge$-elimination (twice), $\to$-suffixing detached (twice), $\to$-prefixing detached, and transitivity. Of course, you may find a different derivation.
  g. *Implications between $\vee$ and $\wedge$*
     (i) $\vdash (A \wedge B) \to (A \vee B)$.
     (ii) $\vdash (A \to B) \to [(A \wedge C) \to (B \vee D)]$.
     (iii) $\vdash [(A \vee B) \to (A \wedge B)] \to (A \to B)$.

Hints for 14.8.g. Results (i) and (ii) follow easily using $\wedge$-elimination, or-introduction, $\to$-prefixing, $\to$-suffixing, and transitivity; we leave the details as an exercise. The proof of 14.8.g(iii) is slightly harder, so we sketch it here; the omitted details are left as an exercise.

(1)	$A \to (A \vee B)$	?

Basic negation

(2)	$(A \wedge B) \to B$	?
(3)	$[A \to (A \wedge B)] \to (A \to B)$	(2), ?
(4)	$[(A \vee B) \to (A \wedge B)] \to [A \to (A \wedge B)]$	(1), ?
(5)	$[(A \vee B) \to (A \wedge B)] \to (A \to B)$	(4), (3), ?

BASIC NEGATION

Next we shall look at some consequences of the *first contrapositive law* and its detachmental corollary:

$$14.1.\text{i} \quad \vdash (A \to \overline{B}) \to (B \to \overline{A}),$$
$$14.1.\text{i}.\delta \quad A \to \overline{B} \vdash B \to \overline{A}.$$

14.9. *Not-introduction.* $\vdash A \to \overline{\overline{A}}$.

  Hints. Apply 14.1.i.δ to $\overline{A} \to \overline{A}$.

14.10. *Second contrapositive law.* $\vdash (A \to B) \to (\overline{B} \to \overline{A})$.

Derivation. The justifications are left as exercises.

(1)	$B \to \overline{\overline{B}}$	?
(2)	$(A \to B) \to (A \to \overline{\overline{B}})$	?
(3)	$(A \to \overline{\overline{B}}) \to (\overline{B} \to \overline{A})$	?
(4)	$(A \to B) \to (\overline{B} \to \overline{A})$	?

Remark. The second contrapositive law could be rewritten as $(A \to B) \to ((\neg B) \to (\neg A))$; thus it is *negation-prefixing* or *not-prefixing* in the sense of 13.2. On the other hand, if we assume the availability of an Ackermann falsehood (as in 5.34), then the second contrapositive law could be rewritten as $(A \to B) \to ((B \to \bot) \to (A \to \bot))$; thus it is *falsehood-suffixing*.

Corollary 14.10.δ. $A \to B \vdash \overline{B} \to \overline{A}$.

14.11. *Brouwer's triple negation law.* $\vdash \overline{\overline{\overline{A}}} \to \overline{A}$.

Sketch of proof. Start from 14.9 and apply 14.10.δ.

Remarks. Theorem 14.9 tells us that we can add two negations to any formula. Theorem 14.11 tells us that we can subtract two negations from any formula, *provided that at least one negation still remains* — i.e., provided that the formula was originally negated at least three times.

Subtracting two negations from a formula that only has two negations requires a slightly stronger logic. It does not follow in general from basic logic, nor even from basic logic plus certain axioms. See 22.17.

14.12. De Morgan's laws
a. $\vdash \overline{A \vee B} \to (\overline{A} \wedge \overline{B})$.
b. $\vdash (\overline{A} \wedge \overline{B}) \to \overline{A \vee B}$.
c. $\vdash (\overline{A} \vee \overline{B}) \to \overline{A \wedge B}$.
d. *Remark.* The fourth De Morgan law, $\overline{A \wedge B} \to (\overline{A} \vee \overline{B})$, is *not* a part of basic or constructive logic. It requires additional assumptions, as in 19.4. The fact that it *cannot* be proved without additional assumptions will follow from 10.4.d and 22.14.

Hints for a. Apply 14.10.δ (to what?) to obtain the theorems $\overline{A \vee B} \to \overline{A}$ and $\overline{A \vee B} \to \overline{B}$. Then what?

Hints for b. Apply 14.1.i.δ (to what?) to obtain the theorems $A \to \neg(\overline{A} \wedge \overline{B})$ and $B \to \neg(\overline{A} \wedge \overline{B})$. Combine those (how?), and then use 14.1.i.δ again.

Hints for c. First prove $\overline{A} \to \overline{A \wedge B}$ and $\overline{B} \to \overline{A \wedge B}$, by applying a contrapositive law (which one?) to some axioms (which ones?).

14.13. *Glivenko's detachment.* $\{\overline{\overline{E}},\ \overline{\overline{E \to F}}\} \vdash \overline{\overline{F}}$. (Postponable — this will only be needed for 22.20.)

Proof. The justifications are left as an *exercise*.

(1)	$(E \to F) \to (\overline{F} \to \overline{E})$	?
(2)	$(\overline{F} \to \overline{E}) \to (\overline{\overline{E}} \to \overline{\overline{F}})$	?
(3)	$(E \to F) \to (\overline{\overline{E}} \to \overline{\overline{F}})$	?
(4)	$\overline{\overline{E}} \to [\overline{\overline{(E \to F)}} \to \overline{\overline{F}}]$	?
(5)	$\overline{\overline{E}}$	?
(6)	$\overline{\overline{(E \to F)}} \to \overline{\overline{F}}$	?

Substitution principles

(7)	$(\neg\neg\neg F) \to \overline{E \to F}$	?
(8)	$\overline{E \to F} \to \neg\neg\neg\neg F$	?
(9)	$(\neg\neg\neg\neg F) \to \overline{\overline{F}}$	?
(10)	$\overline{E \to F} \to \overline{\overline{F}}$	?
(11)	$\overline{E \to F}$	?
(12)	$\overline{\overline{F}}$	?

SUBSTITUTION PRINCIPLES

14.14. *Two-sided substitution principle.* Assume $\vdash$ is some syntactic system that includes at least the assumptions of basic logic. Let $S(Y)$ be a formula scheme in which the metavariable Y appears exactly once. Let $S(A)$ and $S(B)$ be the formula schemes that result from replacing Y with A or B, respectively. If $\vdash A \leftrightarrow B$, then $\vdash S(A) \leftrightarrow S(B)$. That is, if $\vdash A \leftrightarrow B$, then

$$\vdash \left(\begin{array}{c} \text{any formula} \\ \text{involving } A \end{array} \right) \leftrightarrow \left(\begin{array}{c} \text{same formula, except with} \\ \text{at least one } A \text{ replaced by } B \end{array} \right).$$

Proof postponed. We will prove a stronger result in 14.16.

14.15. Before stating and proving our more general substitution principle, we first review some ingredients of that principle. Let A, B, X be any formulas. If we assume $\vdash A \to B$, then we can conclude

prefixing	suffixing
$\vdash (X \to A) \to (X \to B)$,	$\vdash (B \to X) \to (A \to X)$,
$\vdash (X \vee A) \to (X \vee B)$,	$\vdash (A \vee X) \to (B \vee X)$,
$\vdash (X \wedge A) \to (X \wedge B)$,	$\vdash (A \wedge X) \to (B \wedge X)$,
$\vdash (\neg B) \to (\neg A)$.	

Note that in most of these conclusions, A appears before B. The only two exceptions are $\neg$-prefixing and $\to$-suffixing, both of which reverse the order of A and B.

14.16. *One-sided substitution principle.* Assume ⊢ is some syntactic system that includes at least the assumptions of basic logic. Let $S(Y)$ be some formula scheme in which the metavariable Y appears exactly once. Let A and B be some formulas or formula schemes. Let $S(A)$ and $S(B)$ be the formulas or formula schemes obtained from $S(Y)$ by replacing Y with A or with B, respectively. Then:

(i) If ⊢ $A \to B$, and $S(Y)$ is an order preserving function of Y, then ⊢ $S(A) \to S(B)$.

(ii) If ⊢ $A \to B$, and $S(Y)$ is an order reversing function of Y, then ⊢ $S(B) \to S(A)$.

(Review 6.35 for the definitions of order preserving and order reversing functions of formulas.)

Proof. We will use induction on the rank of the formula $S(Y)$.

For the induction initialization, suppose $S(Y)$ has rank 1. Then $S(Y)$ is just the metavariable Y, which is an order preserving subformula of itself. Then the desired conclusion ⊢ $S(A) \to S(B)$ is just the hypothesis ⊢ $A \to B$. This completes the initialization step.

For the induction transition, suppose that $S(Y)$ has rank greater than 1, and the result has been established for all formula schemes of lower rank. Then $S(Y)$ is of one of the forms $Q \to R$, $Q \vee R$, $Q \wedge R$, $\neg Q$, for some formula schemes Q and R. We will carry the proof out in detail only for the case of $S(Y) = Q \to R$; the other three cases can be proved similarly.

Note that Q and R both have rank lower than that of $S(Y)$; hence we may assume that (i) and (ii) both hold for Q and R. We need to prove (i) and (ii) for S.

We will prove only (ii) for S in detail; the proof of (i) is similar. Thus, we may assume Y is an order reversing subformula of $S(Y)$; we want to prove ⊢ $S(B) \to S(A)$.

We now subdivide the proof into two cases, according to whether the metavariable Y appears in Q or in R. (It cannot appear in both, since it only appears once.)

Substitution principles   345

Suppose Y appears in Q. Then we will write Q as $\mathcal{Q}(Y)$, to display its dependence on Y; note that this is an order reversing subformula of $\mathcal{S}(Y)$. Since both Y and $\mathcal{Q}(Y)$ are order reversing subformulas of $\mathcal{S}(Y)$, it follows that Y must be an order preserving subformula of $\mathcal{Q}(Y)$. Since $\mathcal{Q}(Y)$ satisfies (i), we have $\vdash \mathcal{Q}(A) \to \mathcal{Q}(B)$. By $\to$-suffixing, we have $\vdash [\mathcal{Q}(B) \to R] \to [\mathcal{Q}(A) \to R]$. That is, $\vdash \mathcal{S}(B) \to \mathcal{S}(A)$.

On the other hand, suppose Y appears in R. Then we will write R as $\mathcal{R}(Y)$, to display its dependence on Y; note that this is an order preserving subformula of $\mathcal{S}(Y)$.

Since Y is order reversing in $\mathcal{S}(Y)$ and $\mathcal{R}(Y)$ is an order preserving subformula of $\mathcal{S}(Y)$, it follows that $\mathcal{R}(Y)$ is an order reversing function of Y. Since $\mathcal{R}(Y)$ satisfies (ii), we have $\vdash \mathcal{R}(B) \to \mathcal{R}(A)$. By $\to$-prefixing, we have $\vdash [Q \to \mathcal{R}(B)] \to [Q \to \mathcal{R}(A)]$. That is, $\vdash \mathcal{S}(B) \to \mathcal{S}(A)$.

That completes the proof.

Exercise. Now that we have proved 14.16, use it to prove 14.14.

14.17. Remarks. From $A \vdash B$ and the assumption that $\mathcal{S}$ is order preserving, we can *not* conclude that $\mathcal{S}(A) \vdash \mathcal{S}(B)$ in basic logic.

Example. Let $\mathcal{S}(Y) = X \to Y$, for some fixed formula X. Then $\mathcal{S}(Y)$ is an order preserving function of Y. Also, $P \to Q \vdash (R \wedge P) \to (R \wedge Q)$ by 14.7.a. If, as suggested in the remark above, we could conclude that $\mathcal{S}(P \to Q) \vdash \mathcal{S}[(R \wedge P) \to (R \wedge Q)]$, that would tell us

$$X \to (P \to Q) \quad \vdash \quad X \to [(R \wedge P) \to (R \wedge Q)]$$

for every formula X. In particular, taking $X = P \to Q$, we would have

$$(P \to Q) \to (P \to Q) \quad \vdash \quad (P \to Q) \to [(R \wedge P) \to (R \wedge Q)].$$

Since $(P \to Q) \to (P \to Q)$ is a specialization of the identity axiom 13.2.b, we could conclude that $(P \to Q) \to [(R \wedge P) \to (R \wedge Q)]$ is a theorem of basic logic, contradicting 23.18.h.

14.18. *Some applications of the substitution principles.* Hereafter, we will be able to make the following substitutions, in view of 13.23, 14.6, and 14.9–14.12:

a. If $A_1, A_2, \ldots, A_n$ is a sequence of formulas, $B_1, B_2, \ldots, B_n$ is the same sequence in some other order, and R is any formula, then

$$(A_1, A_2, \ldots, A_n) \to R \quad \text{and} \quad (B_1, B_2, \ldots, B_n) \to R$$

(in the notation of 13.21) can be interchanged.

b. $A \to \overline{B}$ and $B \to \overline{A}$ may be interchanged.

c. $\overline{A}$ and $\overline{\overline{\overline{A}}}$ may be interchanged.

d. $\overline{A \vee B}$ and $\overline{A} \wedge \overline{B}$ may be interchanged.

However, we emphasize that the interchange of $\overline{\overline{A}}$ with A, or the interchange of $\overline{A \wedge B}$ with $\overline{A} \vee \overline{B}$, has *NOT* yet been justified, and indeed cannot be justified until we add more assumptions to our logical system. See 14.12.d, 19.2.a, and 19.4 for those and other exchanges. Unjustified interchange of A and $\overline{\overline{A}}$ is one of the most common errors I have seen among beginning students of logic; readers are cautioned to watch out for it.

e. R can be interchanged with $R \vee R$ or with $R \wedge R$.

f. $R \wedge S$ and $S \wedge R$ can be interchanged; $R \vee S$ and $S \vee R$ can be interchanged.

g. $R \wedge (S \wedge T)$ and $(R \wedge S) \wedge T$ can be interchanged; $R \vee (S \vee T)$ and $(R \vee S) \vee T$ can be interchanged. Hence, in any logic that includes basic logic,

> those expressions generally can be written without the parentheses, and we shall do so hereafter.

14.19. *Exercises (optional).* These formulas make use of our newly reduced need for parentheses, mentioned in the preceding paragraph.

a. *Adjunction.* $\{J, K, L\} \vdash P$ if and only if $J \wedge K \wedge L \vdash P$.

b. *Proof by cases.* $\{A \to X, B \to X, C \to X\} \vdash (A \vee B \vee C) \to X$.

c. *Multiple consequences.*
$\{X \to A, X \to B, X \to C\} \vdash X \to (A \wedge B \wedge C)$.

Substitution principles

14.20. *More distributive laws.* The proofs of some of these theorems are made easier by the use of the substitution principles 14.14 and 14.16.
a. $\vdash [R \vee (S \wedge T)] \rightarrow [(R \vee S) \wedge (R \vee T)]$.
b. $\vdash [R \wedge (S \vee T)] \rightarrow [(R \wedge S) \vee (R \wedge T)]$.
c. $\vdash [(R \vee S) \wedge (R \vee T)] \rightarrow [R \vee (S \wedge T)]$.
d. $\vdash [(R \wedge S) \vee (R \wedge T)] \rightarrow [R \wedge (S \vee T)]$.

Derivation of 14.20.a

(1)	$(S \wedge T) \rightarrow S$	14.1.b (left $\wedge$-elimination)
(2)	$(S \wedge T) \rightarrow T$	14.1.c (right $\wedge$-elimination)
(3)	$[R \vee (S \wedge T)] \rightarrow (R \vee S)$	(1), 14.7.b (weak $\vee$-prefix.)
(4)	$[R \vee (S \wedge T)] \rightarrow (R \vee T)$	(2), 14.7.b (weak $\vee$-prefix.)
(5)	$[R \vee (S \wedge T)] \rightarrow$ $[(R \vee S) \wedge (R \vee T)]$	(3), (4), 14.4.a (multiple consequences)

Sketch of 14.20.b: Use transitivity to combine statements (1) through (5), below. Those statements are arrived at separately — i.e., statement $(n+1)$ does *not* follow from statement (n). Start with idempotency, $\vdash R \rightarrow (R \wedge R)$, and apply weak $\wedge$-suffixing, to obtain

(1) $\qquad \vdash [R \wedge \underbrace{(S \vee T)}_{\text{suffix}}] \rightarrow [(R \wedge R) \wedge \underbrace{(S \vee T)}_{\text{suffix}}]$.

(The underbraces are included for emphasis, just to make the proof more readable.) By associativity and commutativity,

(2) $\qquad \vdash [(R \wedge R) \wedge (S \vee T)] \rightarrow [R \wedge (R \wedge (T \vee S))]$.

By the distributivity axiom 14.1.h and substitutions,

(3) $\qquad \vdash [R \wedge \underbrace{(R \wedge (T \vee S))}_{\text{distribute}}] \rightarrow [R \wedge \underbrace{((R \wedge T) \vee S)}_{\text{distribute}}]$.

By associativity and commutativity,

(4) $\qquad \vdash [R \wedge ((R \wedge T) \vee S)] \rightarrow [R \wedge (S \vee (R \wedge T))]$.

By the distributivity axiom 14.1.h again, this time with the substitutions $A = R$, $B = S$, $C = R \wedge T$,

(5) $\quad \vdash [R \wedge (S \vee \underbrace{(R \wedge T)}_{\text{unchanged}})] \to [(R \wedge S) \vee \underbrace{(R \wedge T)}_{\text{unchanged}}].$

Derivations of 14.20.c and 14.20.d are left as exercises. Hint. They are dual (in the sense of 14.2) to 14.20.b and 14.20.a, respectively.

Part D

One-formula extensions

Chapter 15

Contraction

15.1. This chapter considers some extensions of basic logic — i.e., Chapters 13 and 14. Throughout this chapter, we permit the use of the assumptions and results of those chapters.

This chapter extends basic logic by adding one of the formula schemes or inference rules in 15.2 or 15.4. It would have to be as an additional *assumption*, since — as we shall see in 24.3 — none of these added formula schemes or inference rules can be proved using just basic logic, nor can any of the consequences developed in this chapter.

The student is cautioned not to confuse converse (5.25), contrapositive (5.27), contradiction (5.35), contraction (15.4.c), and constructive (2.42).

Weak contraction

15.2. The following results are equivalent over basic logic — i.e., adding any one of them makes the others provable. Formulas 15.2.a–15.2.e have been dubbed "weak" because they are weaker than, respectively, 15.4.a, 15.4.c, 15.5, 23.2.a, and 23.2.b. In fact, they are strictly weaker, since we are not assuming ¬-elimination.

a. *Weak self-distribution.*
$\vdash [A \to (B \to \overline{C})] \to [(A \to B) \to (A \to \overline{C})]$.
b. *Weak contraction.* $\vdash (A \to (A \to \overline{B})) \to (A \to \overline{B})$.
c. *Weak conjunctive detachment.* $\vdash (A \land (A \to \overline{B})) \to \overline{B}$.

d. *Weak reductio.* $\vdash (A \to \overline{A}) \to \overline{A}$.

Also equivalent are these detachmental corollaries.

15.2.a.δ $A \to (B \to \overline{C}) \vdash (A \to B) \to (A \to \overline{C})$.
15.2.a.$\delta\delta$ $\{A \to (B \to \overline{C}), A \to B\} \vdash A \to \overline{C}$.
15.2.b.δ $A \to (A \to \overline{B}) \vdash A \to \overline{B}$.

Optional. The following formulas and inference rules, though possibly less important, are also equivalent.

- **e.** *Weak dichotomy.* $\vdash (\overline{A} \to B) \to ((A \to B) \to \overline{\overline{B}})$.
- **f.** *Weak dichotomy, alternate form.* $\vdash (\overline{A} \to \overline{B}) \to ((A \to \overline{B}) \to \overline{B})$.
- **g.** *Proof by contradiction.* $\vdash (A \to B) \to ((A \to \overline{B}) \to \overline{A})$.

15.2.e.δ $\overline{A} \to B \vdash (A \to B) \to \overline{\overline{B}}$.
15.2.f.δ $\overline{A} \to \overline{B} \vdash (A \to \overline{B}) \to \overline{B}$.
15.2.g.δ $A \to B \vdash (A \to \overline{B}) \to \overline{A}$.

Remarks. Conspicuously absent from our list of equivalents are the detachmental corollaries of weak conjunctive detachment and weak reductio (15.2.c and 15.2.d). They must be omitted because they are not equivalent to those other results. That can be shown using 8.26.g, 8.36, 14.8.a, and soundness.

Proofs of equivalence. Equivalences will be proved in the order shown by the diagram below. The optional parts of this

```
15.2.a   ←  15.2.b                    (15.2.f.δ) ← (15.2.e.δ)
  ↓           ↓    ↘                    ↗    ↑              ↑
15.2.a.δ   15.2.b.δ →  15.2.d        (15.2.f)  ←  (15.2.e)
  ↓     ↗    ↓ ↑          ↑    ↘              ↗
15.2.a.δδ   15.2.c    (15.2.g.δ) ← (15.2.g)
```

section are indicated in parentheses in the diagram. It should be understood that a connection of the form "$x \to y$" in the diagram does not represent "x implies y." Rather, it means "we can go from x to y — that is, if we add x to basic logic then y becomes provable." Of course, all implications of the form $x \to x.\delta$ are just detachmental corollaries, proved as in 13.16–13.17.

Weak contraction

Adding 15.2.b makes 15.2.a provable. The justifications are left as an exercise.

(1)	$(B \to \overline{C}) \to [(A \to B) \to (A \to \overline{C})]$	?
(2)	$[A \to (B \to \overline{C})] \to \{A \to [(A \to B) \to (A \to \overline{C})]\}$	?
(3)	$\{A \to [(A \to B) \to (A \to \overline{C})]\} \to \{(A \to B) \to [A \to (A \to \overline{C})]\}$	?
(4)	$[A \to (A \to \overline{C})] \to (A \to \overline{C})$	?
(5)	$\{(A \to B) \to [A \to (A \to \overline{C})]\} \to [(A \to B) \to (A \to \overline{C})]$	?
(6)	$[A \to (B \to \overline{C})] \to [(A \to B) \to (A \to \overline{C})]$	?

Adding 15.2.a.$\delta\delta$ makes 15.2.b.δ provable. Substituting $B = A$ in 15.2.a.$\delta\delta$ yields $\{A \to (A \to \overline{C}), A \to A\} \vdash A \to \overline{C}$. The hypothesis $A \to A$ is an axiom of basic logic. Now replace C with B to obtain 15.2.b.δ.

Adding 15.2.b.δ makes 15.2.c provable

(1)	$[A \land (A \to \overline{B})] \to A$	?
(2)	$[A \land (A \to \overline{B})] \to (A \to \overline{B})$	?
(3)	$A \to \{[A \land (A \to \overline{B})] \to \overline{B}\}$	?
(4)	$[A \land (A \to \overline{B})] \to \{[A \land (A \to \overline{B})] \to \overline{B}\}$	?
(5)	$[A \land (A \to \overline{B})] \to \overline{B}$	?

Adding 15.2.c makes 15.2.b.δ provable

(1)	$A \to (A \to \overline{B})$	?
(2)	$A \to [(A \to \overline{B}) \to \overline{B})]$	?
(3)	$A \to \{(A \to \overline{B}) \land [(A \to \overline{B}) \to \overline{B})]\}$	?
(4)	$\{(A \to \overline{B}) \land [(A \to \overline{B}) \to \overline{B})]\} \to \overline{B}$	?
(5)	$A \to \overline{B}$	?

Adding 15.2.b.δ makes 15.2.d provable. Most of the justifications are left as an exercise.

(1)	$A \to [(A \to \overline{A}) \to \overline{A}]$	?
(2)	$[(A \to \overline{A}) \to \overline{A}] \to [A \to A \to \overline{A}]$	?
(3)	$A \to [A \to A \to \overline{A}]$	?
(4)	$A \to A \to \overline{\overline{A}}$	(3), 15.2.b.δ

| (5) | $(A \to \overline{A}) \to \overline{A}$ | ? |

Adding 15.2.d makes 15.2.b provable. Most of the justifications are left as an exercise.

(1)	$(A \to \overline{A}) \to \overline{A}$	15.2.d
(2)	$(B \to (A \to \overline{A})) \to (B \to \overline{A})$	?
(3)	$(A \to (B \to \overline{A})) \to (B \to (A \to \overline{A}))$	permutation 13.2.c
(4)	$(A \to (B \to \overline{A})) \to (B \to \overline{A})$	?
(5)	$(B \to \overline{A}) \leftrightarrow (A \to \overline{B})$	?
(6)	$(A \to (A \to \overline{B})) \to (A \to \overline{B})$	?

Proofs of the optional equivalences.

Adding 15.2.d makes 15.2.g provable

(1)	$(A \to B) \to (\overline{B} \to \overline{A})$	14.10 (2nd contrapos.)
(2)	$(\overline{B} \to \overline{A}) \to [(A \to \overline{B}) \to (A \to \overline{A})]$	13.2.d ($\to$-prefixing)
(3)	$(A \to B) \to [(A \to \overline{B}) \to (A \to \overline{A})]$	(1),(2),13.5.b (trans)
(4)	$(A \to \overline{A}) \to \overline{A}$	15.2.d (reductio)
(5)	$[(A \to \overline{B}) \to (A \to \overline{A})] \to [(A \to \overline{B}) \to \overline{A}]$	(4), 13.2.d.δ (pref. det.)
(6)	$(A \to B) \to [(A \to \overline{B}) \to \overline{A}]$	(3),(5),13.5.b (trans)

Adding 15.2.g makes 15.2.e provable

(1)	$(\overline{B} \to \overline{A}) \to [(\overline{B} \to \overline{\overline{A}}) \to \overline{\overline{B}}]$	15.2.g (pf. by contrad.)
(2)	$(A \to B) \to (\overline{B} \to \overline{A})$	14.10 (2nd contrapos.)
(3)	$(A \to B) \to [(\overline{B} \to \overline{\overline{A}}) \to \overline{\overline{B}}]$	(2),(1),13.5.b (trans)
(4)	$(\overline{B} \to \overline{\overline{A}}) \to [(A \to B) \to \overline{\overline{B}}]$	(3), 13.2.c.δ (permut. det.)
(5)	$(\overline{A} \to B) \to (\overline{B} \to \overline{\overline{A}})$	14.10 (2nd contrapos.)
(6)	$(\overline{A} \to B) \to [(A \to B) \to \overline{\overline{B}}]$	(5),(4),13.5.b (trans)

Adding 15.2.e makes 15.2.f provable, and adding 15.2.e.δ makes 15.2.f.δ provable. The proof is left as an exercise. *Hint.* Use 14.11.

Adding 15.2.g.δ or 15.2.f.δ makes 15.2.d provable. Obvious using 13.2.b.

15.3. *Further consequences of weak contraction*
  a. *Specialized contraction.* $\vdash [A \to (A \to \overline{A})] \to (A \to \overline{A})$.
  b. *Inference by contradiction.* $\{A \to B,\ A \to \overline{B}\} \vdash \overline{A}$.
  c. *Inference by dichotomy.* $\{\overline{A} \to B,\ A \to \overline{B}\} \vdash B$.

Contraction

d. *Noncontradiction.* $\vdash \overline{P \wedge \overline{P}}$.   Hint. 15.2.g.δδ.

e. *Weak excluded middle.* $\vdash \overline{\overline{P \vee \overline{P}}}$.   Hint. 15.2.e.δδ.

Remarks. All five of these results are *strictly* weaker than the ones listed in 15.2. That follows from 9.18 and soundness; the details of the proof are left as an exercise.

CONTRACTION

15.4. The following results are equivalent as additions to basic logic. That is, if we add any one of these to basic logic as an additional assumption, then the other results listed here become provable.

a. *Self-distribution.*
$$\vdash [A \to (B \to C)] \to [(A \to B) \to (A \to C)].$$

b. *Self-distribution permuted.*
$$\vdash (A \to B) \to \{[A \to (B \to C)] \to (A \to C)\}.$$

c. *Contraction, or repetition-elimination.*
$$\vdash (D \to (D \to E)) \to (D \to E).$$

Those formulas are also equivalent to these detachmental corollaries:

15.4.a.δ $A \to (B \to C)$ $\vdash$ $(A \to B) \to (A \to C)$.

15.4.b.δ $A \to B$ $\vdash$ $[A \to (B \to C)] \to (A \to C)$.

Proof of equivalence. Obviously 15.4.a.δ, 15.4.b.δ follow from 15.4.a, 15.4.b as detachmental corollaries. It is easy to see that 15.4.a and 15.4.b are equivalent via permutation. To prove 15.4.c from 15.4.a.δ, substitute $A = D$, $B = D \to E$, $C = E$, and use assertion (13.8.a). To prove 15.4.c from 15.4.b.δ substitute $A = B = D$ and $C = E$. To prove 15.4.a from 15.4.c use an argument similar to the proof given in 15.2 for the weak forms of those formulas.

Optional. Here is one more equivalent. (Compare 16.3.g.)

d. *Conjunctive fusion.* $\vdash [A \to (B \to C)] \to [(A \wedge B) \to C]$.

Proof of equivalence. On the one hand, contraction and 14.8.f clearly yield 15.4.d. On the other hand, suppose we add 15.4.d to basic logic, then it includes as a specialization $\vdash [A \to (A \to B)] \to [(A \wedge A) \to B]$. Combine that with idempotency (14.6.a) and one-sided substitution (14.16) to prove contraction.

15.5. *Conjunctive detachment.* $\vdash (A \wedge (A \to B)) \to B$.

Remarks. This formula is a consequence of contraction. In fact, it is equivalent to 15.4.c.δ; that can be proved by reasoning similar to the arguments in 15.2.

15.6. *Iterated contraction.* Let α and β be finite sequences of formulas. Suppose that they are the same list, except that (i) the order of terms in the list has been changed, and (ii) some or all the repetitions in list α have been removed, leaving a shorter list β. Then (using the notation of 13.21)

$$\vdash (\alpha \to Z) \to (\beta \to Z)$$

for any formula Z.

Proof. Use generalized permutation (13.23) to rearrange the lists so that α begins with two copies of one of the terms that is repeated and is to be dropped. Use contraction to drop one of the copies. Repeat this procedure as many times as needed.

Chapter 16

Expansion and positive paradox

16.1. This chapter considers some extensions of basic logic — i.e., Chapters 13 and 14. Throughout this chapter, we permit the use of the assumptions and results of those chapters (but *not* of Chapter 15).

This chapter extends basic logic by appending one of the formula schemes or inference rules in 16.2 or 16.3. It would have to be as an additional *assumption*, since — as we shall see in 23.8–23.10 — none of these added formula schemes or inference rules can be proved using just basic logic, nor can any of the consequences developed in this chapter.

Expansion and Mingle

16.2. The following three schemes are equivalent over basic logic — i.e., adding any one of them makes the others provable. Proofs of equivalence are given after the remarks below.
   a. *Mingle.* $\vdash A \to (A \to A)$.
   b. *Expansion.* $\vdash (A \to B) \to (A \to (A \to B))$.
   c. *Repetition-introduction.* $A \to B \vdash A \to (A \to B)$.

Optional. These five results are also equivalent:
   d. *Prefixed mingle.* $\vdash (A \to B) \to (A \to (B \to B))$.

e. *Collection.* $\vdash (A \to C) \to \{(B \to C) \to [A \to (B \to C)]\}$.
16.2.d.δ. $A \to B \vdash A \to (B \to B)$.
16.2.e.δ. $A \to C \vdash (B \to C) \to [A \to (B \to C)]$.
16.2.e.$\delta\delta$. $\{A \to C, B \to C\} \vdash A \to (B \to C)$.

Remarks: Why is "mingle" called "mingle"? Historically, the name "mingle" was first applied to 16.2.e.$\delta\delta$, which mingles together two partially related implications to produce a third implication. That terminology, at least, was somewhat descriptive. However, with usage, the name "mingle" was shifted to the shorter, simpler formula 16.2.a. I have introduced the name "collection" so that formula 16.2.e can have a name of its own.

The equivalences will be proved in the order shown in the diagram below. (In the diagram, "$x \to y$" does *not* mean "x implies

$$\begin{array}{ccccccc}
16.2.\text{c} & \to & \text{mingle} & \leftarrow & (16.2.\text{e}.\delta\delta) & \leftarrow & (16.2.\text{e}.\delta) \\
\uparrow & \nearrow & \uparrow & \searrow & & & \uparrow \\
\text{expansion} & & (16.2.\text{d}.\delta) & \leftarrow & (\text{prefixed}) & \to & (\text{collection})
\end{array}$$

y." Rather, it means that "we can go from x to y — that is, if we add x to our assumptions, then y becomes provable." Of course, $x \to x.\delta$ is just a detachmental corollary, proved as in 13.16–13.19.) The five optional results are parenthesized.

Using 16.2.c to prove mingle. Substitute $A = B$ and use 13.2.b.

Using mingle to prove expansion

(1)	$(A \to A) \to$ $((A \to B) \to (A \to B))$	13.4 ($\to$-suffixing)
(2)	$A \to (A \to A)$	16.2.a (mingle)
(3)	$A \to ((A \to B) \to (A \to B))$	(2), (1), 13.5.b (trans.)
(4)	$(A \to B) \to (A \to (A \to B))$	(3), 13.2.c.δ (perm. det.)

Proofs of the optional equivalences

Using mingle to prove prefixed mingle. Use 13.2.d.δ.

Using 16.2.d.δ to prove mingle. Substitute $A = B$ and use 13.2.b.

Using 16.2.e.δδ to prove mingle. Substitute $A = B = C$ and use 13.2.b.

Using prefixed mingle to prove collection

(1)	$(C \to C) \to [(B \to C) \to (B \to C)]$	13.2.d ($\to$-prefixing)
(2)	$[A \to (C \to C)] \to \{A \to [(B \to C) \to (B \to C)]\}$	(1), 13.2.d.δ ($\to$-prefixing detached)
(3)	$(A \to C) \to [A \to (C \to C)]$	16.2.d (prefixed mingle)
(4)	$\{A \to [(B \to C) \to (B \to C)]\} \to \{(B \to C) \to [A \to (B \to C)]\}$	13.2.c (permutation)
(5)	$(A \to C) \to \{(B \to C) \to [A \to (B \to C)]\}$	(3), (2), (4), 13.6.a (repeated transitivity)

Positive paradox (strong expansion)

16.3. The term "paradox" is applied here by relevantists, for reasons discussed starting in 5.29. The term "positive" refers to the fact that there are no negations involved, or perhaps to the fact that this formula is one of the simplest and strongest kinds of nonrelevant reasoning.

The following results are equivalent when considered as additions to basic logic. That is, if we append any one of these to basic logic as an additional assumption, then the other results listed here become provable.

a. *Positive paradox.* $\vdash A \to (B \to A)$.
b. *Irrelevant conclusion.* $\vdash B \to (A \to A)$.
c. *Strong adjunction.* $\vdash A \to [B \to (A \land B)]$.
d. *Strong $\lor$-elimination.*
    $\vdash (A \to C) \to \{(B \to C) \to [(A \lor B) \to C]\}$.

Also equivalent are these detachmental corollaries:

16.3.a.δ. $A \vdash B \to A$,
16.3.d.δ. $A \to C \vdash (B \to C) \to [(A \lor B) \to C]$,

and this admissibility rule:

16.3.a.$\delta\alpha$. $\vdash A \;\Rightarrow\; \vdash B{\to}A$.

Remark. The admissibility rule 16.3.a.$\delta\alpha$ says that every theorem A is weaker than any formula B. Thus every theorem is a top element, in the sense of 5.34.

Optional. The following results, though less crucial, are also equivalent; some of them will be used once or twice later in the book.

  e. *Meddling.* $A \to B \vdash A \to (C \to B)$.

  f. *Irrelevant adjunction.* $\vdash A \to [(B {\to} B) \wedge A]$.

  g. *Conjunctive implication* (compare 15.4.d).
$$\vdash [(A \wedge B) \to C] \to [A \to (B \to C)].$$

16.3.c.δ. $A \vdash B \to (A \wedge B)$.

16.3.g.δ. $(A \wedge B) \to C \vdash A \to (B \to C)$.

The equivalences will be proved in the order indicated by the diagram below. The optional items are marked in parentheses.

$$
\begin{array}{ccccccc}
(\text{meddling}) & \leftarrow 16.3.\text{a}.\delta & \to & 16.3.\text{a}.\delta\alpha & \to & \begin{array}{c}\text{irrelevant}\\ \text{conclusion}\end{array} & \leftarrow \left(\begin{array}{c}\text{irrelev.}\\ \text{adjunc.}\end{array}\right) \\
\downarrow & \nearrow & & & & \downarrow & \uparrow \\
\begin{array}{c}\text{positive}\\ \text{paradox}\end{array} & \leftarrow 16.3.\text{d}.\delta & \leftarrow & \begin{array}{c}\text{strong}\\ \text{or-elim.}\end{array} & \leftarrow & \begin{array}{c}\text{strong ad-}\\ \text{junction}\end{array} & \to (16.3.\text{c}.\delta) \\
& \nwarrow & & & \swarrow & & \\
& & (16.3.\text{g}.\delta) \leftarrow & \left(\begin{array}{c}\text{conjunctive}\\ \text{implication}\end{array}\right) & & &
\end{array}
$$

Adding 16.3.a.$\delta\alpha$ yields irrelevant conclusion. As a specialization of 16.3.a.$\delta\alpha$ we have $\vdash (A \to A) \Rightarrow \vdash B \to (A \to A)$.

Adding irrelevant conclusion yields strong adjunction.

(1)	$A \to (B \to B)$	irrelevant conclusion
(2)	$B \to (A \to A)$	irrelevant conclusion
(3)	$A \to (B \to A)$	(2), permutation det.
(4)	$A \to [(B \to A) \wedge (B \to B)]$	(3), (1) 14.4.a
(5)	$[(B \to A) \wedge (B \to B)]$ $\to [B \to (A \wedge B)]$	14.1.f

Positive paradox (strong expansion)

(6)	$A \to [B \to (A \wedge B)]$	(4), (5), transitivity

Adding strong adjunction yields strong ∨-elimination. First observe that

$$(A \to C) \to \{(B \to C) \to [(A \to C) \wedge (B \to C)]\}$$

is a specialization of strong adjunction. Now use 14.1.g and then apply substitution 14.16.

Adding 16.3.d.δ yields positive paradox. Most of the justifications are left as an exercise.

(1)	$B \to [(B \to A) \to A]$	?
(2)	$\{[(B \to A) \to A] \to A\} \to (B \to A)$	?
(3)	$A \to A$	?
(4)	$[(B \to A) \to A] \to \{[A \vee (B \to A)] \to A\}$	(3), 16.3.d.δ
(5)	$[A \vee (B \to A)] \to \{[(B \to A) \to A] \to A\}$	?
(6)	$A \to [A \vee (B \to A)]$	?
(7)	$A \to (B \to A)$	13.6.a, ?

Proofs of the optional equivalences

Adding 16.3.a.δ yields meddling 16.3.e. The inference rule $A \to B \vdash C \to (A \to B)$ is a specialization of 16.3.a.δ; now apply permutation to it.

Adding meddling 16.3.e yields positive paradox. Substitute $B = A$ and use the fact that $A \to A$ is a theorem.

Adding 16.3.g.δ yields positive paradox. Substitute $C = A$.

Adding 16.3.c.δ yields irrelevant adjunction. Observe that 16.3.c.δ has as a specialization $B \to B \vdash A \to ((B \to B) \wedge A)$.

Adding irrelevant adjunction yields irrelevant conclusion. This can be derived in three steps. Use irrelevant adjunction, ∧-elimination, and transitivity (13.5.b). The details are left as an exercise.

Adding strong adjunction yields conjunctive implication. In the following derivation, let the formula scheme $B \to \{[(A \wedge B) \to C] \to C\}$ be abbreviated as Q. As an exercise, fill in the missing justifications, and explain the justification of the last step.

(1)	$(A \wedge B) \to \{[(A \wedge B) \to C] \to C\}$	?
(2)	$[B \to (A \wedge B)] \to Q$	(1), ?
(3)	$A \to [B \to (A \wedge B)]$	?
(4)	$\{[B \to (A \wedge B)] \to Q\} \to (A \to Q)$	?
(5)	$A \to Q$	?
(6)	$[(A \wedge B) \to C] \to [A \to (B \to C)]$	(5), 13.23 *(explain)*

FURTHER CONSEQUENCES OF POSITIVE PARADOX

16.4. *Specialized assertion's converse.* $\vdash A \to [(B \to B) \to A]$. (Compare 13.8.b.)

16.5. *Ackermann falsehood.* $\vdash \overline{A} \leftrightarrow [A \to \overline{B \to B}]$. (Compare 5.34.)

Hints. Use specialized assertion, its converse, permutation, and the first contrapositive law.

16.6. *Weak implicative explosion.* These are equivalent to one another by permutation and the first contrapositive law. (See also 17.1.)
  a. $\vdash B \to (\overline{A} \to \overline{A})$.
  b. $\vdash \overline{A} \to (B \to \overline{A})$.
  c. $\vdash \overline{A} \to (A \to \overline{B})$.
  d. $\vdash A \to (\overline{A} \to \overline{B})$.

16.7. *A lemma.* $X \to (R \to S) \vdash (R \vee S) \to (X \to S)$.

Derivation.

(1)	$X \to (R \to S)$	hypothesis
(2)	$X \to (S \to S)$	16.3.b (irrelevant conclusion)
(3)	$R \to (X \to S)$	(1), 13.2.c.δ (permutation det.)
(4)	$S \to (X \to S)$	(2), 13.2.c.δ (permutation det.)
(5)	$(R \vee S) \to (X \to S)$	(3), (4), 14.4.b (proof by cases)

Further consequences of positive paradox

16.8. *Disjunction implies implication.* We will prove:
a. $\vdash (A \vee B) \rightarrow [(A \rightarrow B) \rightarrow B]$.
b. $\vdash (A \vee \overline{B}) \rightarrow (\overline{A} \rightarrow \overline{B})$.
c. $\vdash (\overline{A} \vee \overline{B}) \rightarrow (A \rightarrow \overline{B})$.

Remark. See related results in 23.4.

Proofs of 16.8. After appropriate substitutions, each of the three desired results is of the form $(R \vee S) \rightarrow (X \rightarrow S)$. That is the conclusion of 16.7. Thus, in each case we merely need a justification for $X \rightarrow (R \rightarrow S)$, the hypothesis of 16.7. The justifications of $X \rightarrow (R \rightarrow S)$ in the three cases follow from (respectively) 13.2.b, 16.6.c, or 16.6.d. The details are left as an exercise.

16.9. *Two characterizations (optional)*
a. $\vdash (A \rightarrow B) \leftrightarrow [(A \vee B) \rightarrow B]$.
b. $\vdash (A \rightarrow B) \leftrightarrow [A \rightarrow (A \wedge B)]$.

Proof. The "←" parts are just 14.8.c and 14.8.d. For 16.9.a, the "→" part follows from 16.8.a by permutation. For 16.9.b, the "→" part can be proved as follows.

(1)	$(A \rightarrow B) \rightarrow (A \rightarrow A)$	irrelevant conclusion
(2)	$(A \rightarrow B) \rightarrow (A \rightarrow B)$	identity
(3)	$(A \rightarrow B) \rightarrow [(A \rightarrow A) \wedge (A \rightarrow B)]$	(1), (2), 14.4.a
(4)	$[(A \rightarrow A) \wedge (A \rightarrow B)] \rightarrow [A \rightarrow (A \wedge B)]$	14.1.f
(5)	$(A \rightarrow B) \rightarrow [A \rightarrow (A \wedge B)]$	(3), (4), transitivity

16.10. ∧-*prefixing.* $\vdash (S \rightarrow T) \rightarrow [(R \wedge S) \rightarrow (R \wedge T)]$. (Compare the slightly weaker results in 14.7, proved in basic logic.)
Some justifications are left as an exercise in the derivation.

(1)	$(R \wedge S) \rightarrow R$	?
(2)	$R \rightarrow [(S \rightarrow T) \rightarrow R]$	16.3.a
(3)	$(R \wedge S) \rightarrow [(S \rightarrow T) \rightarrow R]$	(1),(2),?
(4)	$S \rightarrow [(S \rightarrow T) \rightarrow T]$	?
(5)	$(R \wedge S) \rightarrow S$	?
(6)	$(R \wedge S) \rightarrow [(S \rightarrow T) \rightarrow T]$	(5),(4),?
(7)	$(R \wedge S) \rightarrow$	

(8)	$\{[(S \to T) \to R] \land [(S \to T) \to T]\}$	(3),(6), ?
	$\{[(S \to T) \to R] \land [(S \to T) \to T]\}$ $\to [(S \to T) \to (R \land T)]$	?
(9)	$(R \land S) \to [(S \to T) \to (R \land T)]$	(7), (8), ?
(10)	$(S \to T) \to [(R \land S) \to (R \land T)]$	(9), ?

16.11. ∨-*prefixing.* $\vdash (T \to S) \to [(R \lor T) \to (R \lor S)]$.

The proof, analogous to that of 16.10, is left as an exercise.

16.12. ∧-*suffixing.* $\vdash (S \to T) \to [(S \land R) \to (T \land R)]$.

16.13. ∨-*suffixing.* $\vdash (S \to T) \to [(S \lor R) \to (T \lor R)]$.

Chapter 17

Explosion

17.1. *Explosion* refers to the condition that when we arrive at a contradiction, then the number of theorems in our logic suddenly increases enormously. Several noteworthy versions of explosion and their most important close relatives are shown in the diagram below. These variants are *not* all equivalent to one another as

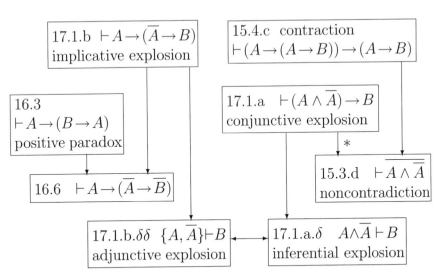

extensions of basic logic. Some formulas in the diagram have been encountered in earlier chapters, some are corollaries, and two are new assumptions to be considered here:

   a. *Conjunctive explosion.* $\vdash (A \wedge \overline{A}) \to B$.
   b. *Implicative explosion.* $\vdash A \to (\overline{A} \to B)$.

Arrows in the diagram indicate which variants of explosion can be proved from which other variants. Most of the proofs are contained in earlier chapters, are obvious, or are easy exercises. Note that adjunctive and inferential explosions are equivalent — i.e., assuming either one makes the other provable.

One implication, marked with an asterisk in the diagram, is less obvious than the others: If we add conjunctive explosion to basic logic, we can prove noncontradiction. Here is a hint for that derivation: Use $(B \to B) \to A \wedge \overline{A}$ as an intermediate step.

Optional. If we assume only basic logic, then the formulas and inference rules in the diagram are not related by any implications except those shown in the diagram. As an exercise, advanced students may wish to verify this claim. Here are some hints:

- The fuzzy logics (8.16) satisfy basic logic plus implicative explosion but not noncontradiction.

- Church's diamond (9.14) satisfies basic logic plus conjunctive explosion but not 16.6.

17.2. *Church's constant.* Assume conjunctive explosion. Then any contradiction (i.e., any formula of the form $A \wedge \overline{A}$) is a Church falsehood, in the sense of 5.34).

17.3. *Ackermann's constant.* Assume implicative explosion. Also suppose that X is some formula for which $\overline{X}$ is a theorem. Then X is an Ackermann falsehood (as discussed in 5.34); that is, $\vdash \overline{A} \leftrightarrow (A \to X)$ for every formula A.

Remark. The condition that $\overline{X}$ be a theorem is satisfied, for instance, by $X = P \to P$. In many later chapters it will also be satisfied by $X = P \wedge \neg P$.

Proof. We have $\vdash \overline{A} \to (A \to X)$ by implicative explosion, regardless of what X is. To prove $\overline{X} \vdash (A \to X) \to \overline{A}$, reason as follows.

| (1) | $\overline{X}$ | known theorem |

(2)	$\overline{X \to [(\overline{X} \to \overline{A}) \to \overline{A}\,]}$	assertion (13.8.a)
(3)	$(\overline{X} \to \overline{A}) \to \overline{A}$	(1), (2), detachment
(4)	$(A \to X) \to (\overline{X} \to \overline{A})$	2nd contrapositive (14.10)
(5)	$(A \to X) \to \overline{A}$	(4), (3), transitivity (13.5.b)

17.4. *Preview of Wajsberg logic.* Łukasiewicz logic is a semantics considered in 8.17. Its three values play a role somewhat analogous to the two values of the classical two-valued interpretation. The corresponding syntactic logic is Wajsberg logic, studied in Chapter 24. For any formula, we always have at least one of three values holding: $\vdash A \vee \overline{A} \vee \widetilde{A}$. But in fact, we cannot have more than one; any two of the three values, when combined, are sufficient for an explosion. Thus, Wajsberg logic satisfies adjunctive explosion 17.1.b.$\delta\delta$, but it also satisfies these two variants: $\{A, \widetilde{A}\} \vdash B$ and $\{\widetilde{A}, \overline{A}\} \vdash B$. See 24.20.b.

17.5. *Definitions.* Let $\mathcal{C}$ be some set of formulas. (Not formula schemes such as $A \to B$, but actual formulas such as $\pi_1 \to \pi_2$.) We shall say that the set $\mathcal{C}$ is *trivializing* (within the context of some particular syntactic logic) if

$$\mathcal{C} \vdash X \qquad \text{for every formula } X$$

— i.e., if adding the members of $\mathcal{C}$ as hypotheses makes all formulas provable, and thus renders the logic trivial. For instance,

- in any logic that includes adjunctive explosion, the set $\{\pi_1, \overline{\pi_1}\}$ is trivializing;
- in Wajsberg logic, the sets $\{\pi_1, \widetilde{\pi_1}\}$ and $\{\widetilde{\pi_1}, \overline{\pi_1}\}$ are also trivializing.

In this book, a set of formulas that does not render the logic trivial will be called *non-trivializing*.[1]

Actually, in a logic that includes basic logic and adjunctive explosion, the following conditions are equivalent:

---

[1] I recommend against omitting the hyphen. "Nontrivializing" conveys to me the meaning of "makes nontrivial," which is a different notion.

- $\mathcal{C}$ is *trivializing* — i.e., $\mathcal{C} \vdash X$ for every X.
- $\mathcal{C}$ is *contradictory*, or *inconsistent* — that is, there exists at least one formula Z such that $\mathcal{C} \vdash Z$ and $\mathcal{C} \vdash \overline{Z}$.
- $\mathcal{C} \vdash Z \wedge \overline{Z}$ for at least one formula Z.

In some contexts the words *inconsistency* or *contradiction* refer to any formula of the form $Z \wedge \overline{Z}$; in other contexts those words refer more generally to any formula W with the property that the singleton $\{W\}$ is inconsistent.

Chapter 18

Fusion

18.1. *Assumption.* This chapter extends basic logic (i.e., Chapters 13 and 14) by adding one new assumption. Throughout this chapter, we permit the use of the assumptions and results of those chapters (but *not* of Chapters 15–17). The one new assumption added now is

> For any two formulas A and B, there exists another formula — denoted by $A \& B$ and called the *fusion* of A and B — with the property that for all formulas X,
>
> $$\vdash \quad [(A \,\&\, B) \to X] \leftrightarrow [A \to (B \to X)].$$

Remarks. For an informal introduction, see 5.33. For examples, the reader may glance ahead to later chapters; some fusion operator & is available in all our major logics.

- In constructive logic, we can use ∧. See 22.10.d.
- All our other major logics include ¬-elimination as an assumption, so we can use ∘. See 19.5.

Our first result, below, will be to show that we are justified in calling this "*the* fusion of A and B," rather than just "*a* fusion of A and B."

18.2. *Uniqueness.* Fusion is determined uniquely, up to syntactic equivalence. That is: If A and B are some particular formulas, and K_1 and K_2 are two formulas satisfying the requirement for $A \,\&\, B$, then $\vdash K_1 \leftrightarrow K_2$.

Remarks. Note that K_1 and K_2 need not be the *same* formula, but the conclusion above says they are interchangeable for most purposes of logic. In particular, they can be substituted for each other, as in 14.14.

Proof of uniqueness. The fusion requirements give us

$$\vdash [K_1 \to X] \leftrightarrow [A \to (B \to X)],$$
$$\vdash [K_2 \to X] \leftrightarrow [A \to (B \to X)]$$

for all formulas X. By transitivity, then, $\vdash [K_1 \to X] \leftrightarrow [K_2 \to X]$ for all formulas X. By weak suffix cancellation 14.8.e, we have $\vdash K_1 \leftrightarrow K_2$.

18.3. *Commutativity.* $\vdash (A \,\&\, B) \leftrightarrow (B \,\&\, A)$.

Hint of proof. Use uniqueness 18.2 and permutation 13.2.c.

Remark. For our semantics we assumed fusion's commutativity and used it to demonstrate permutation (see 11.6); but in our syntactic development (18.3) we find it more convenient to assume permutation and use it to prove fusion's commutativity.

18.4. *Associativity.* $\vdash \bigl[(A \,\&\, B) \,\&\, C\bigr] \leftrightarrow \bigl[A \,\&\, (B \,\&\, C)\bigr]$.

Proof. Four different applications of 18.1 yield

$$\begin{aligned}
\lbrack(P \,\&\, C) \to X\rbrack &\leftrightarrow [P \to (C \to X)], \\
\lbrack(A \,\&\, B) \to R\rbrack &\leftrightarrow [A \to (B \to R)], \\
\lbrack B \to (C \to X)\rbrack &\leftrightarrow [(B \,\&\, C) \to X], \\
\lbrack A \to (Q \to X)\rbrack &\leftrightarrow [(A \,\&\, Q) \to X].
\end{aligned}$$

Now substitute $P = A \,\&\, B$, $Q = B \,\&\, C$, and $R = C \to X$. Also, in

Fusion

the third equivalence above, apply →-prefixing. This transforms the four equivalences above into these four equivalences:

$$[((A \,\&\, B)\,\&\, C) \to X] \leftrightarrow [(A \,\&\, B) \to (C \to X)],$$
$$[(A \,\&\, B) \to (C \,\&\, X)] \leftrightarrow [A \to (B \to (C \,\&\, X))],$$
$$\{A \to [B \to (C \to X)]\} \leftrightarrow \{A \to [(B \,\&\, C) \to X]\},$$
$$[A \to ((B \,\&\, C) \to X)] \leftrightarrow [(A \,\&\, (B \,\&\, C)) \to X].$$

Combine those four lines, using transitivity. That yields

$$[((A \,\&\, B)\,\&\, C) \to X] \leftrightarrow [(A \,\&\, (B \,\&\, C)) \to X].$$

Therefore $\bigl[(A\&B)\&C\bigr] \leftrightarrow \bigl[A\&(B\&C)\bigr]$, by weak suffix cancellation 14.8.e.

Chapter 19

Not-elimination

19.1. This chapter considers some extensions of basic logic — i.e., Chapters 13 and 14. Throughout this chapter, we permit the use of the assumptions and results of those chapters (but *not* of Chapters 15–17). Initially, we do not assume the results of Chapter 18, but those results will become available after we prove (in 19.5) that the assumptions of the present chapter *imply* the assumption of Chapter 18.

NOT-ELIMINATION AND CONTRAPOSITIVES

19.2. This chapter extends basic logic by appending one of the three formula schemes below. It would have to be as an additional *assumption*, since — as we shall see in 22.17 — none of the following three formula schemes can be proved using just basic logic. It does not matter *which* of those three formula schemes we add to our set of assumptions, for — as we shall show below — adding any one of them makes the other two provable.

Note. The resulting logic is called *R-W logic* in some of the literature. The "W" stands for contraction; thus R-W stands for "relevant minus contraction."

a. *Not-elimination.* $\vdash \overline{\overline{A}} \to A$.
b. *Third contrapositive law.* $\vdash (\overline{A} \to B) \to (\overline{B} \to A)$.
c. *Fourth contrapositive law.* $\vdash (\overline{A} \to \overline{B}) \to (B \to A)$.

Adding 19.2.a makes 19.2.b provable

(1)	$(\overline{A} \to B) \to (\overline{B} \to \overline{\overline{A}})$	14.10 (second contrapositive)
(2)	$\overline{\overline{A}} \to A$	19.2.a
(3)	$(\overline{B} \to \overline{\overline{A}}) \to (\overline{B} \to A)$	(2), 13.2.d.δ ($\to$-prefixing det.)
(4)	$(\overline{A} \to B) \to (\overline{B} \to A)$	(1), (3), 13.5.b (transitivity)

Adding 19.2.b makes 19.2.c provable. Justifications are left as an exercise.

(1)	$B \to \overline{\overline{B}}$	?
(2)	$(\overline{\overline{B}} \to A) \to (B \to A)$	?
(3)	$(\overline{A} \to \overline{B}) \to (\overline{\overline{B}} \to A)$	?
(4)	$(\overline{A} \to \overline{B}) \to (B \to A)$	?

Adding 19.2.c makes 19.2.a provable. Hint. Substitute $B = \overline{\overline{A}}$; then what?

INTERCHANGEABILITY RESULTS

19.3. Whenever the assumptions of this chapter are in effect, then (in addition to the interchangeability results already listed in 14.18)

$$A \text{ and } \overline{\overline{A}} \text{ are interchangeable,}$$

in the sense of 14.14. Also,

$$\overline{A \wedge B} \text{ and } \overline{A} \vee \overline{B} \text{ will be interchangeable,}$$

once we prove the next result below.

19.4. *Fourth De Morgan law.* $\vdash \overline{A \wedge B} \to (\overline{A} \vee \overline{B})$.

Hints. Apply the first De Morgan's law (14.12.a) to $\overline{A}$ and $\overline{B}$; this yields $\overline{\overline{A} \vee \overline{B}} \to (\overline{\overline{A}} \wedge \overline{\overline{B}})$. Then apply a contrapositive law (which?), to obtain $\overline{\overline{A} \wedge \overline{\overline{B}}} \rightarrow \left(\overline{A} \vee \overline{B}\right)$. Finally, apply the interchangeability of A and $\overline{\overline{A}}$.

19.5. *Cotenability has the fusion property*

$$\vdash [(A \circ B) \rightarrow X] \leftrightarrow [A \rightarrow (B \rightarrow X)].$$

(Recall that $A \circ B$, cotenability, is an abbreviation for $\overline{A \rightarrow \overline{B}}$.)

Sketch of proof. By contrapositive, permutation, and another contrapositive, each of these formulas is equivalent to the next:

$$A \rightarrow (B \rightarrow X), \quad A \rightarrow (\overline{X} \rightarrow \overline{B}), \quad \overline{X} \rightarrow (A \rightarrow \overline{B}), \quad \overline{A \rightarrow \overline{B}} \rightarrow X.$$

Further observation and notation. As a consequence of the result above, all the results of Chapter 18 are applicable, with the role of & (fusion) played by ∘ (cotenability). Hereafter,

> we will take ∘ to mean both cotenability and fusion, in any logic that includes basic logic plus ¬-elimination.

19.6. *Cotenability is distributive*

$$\vdash [(Q \circ R) \vee (Q \circ S)] \leftrightarrow [Q \circ (R \vee S)].$$

Proof. In 14.5.a, substitute $A = Q$, $B = \overline{R}$, $C = \overline{S}$. This yields

$$\vdash [(Q \rightarrow \overline{R}) \wedge (Q \rightarrow \overline{S})] \leftrightarrow [Q \rightarrow (\overline{R} \wedge \overline{S})].$$

Applying the second contrapositive law yields

$$\vdash \overline{(Q \rightarrow \overline{R}) \wedge (Q \rightarrow \overline{S})} \leftrightarrow \overline{Q \rightarrow (\overline{R} \wedge \overline{S})}.$$

Now substitute, using De Morgan's laws.

Miscellaneous consequences of not-elimination

19.7. *Conjunction fuzzily implies cotenability*

$$\vdash (A \wedge B) \to [(A \wedge B) \to (A \circ B)].$$

More briefly, $(A \wedge B) \rightsquigarrow (A \circ B)$, in the notation of 24.10.

Derivation.

(1)	$A \to [(A \to \overline{B}) \to \overline{B}]$	13.8.a (assertion)
(2)	$(A \wedge B) \to A$	14.1.b (left $\wedge$-elimination)
(3)	$\overline{B} \to (\overline{A} \vee \overline{B})$	14.1.e (right $\vee$-intro.)
(4)	$[(A \to \overline{B}) \to \overline{B}] \to$ $[(A \to \overline{B}) \to (\overline{A} \vee \overline{B})]$	(3), 13.2.d.δ ($\to$-prefixing detached)
(5)	$[\overline{A \circ B} \to \overline{A \wedge B}] \to$ $[(A \wedge B) \to (A \circ B)]$	19.2.c (4th contrapositive)
(6)	$[(A \to \overline{B}) \to (\overline{A} \vee \overline{B})]$ $\to [(A \wedge B) \to (A \circ B)]$	(5), substitutions
(7)	$(A \wedge B) \to$ $[(A \wedge B) \to (A \circ B)]$	(2), (1), (4), (6), 13.6.b (multiple transitivity)

19.8. *Redundancy of an axiom.* Under the assumptions of this chapter, the identity axiom $A \to A$ (13.2.b) is redundant. Indeed, it is proved below using the other axioms (and using 13.6.a, which we proved without using identity).

(1)	$(\neg \neg A) \to A$	19.2.a ($\neg$-elimination)
(2)	$(\neg \neg \neg A) \to \neg A$	19.2.a ($\neg$-elimination)
(3)	$(\neg \neg \neg \neg A) \to \neg \neg A$	19.2.a ($\neg$-elimination)
(4)	$A \to \neg \neg \neg \neg A$	(2), 14.1.i.δ (1st contrapos. det.)
(5)	$A \to A$	(4)(3)(1), 13.6.a (rep. trans.)

19.9. *Chain yields extremes.* If we add

$$\text{chain ordering} \qquad \vdash (A \to B) \vee (B \to A)$$

to the assumptions of this chapter, then we can prove

$$\text{unrelated extremes} \qquad \vdash (A \wedge \overline{A}) \to (B \vee \overline{B}).$$

Sketch of proof. Abbreviate $\alpha = A \wedge \overline{A}$ and $\beta = B \wedge \overline{B}$. Then we have $\overline{\alpha} = A \vee \overline{A}$ and $\overline{\beta} = B \vee \overline{B}$, by De Morgan's laws. What we want to prove is $\alpha \to \overline{\beta}$. For brevity, the derivation below is only an outline, and omits a few minor steps — applying transitivity, commutativity, etc.

(1)	$\alpha \to \overline{\alpha},\ \beta \to \overline{\beta}$	$\wedge$-elim., $\vee$-intro. axioms
(2)	$(\overline{\alpha} \to \beta) \leftrightarrow (\overline{\beta} \to \alpha)$	19.2.b (3rd contrapositive)
(3)	$(\overline{\alpha} \to \beta) \to (\overline{\alpha} \to \overline{\beta})$	(1), 13.2.d.δ ($\to$-pref. det.)
(4)	$\overline{\alpha} \to [(\overline{\alpha} \to \beta) \to \overline{\beta}]$	(3), 13.2.c.δ (perm. det.)
(5)	$\alpha \to [(\overline{\alpha} \to \beta) \to \overline{\beta}]$	(1),(4), 13.5.b (transitivity)
(6)	$\alpha \to [(\overline{\beta} \to \alpha) \to \overline{\beta}]$	(5),(2), substitution 14.14
(7)	$(\overline{\beta} \to \alpha) \to (\alpha \to \overline{\beta})$	(6), 13.2.c.δ (perm. det.)
(8)	$(\alpha \to \overline{\beta}) \to (\alpha \to \overline{\beta})$	13.2.b (identity)
(9)	$(\alpha \to \overline{\beta}) \vee (\overline{\beta} \to \alpha)$	chain order
(10)	$\alpha \to \overline{\beta}$	(7), (8), (9), 14.4.b (cases)

Chapter 20

Relativity

20.1. This chapter considers some extensions of basic logic — i.e., Chapters 13 and 14. Throughout this chapter, we permit the use of the assumptions and results of those chapters, but *not* of Chapters 15–17.

Initially, we do not assume the results of Chapters 18–19. However, those results will become available after we prove (in 20.4) that the assumptions of those two chapters are implied by the assumptions of the present chapter.

20.2. *Motivating remarks (optional).* Most of the logics studied in this book have an Ackermann falsehood $\bot$. Its main property is that $\neg P$ and $P \to \bot$ are equivalent; see 5.34. With that characterization of negation, we can restate some basic formulas about negation as follows. The first three are part of basic logic; the next three are not-elimination and its equivalents.

1st contrapositive	14.1.i	$(A \to (B \to \bot)) \to (B \to (A \to \bot))$
not-introduction	14.9	$A \to ((A \to \bot) \to \bot)$
2nd contrapositive	14.10	$(A \to B) \to ((B \to \bot) \to (A \to \bot))$
not-elimination	19.2.a	$((A \to \bot) \to \bot) \to A$
3rd contrapositive	19.2.b	$((A \to \bot) \to B) \to ((B \to \bot) \to A)$
4th contrapositive	19.2.c	$((A \to \bot) \to (B \to \bot)) \to (B \to A)$

Now, Meyer and Slaney [1989, 2002] asked, why should $\bot$ be special? This is something like Einstein's idea that there is no absolute frame of reference for motion; all motion is relative. What happens to the six formulas above when we "relativize" them by replacing $\bot$ with an arbitrary metavariable? The first three relativizations are familiar theorems of basic logic:

permutation	13.2.c	$(A \to (B \to C)) \to (B \to (A \to C))$
assertion	13.8.a	$A \to ((A \to C) \to C))$
$\to$-suffixing	13.4	$(A \to B) \to ((B \to C) \to (A \to C))$

But the relativizations of 19.2.a, 19.2.b, and 19.2.c are three new formulas, presented in 20.3 below. I would call any of them *relativity* (but Meyer and Slaney applied that name to 20.3.a). Those formulas could also be called *strong not-elimination*, for reasons in 20.4.

See also 2.40 and 20.6.a for further motivation.

We began our motivation with the Ackermann falsehood, but it will not be used in our assumptions about relativity. Rather, the availability of the Ackermann falsehood will be a consequence of our assumptions; see 20.6.c.

20.3. Relativity axioms. It is easy to verify that the following three axiom schemes are not tautological in the two-valued interpretation; hence they are not theorems of classical logic or of basic logic. We will now show that they are equivalent, in the sense that adding any one of them to basic logic makes the other two provable.

a. Coassertion. $\vdash [(A \to B) \to B] \to A$.
b. Meredith's permutation. $\vdash [(P \to Q) \to R] \to [(R \to Q) \to P]$.
c. Suffix cancellation. $\vdash [(Y \to Z) \to (X \to Z)] \to (X \to Y)$.

Hints for suffix cancellation using coassertion. Apply $\to$-prefixing to coassertion (in what form?) to prove $\{X \to [(Y \to Z) \to Z]\} \to (X \to Y)$; then what?

Using suffix cancellation to prove coassertion. Substitute $Y = A$, $Z = B$, $X = (A \to B) \to B$ into suffix cancellation. The right half of the resulting formula is coassertion. The left half is a theorem of basic logic, proved (how?) using identity and permutation.

Using Meredith's permutation to prove coassertion. Substitute $P = A$, $Q = B$, $R = A \to B$.

Derivation of Meredith's permutation, using coassertion

(1)	$[(P \to Q) \to Q] \to P$	(coassertion)
(2)	$\{[(P \to Q) \to R] \to [(P \to Q) \to$	(1), 13.2.d.δ

Relativity 379

	$Q]\} \to \{[(P \to Q) \to R] \to P\}$	($\to$-prefixing detached)
(3)	$(R \to Q) \to \{[(P \to Q) \to R]$ $\to [(P \to Q) \to Q]\}$	13.2.d ($\to$-prefixing)
(4)	$(R \to Q) \to \{[(P \to Q) \to R] \to P\}$	(3),(2), 13.5.b (trans.)
(5)	$[(P \to Q) \to R] \to [(R \to Q) \to P]$	(4), 13.2.c.δ (per.det.)

20.4. *Not-elimination*, $\overline{\overline{A}} \to A$, is a consequence of the above assumptions, as promised a few paragraphs ago.

Derivation

(1)	$(A \to \overline{A}) \to (\overline{\overline{A}} \to \overline{A})$	14.10 (2nd contrapositive)
(2)	$\overline{\overline{A}} \to ((A \to \overline{A}) \to \overline{A})$	(1), 13.2.c.δ (permut.-detach.)
(3)	$((A \to \overline{A}) \to \overline{A}) \to A$	20.3.a (coassertion)
(4)	$\overline{\overline{A}} \to A$	(2), (3), 13.5.b (transitivity)

Remarks. Thus relativity also yields all the results of Chapter 19. In particular, we have all four contrapositive laws and all four De Morgan laws.

20.5. *Unrelated identities.* $\vdash (A \to A) \to (B \to B)$.

Derivation.

(1)	$(A \to A) \to [(B \to A) \to (B \to A)]$	13.2.d ($\to$-prefixing)
(2)	$[(B \to A) \to (B \to A)] \to [B \to ((B \to A) \to A)]$	13.2.c (permutation)
(3)	$\{(A \to A) \to [(B \to A) \to (B \to A)]\} \to \{(A \to A) \to [B \to ((B \to A) \to A)]\}$	(2), 13.2.d.δ ($\to$-pref. det.)
(4)	$(A \to A) \to \{B \to [(B \to A) \to A]\}$	(1), (3), 13.2.a (det.)
(5)	$[(B \to A) \to A] \to B$	20.3.a (coassertion)
(6)	$\{B \to [(B \to A) \to A]\} \to (B \to B)$	(5), 13.2.d.δ (pref.det.)
(7)	$\{(A \to A) \to [B \to ((B \to A) \to A)]\}$	(6), 13.2.d.δ

(8)	→ [(A → A) → (B → B)]	(→-pref.det.)
	(A → A) → (B → B)	(4), (7), 13.2.a (det.)

20.6. *A few corollaries*
  a. *Prefix cancellation.* ⊢ [(C→A)→(C→B)] → (A→B).
  b. *Ackermann truth.* ⊢ B ↔ [(A → A) → B].
  c. *Ackermann falsehood.* ⊢ $\overline{B}$ ↔ (B → $\overline{A → A}$).

Hints. Prefix cancellation follows from suffix cancellation and a few contrapositives. For 20.6.b, use 20.5 and 13.8.b (how?). For 20.6.c, use 20.6.b and some contrapositives.

Part E

Soundness and major logics

Chapter 21

Soundness

Before proceeding further, the reader may find it helpful to review the definitions of "truth-preserving" and "tautology-preserving," in 7.10 and 7.12 respectively.

21.1. *Soundness principles.* Assume

- I is a semantic interpretation of the formal language; and
- A is a syntactic inference system, specified by a collection of axioms and assumed inference rules.

Also assume {axioms of A} $\subseteq$ {tautologies of I}. Then:

a. *Weak soundness principle.* Suppose that every assumed inference rule in A is *tautology-preserving* for I. Then

$$\{\text{theorems of } A\} \subseteq \{\text{tautologies of } I\}.$$

b. *Strong soundness principle.* Suppose that every assumed inference rule in A is *truth-preserving* for I. Then

$$\left\{ \begin{array}{l} \text{syntactic con-} \\ \text{sequences in } A \end{array} \right\} \subseteq \left\{ \begin{array}{l} \text{semantic con-} \\ \text{sequences in } I \end{array} \right\}.$$

That is, $\mathcal{H} \vdash P \Rightarrow \mathcal{H} \vDash P$, for every formula P and every set of formulas $\mathcal{H}$.

(These principles will be used frequently in subsequent chapters.)

Proof. The two principles are proved in almost the same fashion, so we shall treat them simultaneously. Assume that $\mathcal{H} \vdash P$ (where, for the weak principle, we take $\mathcal{H} = \varnothing$). Then $\mathcal{H} \vdash P$ can be given by a lemmaless derivation, as explained in 13.12. Say the lemmaless derivation consists of the sequence of formulas $P_1, P_2, \ldots, P_n$, where $P_n = P$.

For the weak soundness principle, we show that each P_j is a tautology on I. This is established by induction on j; the transition step of the induction argument follows from the fact that each assumed inference rule is tautology preserving. We conclude that $P_n = P$ is a tautology of I.

For the strong soundness principle, we reason a little differently: Fix any valuation $[\![\ ]\!]$ that makes all the members of $\mathcal{H}$ true. We show that each P_j is also true in that valuation — i.e., that $[\![P_j]\!] \in \Sigma_+$. This is established by induction on j; the transition step of the induction argument follows from the fact that each assumed inference rule is truth preserving. We conclude that $P_n = P$ is also true in $[\![\ ]\!]$. Thus $\mathcal{H} \vdash P \Rightarrow \mathcal{H} \vDash P$.

Chapter 22

Constructive axioms: avoiding not-elimination

22.1. By *constructive logic* (also known as *intuitionistic logic* in much of the literature) we shall mean

- basic logic (Chapters 13 and 14), plus
- self-distribution or contraction (15.4),
- positive paradox (16.3), and
- explosion (conjunctive or implicative, 17.1).

We will also

>   replace $\{A,B\} \vdash A \wedge B$ (adjunction)
>   with $\vdash A \to [B \to (A \wedge B)]$ (strong adjunction)

in our list of assumptions. This is not really an additional assumption, since we saw in 16.3.c that adding positive paradox to basic logic makes strong adjunction provable. However, subtracting one item from our list of assumed inference rules and adding one item to our list of axiom schemes will simplify some explanations, particularly in 22.4.b.

Cautionary remark. One of the main things that constructive logic does *not* have, as either an assumption or a derived result, is not-elimination. Unjustified use of not-elimination is the most common error I have seen among beginning students of logic. Not-elimination and its consequences may *not* be used in this chapter.

Constructive implication

22.2. We begin by studying constructive implication — that is, just $\to$, without $\vee, \wedge, \neg$.

Constructive implication adds two axiom schemes — positive paradox and self-distribution — that were not part of basic logic. Interestingly, it is enough to assume two schemes plus detachment. The remaining axiom schemes of basic implication (identity, permutation, and $\to$-prefixing) are no longer needed as assumptions; they become provable. We shall demonstrate that in the next few pages. Thus,

initially we will *NOT* assume the results of basic logic.

Instead, we make only these three assumptions:

a. *Detachment (δ).* $\{A, A \to B\} \vdash B$.
b. *Positive paradox.* $\vdash A \to (B \to A)$.
c. *Self-distribution.* $\vdash [A \to (B \to C)] \to [(A \to B) \to (A \to C)]$.

We also note these detachmental corollaries, which follow by the same techniques as in 13.17:

22.2.b.δ $A \vdash B \to A$
22.2.c.δ $A \to (B \to C) \vdash (A \to B) \to (A \to C)$

22.3. *Some consequences.* Proofs are given below.

a. *Identity.* $\vdash A \to A$.
b. *Transitivity.* $\{S \to T, T \to U\} \vdash S \to U$.
c. *Permutation detached.* $S \to (T \to U) \vdash T \to (S \to U)$.
d. $\to$-*prefixing.* $\vdash (B \to C) \to [(A \to B) \to (A \to C)]$.

Note that 22.3.c is merely the *detachmental corollary* of permutation. Permutation itself, in the undetached form, is harder; its proof is postponed until 22.7.

Derivation of identity

(1)	$A \to ((A \to A) \to A)$	22.2.b (positive paradox)
(2)	$(A \to (A \to A)) \to (A \to A)$	(1), 22.2.c.δ (self dist. det.)
(3)	$A \to (A \to A)$	22.2.b (positive paradox)

| (4) | $A \to A$ | (2), (3), detachment |

Derivation of transitivity. Justifications are left as exercises.

(1)	$T \to U$	?
(2)	$S \to (T \to U)$	?
(3)	$(S \to T) \to (S \to U)$	?
(4)	$S \to T$	?
(5)	$S \to U$	?

Derivation of permutation detached

(1)	$S \to (T \to U)$	hypothesis
(2)	$(S \to T) \to (S \to U)$	(1), 22.2.c.δ (self dist. det.)
(3)	$T \to (S \to T)$	22.2.b (positive paradox)
(4)	$T \to (S \to U)$	(3), (2), 22.3.b (transitivity)

Derivation of $\to$-prefixing. Justifications are left as exercises.

(1)	$(B \to C) \to [A \to (B \to C)]$	?
(2)	$[A \to (B \to C)] \to [(A \to B) \to (A \to C)]$	?
(3)	$(B \to C) \to [(A \to B) \to (A \to C)]$	?

HERBRAND-TARSKI DEDUCTION PRINCIPLE

22.4. *Assumptions.* The deduction principle developed below will apply to constructive logic and classical logic in this book, and a few other logics not in this book. It will apply to logics satisfying these assumptions:

a. Our list of theorems includes at least the axioms of constructive implication — i.e., self-distribution and positive paradox.

(We emphasize that this restriction does not prohibit us from adding *more* axioms or theorems. Thus, the same principle will also be applicable in classical logic, Chapter 25.)

b. Our only assumed inference rule is detachment.

(For instance, we do not take adjunction as an assumed inference rule; see the discussion in 22.1. However, we emphasize that this restriction does not prohibit us from using *derived* inference rules such as transitivity and permutation detached.)

22.5. *Classical deduction principle.*[1] With the assumptions noted above, suppose that F and G are some formulas such that $F \vdash G$. Then $F \to G$ is a theorem; that is, $\vdash F \to G$.

More generally, suppose F and G are some formulas and $\mathcal{H}$ is some set of formulas, such that $\mathcal{H} \cup \{F\} \vdash G$. Then $\mathcal{H} \vdash F \to G$.

Demonstration. Since we could take $\mathcal{H} = \varnothing$, clearly the assertion of the first paragraph is a special case of the assertion in the second paragraph. Thus it suffices to prove the latter (i.e., with $\mathcal{H}$ not necessarily empty). Note that the assumption $\mathcal{H} \cup \{F\} \vdash G$ says that there exists a derivation of G from $\mathcal{H} \cup \{F\}$. But then there also exists a *lemmaless* derivation of G from $\mathcal{H} \cup \{F\}$, as explained in 13.12; we will use that fact momentarily.

Any such lemmaless derivation has only finitely many steps. Our demonstration will be by induction on the number of steps in that given derivation. Thus, it suffices to prove the following implication for each positive integer n:

$P(n)$ $\begin{cases} \text{If } F \text{ and } G_n \text{ are any formulas and } \mathcal{H} \text{ is} \\ \text{any set of formulas, and there exists a} \\ \text{lemmaless derivation of } \mathcal{H} \cup \{F\} \vdash G_n \\ \text{in fewer than } n \text{ steps, then } \mathcal{H} \vdash F \to G_n. \end{cases}$

We have worded $P(n)$ very carefully — perhaps more carefully than the student might realize — in order to avoid certain subtle difficulties in the subsequent argument. Before proceeding further, we point out some subtleties in the wording of $P(n)$:

- The statement $P(n)$ assumes that a derivation of $\mathcal{H} \cup \{F\} \vdash G_n$ is given that is both lemmaless and of a certain length. The statement $P(n)$ concludes that a derivation of $\mathcal{H} \vdash F \to G_n$ exists, but $P(n)$ does *not* assert that this derivation is lemmaless or of a certain length.

---

[1] Known as the "Deduction Theorem" in most of the literature; see 2.18.

Herbrand-Tarski Deduction Principle

- We have subscripted the G's. The formulas G_n and G_{n+1} used in $P(n)$ and $P(n+1)$ may be different.[2] When we are trying to prove $P(n) \Rightarrow P(n+1)$, we may assume that $P(n)$ is already known for *all* choices of G_n.

- Note that $P(n)$ does not say that the given derivation is "in n steps." Rather, it says "in fewer than n steps." At first glance, that may look like an error or an unnecessary complication, but it actually will make the proof easier in two ways: It allows us to apply the induction hypothesis to derivations of many different lengths, and it gives us $P(1)$ for free — it makes $P(1)$ vacuously true, since there cannot be a lemmaless derivation in less than 1 step. See the remarks in 5.60.

Now assume $P(n)$ for some $n \geq 1$; we wish to prove $P(n+1)$. Assume that some particular $F, G_{n+1}, \mathcal{H}$ are given, and some lemmaless derivation of $\mathcal{H} \cup \{F\} \vdash G_{n+1}$ is given that takes n or fewer steps. We shall transform that given derivation into the desired derivation of $\mathcal{H} \vdash F \to G_{n+1}$. All of n, F, G_{n+1}, $\mathcal{H}$, and the given lemmaless derivation will be held fixed for the remainder of this proof.

The last step of that given derivation is the formula G_{n+1}. That step has some justification, which falls into one of these four categories:

(a) G_{n+1} is an axiom (i.e., an assumed rule with no hypotheses) in our logical system,

(b) G_{n+1} is one of the hypotheses in the set $\mathcal{H}$,

(c) G_{n+1} is the hypothesis F, or

(d) G_{n+1} follows by detachment from two of the earlier steps in the given lemmaless derivation.

In each of those four[3] cases, by different methods, we shall describe a derivation of $\mathcal{H} \vdash F \to G_{n+1}$. We've already done most of

---

[2] Actually, we can also put subscripts on F and $\mathcal{H}$ if we wish, but it turns out that we won't need to.

[3] *Remarks.* The number of cases is kept down to four, in part because we're working from a *lemmaless* derivation, and in part because we only need one assumed inference rule. In 23.6 we will prove an analogous deduction

the work for the first three cases; only the fourth case will involve much further effort.

In cases (a) or (b). The formula G_{n+1} is either an axiom or a member of $\mathcal{H}$ (i.e., a hypothesis). In either of those cases, we use this three-step derivation:

(1)	G_{n+1}	(axiom or hypothesis)
(2)	$G_{n+1} \to (F \to G_{n+1})$	22.2.b (pos. pdx.)
(3)	$F \to G_{n+1}$	(1), (2). δ

In case (c). There is a one-step derivation, because the desired formula $F \to G_{n+1}$ is an instance of the identity theorem 22.3.a.

In case (d): The justification of the last step in the given derivation is detachment. In other words, one of the earlier steps in the derivation (say step p) was some formula X, and another of the earlier steps (say step q) was the formula $X \to G_{n+1}$. Thus, the given lemmaless derivation looks something like this:

step number	formula	justification
(1) (2) (3)	(some formulas)	(some reasons)
$\vdots$	$\vdots$	$\vdots$
(p)	X	(some reason)
$\vdots$	$\vdots$	$\vdots$
(q)	$X \to G_{n+1}$	(some reason)
$\vdots$	$\vdots$	$\vdots$
(n or less)	G_{n+1}	(p), (q), Detachment

(I say it looks "*something* like this" only because we don't know whether $p < q$ or $p > q$. But that's a very superficial matter which has no effect on our argument.)

---

principle for relevant logic, but it requires two assumed inference rules, so it requires five cases. It is in anticipation of proofs like this that we use as few assumed inference rules as possible.

Now, what if we use just *part* of that derivation? If we stop after p steps or after q steps, respectively, we still haven't used any lemmas, and we still haven't used any hypotheses except members of $\mathcal{H} \cup \{F\}$. Thus we obtain lemmaless derivations, in p or q steps, of these two results:

$$\mathcal{H} \cup \{F\} \vdash X, \qquad \mathcal{H} \cup \{F\} \vdash X \to G_{n+1}.$$

Since $p < n$ and $q < n$, each of those lemmaless derivations takes fewer than n steps. Apply the induction hypothesis $P(n)$ to each of those lemmaless derivations — once with $G_n = X$ and once with $G_n = X \to G_{n+1}$. The conclusions thereby obtained from $P(n)$ are

(*) $\qquad \mathcal{H} \vdash F \to X, \qquad \mathcal{H} \vdash F \to (X \to G_{n+1}).$

Finally, we use those derivations as ingredients in the derivation we're looking for; the only hypotheses used are members of $\mathcal{H}$:

step	formula	justification
$\vdots$	$\vdots$	$\vdots$
(*)	various steps	
$\vdots$	$\vdots$	$\vdots$
(r)	$F \to X$	from (*)
$(r+1)$	$F \to (X \to G_{n+1})$	from (*)
$(r+2)$	$[F \to (X \to G_{n+1})] \to$ $[(F \to X) \to (F \to G_{n+1})]$	22.2.c (self-distrib'n)
$(r+3)$	$(F \to X) \to (F \to G_{n+1})$	$(r+1)$, $(r+2)$, detach.
$(r+4)$	$F \to G_{n+1}$	(r), $(r+3)$, detach.

Thus we have established $\mathcal{H} \vdash F \to G_{n+1}$ as required. This completes the induction argument, and hence the proof of the deduction principle.

22.6. *Converse to the deduction principle.* If we combine the classical deduction principle ($\Rightarrow$) with the detachmental corollary procedure ($\Leftarrow$), we obtain this equivalence:

(*) $\qquad \mathcal{H} \cup \{F\} \vdash G \qquad \Longleftrightarrow \qquad \mathcal{H} \vdash F \to G.$

We proved this equivalence using detachment, self-distribution, and positive paradox. We now will prove a converse:

> Any syntactic system (as described in Chapter 12) that satisfies (∗), must also satisfy detachment, self-distribution, and positive paradox.

Proof of detachment. Let $\mathcal{H}$ be the set $\{A \to B\}$. Then clearly $\mathcal{H} \vdash A \to B$ (with a one-line derivation whose justification is "hypothesis"). By (∗), we have $\mathcal{H} \cup \{A\} \vdash B$. But that is $\{A, A \to B\} \vdash B$, which is detachment.

Proof of self-distribution. First we shall show

$$\{A, A \to B, A \to (B \to C)\} \vdash C,$$

using nothing more than three applications of detachment (δ).

(1)	A	hypothesis
(2)	$A \to B$	hypothesis
(3)	$A \to (B \to C)$	hypothesis
(4)	B	(1), (2), δ
(5)	$B \to C$	(1), (3), δ
(6)	C	(4), (5), δ

Now apply the deduction principle three times, to successively move hypotheses from left to right past the turnstile. The inference rule that we have just derived yields, successively,

$$\{A \to B, A \to (B \to C)\} \vdash A \to C;$$
$$A \to (B \to C) \vdash (A \to B) \to (A \to C);$$

and finally $\vdash [A \to (B \to C)] \to [(A \to B) \to (A \to C)]$.

Proof of positive paradox. A one-line derivation yields $\{A, B\} \vdash A$. Apply the deduction principle to obtain $A \vdash B \to A$; apply it again to obtain $\vdash A \to (B \to A)$.

22.7. *Permutation.* $\vdash [S \to (T \to U)] \to [T \to (S \to U)]$.

Proof. Immediate from 22.3.c and the deduction principle.

Exercise for advanced students. The argument given above tells us that there exists a derivation of permutation, using just the three assumptions in 22.2, but the argument given above does not explicitly produce that derivation. Find the derivation and write it out, in the style of 22.3. (You may get some hints for your derivation by studying the proof of the deduction principle; that proof actually does give an algorithm for producing derivations.)

An even harder exercise. Give a *lemmaless* derivation of permutation. *Remark.* Norm Megill has informed me that, inspired by my exercise, he carried out an exhaustive computer search. It turns out that a minimum of 19 steps is required for the lemmaless derivation. Also, some of the steps are very long; in the 19-step derivation that Norm sent me, two of the steps were formulas of rank 53.

Basic logic revisited

22.8. *Basic implication revisited.* We have now established that the assumptions of basic implication, given in 13.2, are among the results of constructive implication. Hence

$$\left\{ \begin{array}{c} \text{theorems of} \\ \text{basic} \\ \text{implication} \end{array} \right\} \subseteq \left\{ \begin{array}{c} \text{theorems of} \\ \text{constructive} \\ \text{implication} \end{array} \right\}.$$

Therefore we may use all the results of Chapter 13 throughout the remainder of the present chapter.

22.9. *Reformulating some axioms.* Throughout the remainder of this chapter we shall permit the use of the deduction principle 22.5, plus all the assumptions of constructive logic, as listed in 22.1. However, some of those assumptions can now be reformulated in ways that are either simpler or in some other fashion more convenient.

22.10. *Constructive "and."* Basic logic's assumptions regarding conjunction (and implication) are

adjunction $\{A, B\} \vdash A \wedge B$
$\wedge$-elimination $\vdash (A \wedge B) \to A$, $\vdash (A \wedge B) \to B$
$\wedge$-introduction $\vdash [(A \to B) \wedge (A \to C)] \to [A \to (B \wedge C)]$

We can reformulate those in constructive logic:

a. First, as we already remarked in 22.1, for constructive logic, adjunction must be replaced with *strong adjunction*:

$$\vdash A \to [B \to (A \wedge B)].$$

b. Second, $\wedge$-introduction (14.1.f) is redundant in constructive logic, and may be dropped from our list of assumptions. Indeed, to prove the theorem 14.1.f, it suffices to first prove the inference rule

$$\{(A \to B) \wedge (A \to C),\ A\} \vdash (B \wedge C)$$

and then apply the deduction principle twice. That inference rule can be proved using adjunction, as follows:

(1)	$(A{\to}B){\wedge}(A{\to}C)$	hypothesis
(2)	A	hypothesis
(3)	$A \to B$	(1), 14.1.b.δ ($\wedge$-elim. det.)
(4)	$A \to C$	(1), 14.1.c.δ ($\wedge$-elim. det.)
(5)	B	(2), (3), δ
(6)	C	(2), (4), δ
(7)	$B \wedge C$	(5), (6), 14.1.a (adjunction)

c. Actually, all of our rules for conjunction can be replaced by (i.e., are equivalent to) this one fairly simple principle. Let any formulas J, K, L be given. Then

(i) $L \vdash J \wedge K$ holds

if and only if (ii) $L \vdash J$ and $L \vdash K$ both hold.

Hints for the proof. Assume constructive implication, including the classical deduction principle. Show that adjunction is equivalent to (ii) $\Rightarrow$ (i), and the two $\wedge$-elimination axioms together are equivalent to (i) $\Rightarrow$ (ii).

Basic logic revisited

d. From 16.3.g and 15.4.d we have

$$\vdash [(A \wedge B) \to X] \leftrightarrow [A \to (B \to X)].$$

Thus, in constructive logic, $\wedge$ plays the role of *fusion* — i.e., it satisfies the requirement of & in 18.1.

22.11. *Constructive "or."* Basic logic's assumptions regarding disjunction (and implication and conjunction) are

$\vee$-introduction $\quad \vdash A \to (A \vee B), \quad \vdash B \to (A \vee B)$
$\vee$-elimination $\quad \vdash [(B \to A) \wedge (C \to A)] \to [(B \vee C) \to A]$
distribution $\quad \vdash [A \wedge (B \vee C)] \to [(A \wedge B) \vee C]$

Those, too, can be reformulated in certain ways:

a. In constructive logic, $\vee$-elimination can be reformulated in various other ways, such as *strong $\vee$-elimination*:

(16.3.d) $\quad \vdash (B \to A) \to \{(C \to A) \to [(B \vee C) \to A]\}.$

(The proof is left as an exercise.) Some readers may prefer that formula because it only involves $\vee$ and $\to$; it does not involve $\wedge$. However, this replacement cannot be made in basic logic or in some other logics; for instance, see 23.17.

b. In constructive logic, the distribution axiom 14.1.h becomes redundant. It can be proved from the other axioms. Indeed, it follows by the classical deduction principle from

$$A \wedge (B \vee C) \vdash (A \wedge B) \vee C,$$

which we prove as follows:

(1)	$A \wedge (B \vee C)$	hypothesis
(2)	A	(1), $\wedge$-elimination.δ
(3)	$B \vee C$	(1), $\wedge$-elimination.δ
(4)	$A \to [B \to (A \wedge B)]$	strong adjunction
(5)	$B \to (A \wedge B)$	(2), (4), δ
(6)	$(A \wedge B) \to [(A \wedge B) \vee C]$	$\vee$-introduction
(7)	$B \to [(A \wedge B) \vee C]$	(5), (6), transitivity
(8)	$C \to [(A \wedge B) \vee C]$	$\vee$-introduction

| (9) | $(B \vee C) \to [(A \wedge B) \vee C]$ | (7), (8), cases 14.4.b |
| (10) | $(A \wedge B) \vee C$ | (3), (9), δ |

c. The ∨-elimination and ∨-introduction formulas can be replaced by a single principle. Let any formulas J, K, L be given. Then

$$\text{if and only if} \quad \begin{array}{l} \text{(i)} \quad J \vee K \vdash L \text{ holds} \\ \text{(ii)} \quad J \vdash L \text{ and } K \vdash L \text{ both hold}. \end{array}$$

Hints for the proof. Assume constructive implication, including the classical deduction principle. Show that ∨-elimination is equivalent to (ii) ⇒ (i), and the two ∨-introduction axioms together are equivalent to (i) ⇒ (ii).

22.12. *Explosion.* Some form of explosion is assumed as part of constructive logic. It doesn't greatly matter which version of explosion we assume; several versions are equivalent once we have basic logic and the deduction principle. In particular, we can assume any of these:

- $\vdash A \to (\overline{A} \to B)$ or $\vdash \overline{A} \to (A \to B)$ (implicative explosion)
- $\vdash (A \wedge \overline{A}) \to B$ (conjunctive explosion)
- $A \wedge \overline{A} \vdash B$ (inferential explosion)
- $\{A, \overline{A}\} \vdash B$ (adjunctive explosion)

Exercise. Show that assuming any of those makes the others provable.

22.13. In this section we consider logics that include *at least* constructive logic. Also, recall from 17.5 the definition of "trivializing."

Let $\mathcal{C}$ be any collection of formulas, and let A be any formula. Then:

a. $\mathcal{C} \vdash \overline{A}$ if and only if $\mathcal{C} \cup \{A\}$ is trivializing.

b. If $\mathcal{C} \vdash A$, then $\mathcal{C} \cup \{\overline{A}\}$ is trivializing.

If we also assume that the logic includes not-elimination (a condition satisfied in classical logic), then:

c. $\mathcal{C} \vdash A$ if and only if $\mathcal{C} \cup \{\overline{A}\}$ is trivializing.

Proofs. For 22.13.a, the "only if" part is easy, and is left as an exercise. For the "if" part, assume that $\mathcal{C} \cup \{A\}$ is trivializing. In particular $\mathcal{C} \cup \{A\} \vdash \overline{A}$. Therefore $\mathcal{C} \vdash A \to \overline{A}$ by the deduction principle. From 15.2.d (weak reductio) we have $\mathcal{C} \vdash (A \to \overline{A}) \to \overline{A}$. Combining those two results by detachment, we obtain $\mathcal{C} \vdash \overline{A}$. This completes our proof of 22.13.a.

Substituting $\overline{A}$ for A in 22.13.a yields this specialization:

$$\mathcal{C} \vdash \overline{\overline{A}} \quad \text{if and only if} \quad \mathcal{C} \cup \{\overline{A}\} \text{ is trivializing.}$$

In constructive logic, we have $\vdash A \to \overline{\overline{A}}$, hence $\mathcal{C} \vdash A \Rightarrow \mathcal{C} \vdash \overline{\overline{A}}$, hence 22.13.b. If we also assume not-elimination, $\vdash \overline{\overline{A}} \to A$, then we have $\mathcal{C} \vdash \overline{\overline{A}} \Rightarrow \mathcal{C} \vdash A$, hence 22.13.c.

Remark. The "if" part of 22.13.c is *not* true for constructive logic. For instance, let $A = \pi_1 \vee \overline{\pi_1}$. We know $\overline{\overline{A}}$ is a theorem of constructive logic, by 15.3.e; but we shall see in 22.15 that $A = \pi_1 \vee \overline{\pi_1}$ itself is not a theorem of constructive logic. From these results it follows that $\varnothing \cup \{\overline{A}\}$ is trivializing but $\varnothing \nvdash A$.

The remainder of this chapter is *not* part of constructive syntactic logic.

SOUNDNESS

22.14. We saw in 10.8 that every axiom of constructive logic (as listed in 22.1) is a tautology for every topology. It is also easy to verify that detachment is truth-preserving — i.e., that whenever $[\![A]\!] = \Omega$ and $[\![A \to B]\!] = \Omega$, then $[\![B]\!] = \Omega$. Hence, by the soundness principles 21.1,

> every theorem of constructive logic is a tautology of every topological interpretation

and, moreover,

> every valid inference rule of constructive logic is a valid inference rule of every topological interpretation — that is, $\mathcal{C} \vdash A \Rightarrow \mathcal{C} \vDash A$.

(A converse is also true, but proving it will require much more advanced techniques; see 29.29.)

22.15. *A few examples of nonconstructive formulas.* In Chapter 10 the following formulas were shown to be nontautological for one topology or another. By the preceding soundness result, it follows that these formulas are not theorems of constructive logic.

$$(A \to B) \lor (B \to A), \qquad A \lor \overline{A}, \qquad \overline{A} \lor \overline{\overline{A}}, \qquad \overline{\overline{A}} \to A.$$

22.16. *Gödel's finite interpretation principle.* There does not exist a finite functional interpretation I (topological or otherwise) whose set of tautologies is the same as the set of theorems of constructive logic.

Outline of proof. Suppose that I were such an interpretation; we shall obtain a contradiction. Say I has n semantic values, where n is a positive integer.

Since some formulas are theorems of constructive logic, I has some tautologies; thus I must have at least one true semantic value. Since some formulas are not theorems of constructive logic, they are not tautologies; therefore I must have at least one false semantic value. Thus, $n \geq 2$.

The formula scheme $(S \to S) \land (S \to S)$ is a theorem scheme of constructive logic. Also, if at least one of the formulas S, T is a theorem of constructive logic, then $S \lor T$ is too. Thus, the hypotheses of 11.14 are satisfied, with $m = n + 1$.

Therefore $\mathcal{D}_{n+1}$ is a tautology for I, and so it is a theorem of constructive logic. By 22.14, therefore, it is a tautology of every topological interpretation.

However, let us equip $\Omega = \{1, 2, 3, \ldots, n\}$ with the topology

$$\Sigma = \Big\{ \varnothing, \{1\}, \{1,2\}, \{1,2,3\}, \ldots, \{1,2,\ldots,n-1\}, \Omega \Big\}.$$

It is easy to verify that the resulting topological interpretation satisfies the assumptions of 11.16 with $m = n+1$. Hence $\mathcal{D}_{n+1}$ is not tautological for this topology. This contradiction completes the proof.

NONCONSTRUCTIVE AXIOMS AND CLASSICAL LOGIC

22.17. *Some nonconstructive formulas and rules.* The formulas and inference rules listed below can *not* be proved using just the assumptions of constructive logic. For some of these formulas, we already established nonconstructiveness in 22.15. We shall now show that items on the following list are all *equally* nonconstructive — i.e., adding any one of these formulas or inference rules to constructive logic, as an additional assumption, makes all the other items on the list provable. (The resulting logic is then *classical logic*, described further in Chapter 25.)

- a. *Switch* $\vdash [(A \to B) \to B] \to [(B \to A) \to A]$
- b. *Peirce's law* $\vdash [(A \to B) \to A] \to A$
- c. *Specialized Peirce* $\vdash [(A \to \overline{A}) \to A] \to A$
- d. *Not-elimination* $\vdash \overline{\overline{A}} \to A$
- e. *3rd contrapositive* $\overline{A} \to B \vdash \overline{B} \to A$
- f. *4th contrapositive* $\overline{A} \to \overline{B} \vdash B \to A$
- g. *Excluded middle* $\vdash A \lor \overline{A}$
- h. *Proof by dichotomy* $\{A \to B, \overline{A} \to B\} \vdash B$
- i. *Strong reductio* $\vdash (\overline{A} \to A) \to A$

plus two more equivalents that I would call "optional":

- j. *Disjunctive implication* $\vdash \overline{A \land \overline{B}} \to (A \to B)$
- k. *Long proof by contradiction* $(A \land \overline{B}) \to (C \land \overline{C}) \vdash A \to B$

Proof of equivalence. The equivalence of 22.17.d, 22.17.e, 22.17.f was already proved in section 19.2. The remaining proofs will be given in the order shown by the diagram below.

400   Chapter 22. Constructive axioms: avoiding not-elimination

$$22.17.\text{a} \leftrightarrow 22.17.\text{b} \to 22.17.\text{c} \to \boxed{\begin{array}{c} 22.17.\text{d}, \\ 22.17.\text{e}, \end{array}} \to 22.17.\text{j}$$
$$\nearrow \qquad \qquad \qquad \qquad \qquad \downarrow$$
$$22.17.\text{i} \leftarrow 22.17.\text{h} \leftarrow 22.17.\text{g} \leftarrow \boxed{22.17.\text{f}} \leftarrow 22.17.\text{k}$$

Switch 22.17.a yields Peirce's law 22.17.b

(1)	$[A \to (A \to B)] \to (A \to B)$	contraction 15.4.c
(2)	$\{[A \to (A \to B)] \to (A \to B)\}$	
	$\to \{[(A \to B) \to A] \to A\}$	switch
(3)	$[(A \to B) \to A] \to A$	(1), (2), δ

Peirce's law 22.17.b yields switch 22.17.a. This will follow from two applications of the deduction principle, after we demonstrate $\{(A \to B) \to B, B \to A\} \vdash A$ as follows. (The justifications are left as exercises.)

(1)	$(A \to B) \to B$	?
(2)	$B \to A$	?
(3)	$(A \to B) \to A$	?
(4)	$[(A \to B) \to A] \to A$	?
(5)	A	?

Peirce's law 22.17.b yields specialized Peirce 22.17.c. Obvious.

Specialized Peirce 22.17.c yields not-elimination 22.17.d. It suffices to apply the deduction principle after proving $\overline{\overline{A}} \vdash A$. We prove that inference as follows:

(1)	$\overline{\overline{A}}$	hypothesis
(2)	$\overline{\overline{A}} \to A$	(1), 22.12.δ
(3)	$(A \to \overline{A}) \to \overline{A}$	weak reductio 15.2.d
(4)	$(A \to \overline{A}) \to A$	(3), (2), transitivity
(5)	$[(A \to \overline{A}) \to A] \to A$	specialized Peirce
(6)	A	(4), (5), detachment

Not-elimination 22.17.d yields excluded middle 22.17.g. Immediate from weak excluded middle 15.3.e.

Nonconstructive axioms and classical logic

Excluded middle 22.17.g yields proof by dichotomy 22.17.h. Immediate from 14.4.b.δ.

Proof by dichotomy 22.17.h yields strong reductio 22.17.i. The formula $\overline{A} \to [(\overline{A} \to A) \to A]$ is an instance of assertion 13.8.a, and the formula $A \to [(\overline{A} \to A) \to A]$ is an instance of positive paradox 16.3.a. Combine those using proof by dichotomy.

Strong reductio 22.17.i yields Peirce 22.17.b. The justifications are left as exercises.

(1)	$\overline{A} \to (A \to B)$	?
(2)	$[(A \to B) \to A] \to (\overline{A} \to A)$	?
(3)	$(\overline{A} \to A) \to A$	?
(4)	$[(A \to B) \to A] \to A$	?

Proof of the optional equivalences.

Fourth contrapositive 22.17.f yields disjunctive implication 22.17.j. By the fourth contrapositive we have $[\overline{B} \to \overline{(A \wedge \overline{B})}] \to [A \wedge \overline{B} \to B]$. Also, we have $A \to [\overline{B} \to \overline{(A \wedge \overline{B})}]$ by strong adjunction. Combine those two formulas using transitivity; then apply permutation.

Disjunctive implication 22.17.j yields the long form of proof by contradiction 22.17.k. Use noncontradiction (15.3.d), the detachmental corollary of assertion (13.8.a.δ), disjunctive implication (22.17.j), and transitivity. The details are left as an exercise.

Proof by contradiction 22.17.k yields 4th contrapositive 22.17.f.

(1)	$\overline{A} \to \overline{B}$	hypothesis
(2)	$(B \wedge \overline{A}) \to (B \wedge \overline{B})$	(1), 14.7.a (wk. ∧-prefixing)
(3)	$B \to A$	(2), 22.17.k

22.18. Slightly nonconstructive logics. In much of the literature, an *intermediate logic* means any logic between constructive and the classical two-valued.[4] For instance,

$$\left\{\begin{array}{c}\text{topological}\\ \text{tautols.}\\ (22.14)\end{array}\right\} \subsetneq \left\{\begin{array}{c}\text{upper set}\\ \text{tautols.}\\ (4.6.\text{h})\end{array}\right\} \subsetneq \left\{\begin{array}{c}\text{3-valued}\\ \text{tautols.}\\ (10.5)\end{array}\right\} \subsetneq \left\{\begin{array}{c}\text{classical}\\ \text{tautols.}\\ (29.12)\end{array}\right\}.$$

The inclusions ($\subseteq$) follow from the restriction principle 11.8. That the inclusions are strict ($\subsetneq$) can be explained as follows:

- The chain axiom $(A \to B) \vee (B \to A)$ and the fourth De Morgan's law are tautological for the upper set topology (4.6.h) but not for some other topologies.

- Stone's formula, $\overline{A} \vee \overline{\overline{A}}$, is tautological for the three-valued interpretation (10.5) but not for the upper set topology.

- $A \vee \overline{A}$ is tautological in classical logic but not in the 3-valued interpretation.

Remark. It can be shown that the tautologies of the upper set topology (4.6.h) are exactly the same as the theorems of the axiomatic system known as *Dummett's LC*, which consists of constructive logic plus the axiom scheme $(A \to B) \vee (B \to A)$. The proof of this completeness pairing is more difficult and will not be given here; it can be found in Dummett [1959].

Further note. We have restated Dummett's result to make it more compatible with the rest of this book. Actually, Dummett used the set $\Sigma = \{0, 1, 2, 3, \ldots, +\infty\}$, which is not a topology. But his techniques translate readily to the upper set topology, with his number n corresponding to our open set $\{n+1, n+2, n+3, \ldots\}$.

GLIVENKO'S PRINCIPLE

22.19. *Glivenko's not-elimination.* $\vdash \overline{\overline{A}} \to A$.

---

[4] That terminology seems a bit slanted, when seen from this book's viewpoint. Wouldn't the extensions of relevant or fuzzy logic be just as deserving of the name "intermediate logic"? But historically, constructive logic was studied earlier and more extensively than other nonclassical logics, so it got hold of some of the best names.

Proof

(1)	$\overline{\overline{A}} \to (\overline{\overline{A}} \to A)$	explosion 22.12
(2)	$A \to (\overline{\overline{A}} \to A)$	positive paradox 16.3.a
(3)	$\overline{\overline{A}} \to A$	(2), (1), 15.2.e.$\delta\delta$

22.20. *Glivenko's principle.* Let $\mathcal{H}$ be some collection of formulas, and let F be a formula. Then the following conditions are equivalent:

(i) $\mathcal{H} \vdash F$ classically. That is, the formula F can be proved using just the hypotheses $\mathcal{H}$, constructive logic (22.1–22.13), and not-elimination.

(ii) $\mathcal{H} \vdash \overline{\overline{F}}$ classically.

(iii) $\mathcal{H} \vdash \overline{\overline{F}}$ constructively. That is, the formula $\overline{\overline{F}}$ can be proved using just the hypotheses $\mathcal{H}$ and the constructive inference system developed in 22.1–22.13.

Demonstration. The implication (iii) $\Rightarrow$ (ii) is obvious, since any constructive proof is also a classical proof — i.e., even if the nonconstructive axiom $\overline{\overline{F}} \to F$ is made available to us, we are not actually *required* to use it.

The implication (ii) $\Rightarrow$ (i) is also fairly easy: Take any classical derivation of the formula $\overline{\overline{F}}$. Add to it two more steps: $\overline{\overline{F}} \to F$ (an instance of axiom scheme 22.17.d) and F (justified by detachment).

Thus, we just have to demonstrate (i) $\Rightarrow$ (iii). The proof is similar to the proofs of the deduction principle in 22.5; we shall sketch a few of the details and leave the remaining details as an exercise for the reader. It suffices to prove this statement for every positive integer n:

$P(n)$ $\begin{cases} \text{Let } F_n \text{ be a formula, let } \mathcal{H} \text{ be a set of formulas,} \\ \text{and suppose } \mathcal{H} \vdash F_n \text{ has a lemmaless classical} \\ \text{derivation of fewer than } n \text{ steps. Then there} \\ \text{exists a constructive derivation of } \mathcal{H} \vdash \overline{\overline{F_n}}. \end{cases}$

As in 22.5, $P(1)$ is vacuously true. Now assume $P(n)$ is true, for some integer $n \geq 1$; we wish to prove $P(n+1)$. Assume that some particular $\mathcal{H}$ and F_{n+1} are given, and some classical lemmaless derivation of $\mathcal{H} \vdash F_{n+1}$ is given that takes fewer than $n+1$ steps — i.e., it takes at most n steps. The last step of the derivation is the formula F_{n+1}. The justification for that last step is one of the following:

(a) F_{n+1} is a hypothesis — that is, $F_{n+1} \in \mathcal{H}$.

(b) F_{n+1} is an instance of one of the axioms in the constructive inference system developed in 22.1–22.13.

(c) F_{n+1} is an instance of the nonconstructive axiom $\overline{\overline{A}} \to A$.

(d) The last step, formula F_{n+1}, is obtained from two earlier steps by detachment.

In each of these four cases, we can derive $\mathcal{H} \vdash \overline{\overline{F_{n+1}}}$ in constructive logic. *Hints:* You may use 14.9, 14.13, and 22.19.

22.21. *Corollary.* Let $\mathcal{H}$ be any set of formulas, and let F be any formula. Then $\mathcal{H} \vdash F$ classically if and only if $\mathcal{H} \vdash \overline{\overline{F}}$ constructively.

22.22. *Remarks.* Glivenko's principle gives constructivists a complete and simple description of classical syntactics. Indeed, let us imagine a constructivist mathematician who is unwilling or unable to carry out classical proofs. The constructivist can nevertheless determine which formulas are considered theorems by his or her classicist colleagues — they are precisely those formulas F with the property that $\overline{\overline{F}}$ is a theorem in the eyes of the constructivist.

The converse fails — i.e., Glivenko's principle does not give classicists a description of constructive logic. In fact, Glivenko's principle tells us nothing about the constructive provability of formulas that do not begin with $\neg$.

Chapter 23

Relevant axioms: avoiding expansion

23.1. By *relevant implication*, we shall mean basic implication, plus contraction or self-distribution (Chapters 13 and 15).

By *relevant logic* (sometimes denoted by R), we shall mean

- basic logic and its consequences (Chapters 13 and 14), plus
- contraction or self-distribution (Chapter 15) and
- not-elimination (Chapter 19).

Relevant logic takes its name largely from the *Belnap relevance property*, presented in 23.9.b; for motivation the reader may wish to glance ahead to that result.

Many logics lie between relevant and classical logic. Probably the most interesting of those is RM (relevant plus mingle), which is also studied later in this chapter. We might describe that as "mildly irrelevant" — it adds to R some slightly irrelevant formulas such as expansion and mingle. But RM still excludes strongly irrelevant formulas such as positive paradox, which can be found in classical logic.

Some syntactic results

The following theorems and inference rules are proved using just the assumptions of relevant logic.

23.2. *Substituting with not-elimination.* We can apply the substitution 19.3 to the consequences of contraction. In particular, we obtain

  a. *Reductio.* $\vdash (\overline{A} \to A) \to A$.
  b. *Proof by dichotomy.* $\vdash (\overline{A} \to B) \to [(A \to B) \to B]$.
  c. *Excluded middle.* $\vdash A \vee \overline{A}$.

These are immediate from the "weak" versions in 15.2.d, 15.2.e, and 15.3.e.

23.3. *Mixing.* $\{B \to C,\ \overline{A} \to C\} \vdash (A \to B) \to C$.

Derivation

(1)	$B \to C$	hypothesis
(2)	$\overline{A} \to C$	hypothesis
(3)	$\overline{C} \to A$	(2), contrapositive 19.2.b.δ
(4)	$(A \to B) \to (A \to C)$	(1), $\to$-prefixing 13.2.d.δ
(5)	$(A \to C) \to (\overline{C} \to C)$	(3), suffixing 13.4.δ
(6)	$(\overline{C} \to C) \to C$	reductio 23.2.a
(7)	$(A \to B) \to C$	(4), (5), (6), rep. trans. 13.6.a

23.4. *Disjunctive consequence*
$\vdash (A \to B) \to (B \vee \overline{A})$, $\vdash (\overline{A} \to B) \to (B \vee A)$,
$\vdash (A \to \overline{B}) \to (\overline{B} \vee \overline{A})$, $\vdash (\overline{A} \to \overline{B}) \to (\overline{B} \vee A)$.

Proof. The first of these follows from 23.3 with $C = B \vee \overline{A}$. For the remaining three, replace A with $\overline{A}$, or B with $\overline{B}$, or both, and apply 19.3 as needed.

Remark. In classical logic, $A \to B$ and $B \vee \overline{A}$ are equivalent. In relevant logic, half of that equivalence is a theorem, as we have just shown. The other half is an admissible inference rule, but not a theorem; see 23.18.e and 27.17, and related results in 16.8.

Relevant deduction principle
(optional)

We will now prove a deduction principle for relevant logic, analogous to (but necessarily different from) the principle in 22.5. The following result will not be used later in this book.

23.5. The following assumptions are satisfied by relevant logic and RM (see 23.13), as well as some other logics not studied in this book.

Assumption. We assume that $\vdash$ is the syntactic consequence of a logical system satisfying these conditions:

    a. Our list of axioms or theorems includes at least the axioms of relevant logic.

    b. Our only *assumed* inference rules are Detachment (13.2.a) and Adjunction (14.1.a).

We emphasize that the assumptions above do not prohibit the addition of other axioms or other *derived* inference rules.

23.6. *Relevant deduction principle.* With the assumptions noted above, suppose that F and G are some formulas such that $F \vdash G$. Then $\overline{F} \vee G$ is a theorem; that is, $\vdash \overline{F} \vee G$.

More generally, suppose F and G are some formulas and $\mathcal{H}$ is some set of formulas, such that $\mathcal{H} \cup \{F\} \vdash G$. Then $\mathcal{H} \vdash \overline{F} \vee G$.

Demonstration. The proof is similar to the proof given in 22.5. We shall just sketch some of the proof, and leave the remaining details as an exercise for the reader.

It suffices to prove the following implication for each positive integer n:

$$P(n) \begin{cases} \text{If } F \text{ and } G_n \text{ are any formulas and } \mathcal{H} \text{ is} \\ \text{any set of formulas, and there exists a} \\ \text{lemmaless derivation of } \mathcal{H} \cup \{F\} \vdash G_n \\ \text{in fewer than } n \text{ steps, then } \mathcal{H} \vdash \overline{F} \vee G_n. \end{cases}$$

As in 22.5, we obtain $P(1)$ for free. Now assume $P(n)$ for some $n \geq 1$; we wish to prove $P(n+1)$. Assume that some particular $F, G_{n+1}, \mathcal{H}$ are given, and some lemmaless derivation of $\mathcal{H} \cup \{F\} \vdash G_{n+1}$ is given that takes n or fewer steps. The last step of that derivation is the formula G_{n+1}, which is justified in one of five ways:

    (a) G_{n+1} is an axiom (i.e., an assumed rule with no hypotheses) in our logical system,

    (b) G_{n+1} is one of the hypotheses in $\mathcal{H}$,

(c) G_{n+1} is the hypothesis F,

(d) G_{n+1} follows by Detachment from two of the earlier steps in the given lemmaless derivation, or

(e) G_{n+1} follows by Adjunction from two of the earlier steps in the given lemmaless derivation.

(There were only four cases in 22.5, but for relevant logic we must add adjunction as a fifth possible justification.) In each of those five cases, we must describe a derivation of $\mathcal{H} \vdash \overline{F} \vee G_{n+1}$. The methods are a bit different from those in 22.5, for a couple of reasons:

- We have changed the goal: We are now trying to prove $\mathcal{H} \vdash \overline{F} \vee G_{n+1}$ instead of $\mathcal{H} \vdash F \to G_{n+1}$.

- We have changed the available tools: positive paradox (16.3.a) is a theorem of constructive logic but not of relevant logic; excluded middle (23.2.c) is a theorem of relevant logic but not of constructive logic.

We leave it as an exercise for the reader to fill in the details in the five cases. *Hint.* There may be more than one proof possible. My proof used 14.20 and 15.5, among other things.

23.7. *Remarks.*

a. The converse of the relevant deduction principle would state that

$$\vdash \overline{F} \vee G \quad \Rightarrow \quad F \vdash G.$$

That is true for classical logic, but it fails for relevant logic — for instance, when $F = A \wedge (\overline{A} \vee B)$ and $G = B$. *Exercise:* Fill in the details. *Hint:* You may use the results that are stated without proof in 27.17.

b. The classical deduction principle, $F \vdash G \Rightarrow \vdash F \to G$, stated in 22.5, fails for relevant logic — for instance, with $G = F \vee \overline{F}$. *Exercise:* Fill in the details.

The remainder of this chapter is *not* part of syntactic relevant logic.

SOUNDNESS

23.8. Relevant logic (R) can be characterized by a semantic interpretation, but it is too complicated to present in this book. See Anderson and Belnap [1975] or Dunn [1986] for details.

Soundness   409

Although completeness of R is beyond the scope of this book, *soundness* (the easy half of completeness) is not. Indeed, we have already verified the assumptions of R in four different functional interpretations:

- Sugihara (8.38); verifications in 8.42.
- Church's diamond (9.14); verifications in 9.17 and 9.18.a.
- Crystal (9.7); verifications in 9.11.
- Church's chain (9.13, including verifications).

By the strong soundness principle 21.1.b, it follows that any theorem of R is a tautology of all four of those interpretations, and any valid inference rule of R is truth-preserving in those interpretations.

Likewise, since mingle is tautological in Sugihara's interpretation (see 8.39.b(i)), any theorem or inference rule of RM (relevant plus mingle) is also tautological in Sugihara's interpretation. Since {theorems of R} ⊆ {theorems of RM}, this leads us to information about R as well.

23.9. Here are some consequences of the preceding soundness observations.

  a. No Dugundji formula is a theorem of R or RM. (Immediate from 11.17.b.)

  b. If $A \to B$ is a theorem of relevant logic,[1] then the subformulas A and B must share at least one propositional variable symbol. (Immediate from soundness and 9.12.) This is sometimes known as the *Belnap relevance property*, since it was first proved by Belnap (by a more complicated proof using an eight-valued interpretation; see Belnap [1960]).

  c. If $A \to B$ is a theorem of RM that does not involve negation, then A and B share at least one propositional variable symbol. (Immediate from 8.43.)

  d. If X is a theorem of RM that only involves implication (i.e.,

---

[1] Or more generally, a tautology of the crystal interpretation.

that does not involve $\vee, \wedge, \neg$), then every propositional variable symbol that appears in X must appear at least twice. (Immediate from 8.44.)

e. None of the formulas listed in 8.39.c, 8.39.c(iii), 8.45, or 11.17.b are theorems of R or RM.

f. Neither R nor RM possesses either of Church's constants, described in 5.34. (Immediate from 8.32.)

23.10. The four interpretations listed above lie between relevant logic and the two-valued interpretation. We saw in 9.13 that {crystal tautologies} ⊆ {Church's chain tautologies}.

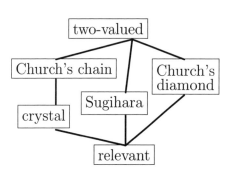

We now claim that there are no other inclusions among those six logics. Thus, they are related as in the diagram shown here.

Proof. It suffices to exhibit formulas that are tautological in one interpretation but not another. Those formulas are given in the table below. For in-

tautology?		↓ logics ↓					
↓ formulas ↓		relev.	crystal	C.chain	C.dia	Sugi.	2-val.
mingle $A \to (A \to A)$		No	No	No	No	✓	✓
chain $(A \to B) \vee (B \to A)$		No	No	✓	No	✓	✓
Brady $A \vee (A \to B)$		No	✓	✓	No	No	✓
conjunc. explos. $(A \wedge \overline{A}) \to B$		No	No	No	✓	No	✓

stance, the table indicates that the chain formula is tautological in Church's chain but not in Church's diamond. Those facts are both easy to verify. It follows that the set of tautologies of Church's chain is not a subset of the set of tautologies of Church's diamond. In other words, our diagram should not have a line segment running upward from Church's chain to Church's diamond (and indeed it does not).

23.11. *Some completeness results.*

a. No *finite* functional interpretation can characterize RM or relevant logic.

Proof. If either of those logics could be characterized by a finite functional interpretation, then by 11.14 that interpretation would make tautologous at least one Dugundji formula. But by 23.9.a, no Dugundji formula is a theorem of either of those logics.

Optional. The next two results are stated here without proof. The results are too interesting not to mention, but the proofs are too difficult for this book. The results will not be used for proofs elsewhere in this book, though they may be used for motivation and discovery.

b. The crystal interpretation (9.7) corresponds exactly to relevant logic plus these two axiom schemes:

$$\vdash A \vee (A \to B), \qquad \vdash (\overline{A} \wedge B) \to [(\overline{A} \to A) \vee (A \to B)].$$

This result, due to Brady, appears in Theorem 9.8.3 of Brady [2003].

c. The Sugihara interpretation (8.38) corresponds exactly to RM logic (i.e., relevant plus mingle). This completeness result is proved in Anderson and Belnap [1975]. Some further properties of RM are developed in the next few pages (but will not rely on the completeness result).

MINGLE: SLIGHTLY IRRELEVANT

23.12. "Mingle" is the formula scheme $A \to (A \to A)$. Though it is a theorem of classical logic, it is not provable in relevant logic; that fact can be demonstrated easily using the soundness of Church's diamond or the crystal logic or Church's chain.

Moreover, when we add mingle to relevant logic, we lose the

Belnap relevance property 23.9.b. For instance, these formulas become theorems:

"unrelated extremes" $(A \wedge \overline{A}) \to (B \vee \overline{B})$,
and "unrelated identities" $\overline{A \to A} \to (B \to B)$.

These formulas are contrary to the philosophy of relevantists, since they assert that one formula (involving only A) implies an entirely unrelated formula (involving only B).

Thus mingle is an irrelevant formula. But it is only mildly irrelevant: It does not violate the symbol-sharing criterion directly in its own right, and it does not produce quite as many non-relevant results as the "positive paradox" axiom explored later in this chapter.

Aside from its indirect production of unrelated implications, relevantists may also find mingle philosophically distasteful for more direct reasons. The formula $A \to A$ is not a statement about A, but rather a statement about implication, i.e., about the relationship between A and A. For instance, "if today is Tuesday then today is Tuesday" says something about the relationship between the day and the day, without telling us what the day actually is. On the other hand, "today is Tuesday" tells what the day is, without addressing the issue of how the day relates to the day. Thus, the two statements A and $A \to A$ are really about different topics; they are not directly relevant to one another. This may seem like a rather mild irrelevance, but perhaps it explains why mingle leads to more blatant irrelevancies such as unrelated extremes. Once the camel gets its nose inside the tent, we find it difficult to keep the rest of the camel from coming in.

23.13. *Equivalents of mingle.* We will now list some formulas and inference rules that are irrelevant *to the same extent as* mingle — i.e., adding any one of these to relevant logic makes all the others provable. The resulting logic is then called RM, or *relevant logic plus mingle*. (The formulas in 16.2 could also be added to this list.)

a. *Negated mingle.* $\vdash \overline{A} \to (\overline{A} \to \overline{A})$.

b. *Prenegated mingle.* $\vdash \overline{A} \to (A \to A)$.
c. *Permuted mingle.* $\vdash A \to (\overline{A} \to A)$.
d. *Implicative conjunction.* $\vdash (A \land B) \to (\overline{A} \to B)$.
e. *Unrelated identities.* $\vdash \overline{A \to A} \to (B \to B)$.
f. *Reversed implications.* $\vdash \overline{A \to B} \to (B \to A)$.
g. *Negated identities.* $\vdash \overline{A \to A} \to (A \to A)$.

Following are proofs of equivalence.

Mingle and negated mingle are equivalent. Substitute $\overline{A}$ for A.

Negated and prenegated are equivalent. Contrapositive laws.

Prenegated and permuted are equivalent. Permutation.

Permuted mingle yields implicative conjunction.

(1)	$(A \land B) \to A$	axiom 14.1.b
(2)	$(A \land B) \to B$	axiom 14.1.c
(3)	$(A \land B) \to [\overline{A \land B} \to (A \land B)]$	perm. mingle 23.13.c
(4)	$[\overline{A \land B} \to (A \land B)]$ $\to [\overline{A \land B} \to A]$	(1), $\to$-pref. 13.2.d.δ
(5)	$[\overline{A \land B} \to A] \to [\overline{A} \to (A \land B)]$	3rd contrapos 19.2.b
(6)	$[\overline{A} \to (A \land B)] \to (\overline{A} \to B)$	(2), $\to$-pref. 13.2.d.δ
(7)	$(A \land B) \to (\overline{A} \to B)$	(3)(4)(5)(6) tr 13.6.a

Implicative conjunction yields unrelated identities. We have $A \to A$ and $B \to B$ as theorems, and

$$[(A \to A) \land (B \to B)] \to [\overline{A \to A} \to (B \to B)]$$

as an instance of implicative conjunction.

Unrelated identities yields reversed implications. By substitution, switch the roles of A and B; thus we obtain $\overline{B \to B} \to (A \to A)$. By permutation, $A \to \left(\overline{B \to B} \to A\right)$. By contraposition, $A \to [\overline{A} \to (B \to B)]$. Several more permutations yields $B \to [\overline{A} \to (A \to B)]$ (the details of those permutations are left as an exercise). Then contrapositive again: $B \to (\overline{A \to B} \to A)$. Finally, one more permutation yields reversed implications.

Reversed implications yields negated identities. Let $A = B$.

Negated identities yields prenegated mingle. By permutation we have $A \to (\overline{A \to A} \to A)$. Then $A \to [\overline{A} \to (A \to A)]$ by contraposition. By another permutation, $\overline{A} \to [A \to (A \to A)]$. On the other hand, we have $[A \to (A \to A)] \to (A \to A)$ by contraction. Combine those last two results using transitivity; thus we obtain prenegated mingle.

23.14. *Weaker corollaries.* The following formulas are also theorems of RM. However, they are not equivalent to mingle, as an extension of relevant logic. Chain order is strictly weaker than mingle, and unrelated extremes is strictly weaker than chain order.
  a. *Chain order.* $\vdash (Q \to P) \lor (P \to Q)$.
  b. *Unrelated extremes.* $\vdash (A \land \overline{A}) \to (B \lor \overline{B})$.

Proofs. To prove chain order, apply 23.4.δ to 23.13.f. To prove unrelated extremes using chain order, follow the argument given in 19.9. To show that relevant plus chain does not imply mingle, use soundness and Church's chain 9.13. To show that relevant plus unrelated extremes does not imply chain, use soundness and Church's diamond 9.18.

23.15. *Additional mingles (exercise).* The following formulas and inference rules are also equivalent to mingle. For how many of them can you prove equivalence?
  a. $A \to B \vdash A \to (A \to B)$.
  b. $\vdash A \to (A \to (A \to A))$.
  c. $\{\overline{A}, B\} \vdash A \to B$.
  d. $\overline{A \to B} \vdash B \to A$.
  e. $A \vdash \overline{A} \to A$.
  f. $\vdash \overline{A} \to [A \to (A \to A)]$.

> The remainder of this chapter is *not* part of RM.

Positive paradox and classical logic

23.16. "Positive paradox" is the formula scheme $A \to (B \to A)$. Though it is a theorem of classical logic, it is not provable in relevant logic, nor even in the logic RM; that follows from the soundness of the Sugihara interpretation. Thus it is an "irrelevant" formula, and strongly so.

23.17. *Equivalents of positive paradox.* We will now list some formulas and inference rules that are irrelevant *to the same extent as* positive paradox — i.e., adding any one of these to relevant logic makes all the others provable. (The formulas in 16.3 could also be added to this list.) The resulting logic is actually *classical logic*, discussed in Chapter 25.

a. *Modified paradox.* $\vdash A \to (\overline{B} \to A)$.
b. *Implicative explosion.* $\vdash A \to (\overline{A} \to B)$.
c. *Permuted explosion.* $\vdash \overline{A} \to (A \to B)$.
d. *Peirce's law.* $\vdash ((A \to B) \to A) \to A$.

Sketch of proof of equivalence. In the diagram below, the implications indicated by the labeled, doubleheaded arrows are fairly

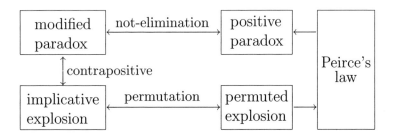

obvious. The two unlabeled singleheaded arrows are given by the derivations below.

Permuted explosion yields Peirce.

(1)	$\overline{A} \to (A \to B)$	perm. explosion
(2)	$[(A \to B) \to A] \to (\overline{A} \to A)$	(1), $\to$-suffixing 13.4.δ
(3)	$(\overline{A} \to A) \to A$	reductio 23.2.a
(4)	$[(A \to B) \to A] \to A$	(2), (3), 13.5.b (trans)

Peirce yields positive paradox. (Proof from Meyer [1990].)

(1)	$\{[A \to (B \to A)] \to A\} \to A$	Peirce
(2)	$[A \to (B \to A)] \to [(A \circ B) \to A]$	19.5
(3)	$(\{[A \to (B \to A)] \to A\} \to A)$ $\to (\{[(A \circ B) \to A] \to A\} \to A)$	(2), $\to$-suffixing 13.4.δ
(4)	$\{[(A \circ B) \to A] \to A\} \to A$	(1), (3), detach.
(5)	$(A \circ B) \to [((A \circ B) \to A) \to A]$	assertion
(6)	$(A \circ B) \to A$	trans., (5), (4)
(7)	$A \to (B \to A)$	(6), 19.5.δ

23.18. Exercise: More equivalents. The following formulas and inference rules can be added to the preceding list — i.e., they are equivalent to positive paradox, as extensions of relevant logic. Prove the equivalence of as many as you can.

a. $\overline{A} \vdash A \to B$.
b. $\vdash (A \to B) \to (A \to A)$.
c. $\vdash (A \to B) \to (B \to B)$.
d. $\overline{A} \lor B \vdash A \to B$.
e. $\vdash (\overline{A} \lor B) \to (A \to B)$.
f. $\vdash (A \to B) \to [A \to (A \land B)]$.
g. $\vdash (A \to B) \to [(A \lor B) \to B]$.
h. $\vdash (A \to B) \to [(C \land A) \to (C \land B)]$.
i. $\vdash (A \to B) \to [(C \lor A) \to (C \lor B)]$.

Chapter 24

Fuzzy axioms: avoiding contraction

AXIOMS

24.1. By *Rose-Rosser logic* we shall mean

- basic logic (investigated in Chapters 13 and 14), plus
- positive paradox (Chapter 16),
- not-elimination (Chapter 19), and
- *implicative disjunction*, $((A \to B) \to B) \to (A \vee B)$, which was mentioned in 8.39.c(iv).

By *Wajsberg logic* we shall mean those assumptions plus one more axiom scheme,

- *specialized contraction*, $[A \to (A \to \overline{A})] \to (A \to \overline{A})$, which was introduced in 15.3.a.

(The axiom systems presented here are not actually identical to those developed by Rose and Rosser or by Wajsberg, but they are in some sense equivalent, and close enough to justify the names.)

24.2. *Soundness and preview of completeness.* We have already shown in 8.22–8.23 that the assumptions of Rose-Rosser logic and of Wajsberg logic are satisfied, respectively, by the $[0,1]$-valued Zadeh logic (8.16–8.26) and the $\{0, \frac{1}{2}, 1\}$-valued Łukasiewicz logic

(8.17). By the soundness principles 21.1, therefore, we now know

{Rose-Rosser theorems} ⊆ {Zadeh tautologies} and
{Wajsberg theorems} ⊆ {Łukasiewicz tautologies}.

Both of those inclusions (⊆) are actually equalities (=). The completeness proof for Rose-Rosser and Zadeh is too difficult for this book, and is omitted; it can be found in Rose and Rosser [1958]. The completeness proof for Wajsberg and Łukasiewicz will be given in 29.20 in this book, using results developed in the present chapter. Among those results, some highlights are chain ordering (24.9) and a deduction principle (24.26).

24.3. *Absence of contraction.* In 8.21.e we saw that contraction is not tautologous in the fuzzy semantics. Therefore (by soundness) it is not a theorem of Rose-Rosser or Wajsberg logic. Hence we cannot expect any of the results of Chapter 15 to hold in Rose-Rosser or Wajsberg logic.

However, we can certainly watch for possible variants of those results. For instance, excluded middle is not valid in fuzzy logics (see 8.21.c), nor is the variant given in 15.3.e. But a three-valued variant is valid in Wajsberg logic; it is given in 24.16.

24.4. *Quick consequences of familiar axioms.* Rose-Rosser logic includes among its assumptions both positive paradox and not-elimination. We have already studied those axioms separately (Chapters 16 and 19). Combining those axioms has a few immediate consequences that we note here:
a. $\vdash A \to (\overline{A} \to B)$.
b. $\vdash \overline{A} \to (A \to B)$.
c. $\vdash (A \lor B) \to (\overline{A} \to B)$.
d. $\vdash (\overline{A} \lor B) \to (A \to B)$.
e. $\vdash (A \circ B) \to (A \land B)$. (*Hint:* Use 16.8.c.)

24.5. *Quick consequences of our new axiom.* In addition to a new combination of old axioms, Rose-Rosser logic also has one axiom that we have not studied before,

$$\vdash [(A \to B) \to B] \to (A \lor B),$$

called "*implicative disjunction*" in this book. Here are some immediate consequences:
  a. ⊢ $[(A \to B) \to B] \leftrightarrow (A \lor B)$.
  b. ⊢ $[(A \to B) \to B] \leftrightarrow [(B \to A) \to A]$.

Hereafter, the expressions on the two sides of the "↔" can be substituted for each other, in the sense of 14.14.

Proofs. The converse of implicative disjunction was proved in 16.8.a; combining that with implicative disjunction yields 24.5.a. Commutativity of ∨ was proved in 14.6.d; combining that with 24.5.a yields 24.5.b.

24.6. *Preview: Expressions of three metavariables.* Since implication (→) is neither commutative nor associative, the six expressions listed below are all different — but they are similar enough in appearance to be easily confused with one another. Most or all of them play substantial roles in this chapter. To reduce confusion, we list them together here for comparison and convenient reference.

$A \to (B \to A)$ is a theorem, by 16.3.a.
$B \to (A \to A)$ is a theorem, by 16.3.b.
$A \to (A \to B)$ is abbreviated "$A \rightsquigarrow B$," starting in 24.10.
$(A \to A) \to B$ is equivalent to B by 13.8.b and 16.4.
$(B \to A) \to A$ is equivalent to $A \lor B$ by 24.5.a.
$(A \to B) \to A$ unfortunately does not simplify.

Meredith's chain proof

24.7. *A few technical lemmas.*
  a. ⊢ $[(P \to Q) \to (R \to Q)] \to [R \to ((P \to Q) \to Q)]$.
  b. ⊢ $[R \to ((Q \to P) \to P)] \to [(Q \to P) \to (R \to P)]$.
  c. ⊢ $[(P \to Q) \to (R \to Q)] \to [R \to ((Q \to P) \to P)]$.

d. ⊢ $[(P \to Q) \to (R \to Q)] \to [(Q \to P) \to (R \to P)]$.
e. ⊢ $[(\overline{Y} \to \overline{X}) \to (\overline{Z} \to \overline{X})] \to [(\overline{X} \to \overline{Y}) \to (\overline{Z} \to \overline{Y})]$.
f. ⊢ $[(X \to Y) \to (X \to Z)] \to [(Y \to X) \to (Y \to Z)]$.

Proofs. Formulas a and b are instances of permutation (13.2.c). Apply 24.5.b at the right end of formula a to obtain formula c. Combine b and c via transitivity to get formula d. In formula d, make the substitutions $P = \overline{Y}, Q = \overline{X}$, and $R = \overline{Z}$ to obtain e. Then contrapositives yield f.

24.8. ⊢ $\Big((P \to Q) \to (Q \to P)\Big) \to (Q \to P)$.

Proof. Begin with a specialization of 13.7, with the substitutions
$A = Q \to P$, $B = P \to Q$, $C = P$, $D = [Q \to (P \to Q)] \to (Q \to P)$.
This yields

$$\Big\langle \{[(Q \to P) \to P] \to [(P \to Q) \to P]\} $$
$$\to \{[Q \to (P \to Q)] \to (Q \to P)\} \Big\rangle \to \Big\langle [(P \to Q) $$
$$\to (Q \to P)] \to \{[Q \to (P \to Q)] \to (Q \to P)\} \Big\rangle.$$

Next, at the left end of that formula, replace $[(Q \to P) \to P]$ with $[(P \to Q) \to Q]$, using 24.5.b. This yields

$$\Big\langle \{[(P \to Q) \to Q] \to [(P \to Q) \to P]\} $$
$$\to \{[Q \to (P \to Q)] \to (Q \to P)\} \Big\rangle \to \Big\langle [(P \to Q) $$
$$\to (Q \to P)] \to \{[Q \to (P \to Q)] \to (Q \to P)\} \Big\rangle.$$

Now, substituting $X = P \to Q$, $Y = Q$, $Z = P$ into 24.7.f yields the left half of the formula above. Hence, by detachment, we have the right half of the formula above. That is,

$$[(P \to Q) \to (Q \to P)] \to \Big\{[Q \to (P \to Q)] \to (Q \to P)\Big\}.$$

Then the detachmental corollary of permutation yields

$$[Q \to (P \to Q)] \to \Big\{[(P \to Q) \to (Q \to P)] \to (Q \to P)\Big\}.$$

Additional notations   421

The left side of that is positive paradox. So we can obtain the right side via detachment; but that is just 24.8.

24.9. *Chain order.* $\vdash (P \to Q) \lor (Q \to P)$.

Proof. Immediate from 24.5.a and 24.8.

ADDITIONAL NOTATIONS

24.10. We now introduce a couple of abbreviations that will be useful in the remainder of this chapter:

$\widetilde{A}$ will stand for $(A \to \overline{A}) \land (\overline{A} \to A)$;
$A \rightsquigarrow B$ will stand for $A \to (A \to B)$, or $(A \circ A) \to B$.

Remarks. Intuitively, it may help to think of $\widetilde{A}$ as "A is indeterminate," something other than true or false. For motivation, the reader might find it helpful to glance ahead to 24.11.b and 29.19. Cotenability ($\circ$) was introduced in 5.32, 11.3, 18.1, and 19.5.

24.11. *Some immediate consequences.* Applying the new notation to some past results, we obtain the following
  a. $\vdash \widetilde{\neg A} \leftrightarrow \widetilde{A}$.
  b. $\vdash (A \to B) \to (A \rightsquigarrow B)$. (Restatement of 16.2.b.)
  c. $\vdash (A \land B) \rightsquigarrow (A \circ B)$. (Restatement of 19.7.)
  d. $\vdash (A \to \overline{A}) \to (A \rightsquigarrow B)$. (*Proof.* $\to$-prefixing and 24.4.b.)
  e. ($\rightsquigarrow$-prefixing) $(B \to C) \to [(A \rightsquigarrow B) \to (A \rightsquigarrow C)]$.
    Hint. Write the formula without the $\rightsquigarrow$ abbreviation; you'll see that this is a special case of 13.22. Alternatively, prove this using $\to$-prefixing twice.

24.12. *Fuzzy detachment.* If $\mathcal{H} \vdash A \rightsquigarrow B$ then $\mathcal{H} \cup \{A\} \vdash B$. (Compare with 24.26.)

Proof. Assume $\mathcal{H} \vdash A \rightsquigarrow B$ as a given theorem; then we have this derivation of $\mathcal{H} \cup \{A\} \vdash B$:

(1)	A	hypothesis
(2)	$A \rightarrow (A \rightarrow B)$	given theorem
(3)	$A \rightarrow B$	(1), (2), detachment
(4)	B	(1), (3), detachment

24.13. *Fuzzy proof by cases.* (Compare 14.4.b.δ.)

$$\{S \rightsquigarrow Z,\ T \rightsquigarrow Z,\ S \vee T\} \ \vdash \ Z$$

Proof.

(1)	$S \rightarrow (S \rightarrow Z)$	hypothesis
(2)	$T \rightarrow (T \rightarrow Z)$	hypothesis
(3)	$T \rightarrow (S \rightarrow (S \rightarrow Z))$	(1), 16.3.a.δ (pos. pdx.-det.)
(4)	$S \rightarrow (T \rightarrow (S \rightarrow Z))$	(3), 13.2.c.δ (permut-det)
(5)	$S \rightarrow (S \rightarrow (T \rightarrow Z))$	(3), 13.23.δ (gen.permut-det)
(6)	$S \rightarrow (T \rightarrow (T \rightarrow Z))$	(2), 16.3.a.δ (pos. pdx.-det.)
(7)	$T \rightarrow (S \rightarrow (T \rightarrow Z))$	(6), 13.2.c.δ (permut-det)
(8)	$T \rightarrow (T \rightarrow (S \rightarrow Z))$	(6), 13.23.δ (gen.permut-det)
(9)	$S \vee T$	hypothesis
(10)	$S \rightarrow (T \rightarrow Z)$	(5),(7),(9), 14.4.b.δ (cases)
(11)	$T \rightarrow (S \rightarrow Z)$	(4),(8),(9), 14.4.b.δ (cases)
(12)	$S \rightarrow Z$	(1),(11),(9), 14.4.b.δ (cases)
(13)	$T \rightarrow Z$	(2),(10),(9), 14.4.b.δ (cases)
(14)	Z	(12),(13),(9), 14.4.b.δ (cases)

The remainder of this chapter is *not* part of Rose-Rosser logic.

Wajsberg logic

24.14. We now add one more axiom scheme; the resulting logic will be called *Wajsberg logic* in this book. The additional axiom

Wajsberg logic

can be any of the following, as they are all equivalent — i.e., adding any one of these formulas to Rose-Rosser logic makes the others provable. The equivalences follow easily from 14.6.d, 24.5.a, 24.10, and 24.11.b.

a. $\vdash (A \leadsto \overline{A}) \leftrightarrow (A \to \overline{A})$.
b. $\vdash (A \leadsto \overline{A}) \to (A \to \overline{A})$.
c. $\vdash [A \to (A \to \overline{A})] \to (A \to \overline{A})$ (same as 15.3.a)
d. $\vdash A \vee (A \to \overline{A})$.
e. $\vdash (A \to \overline{A}) \vee A$.
f. $\vdash [(A \to \overline{A}) \to A] \to A$ (specialized Peirce).

24.15. *Specialized conjunctive detachment.* (Compare with 15.5.) This formula is also equivalent to the ones listed in 24.14, but the proof of equivalence is not quite so obvious.

$$\vdash \left[A \wedge (A \to \overline{A})\right] \to \overline{A}.$$

Proof of equivalence. We shall show that the formula above is equivalent to 24.14.c. Each of the following steps is reversible.

$(A \to (A \to \overline{A})) \to (A \to \overline{A})$		
$\Updownarrow$	13.2.c.δ	permutation-det
$A \to [(A \to (A \to \overline{A})) \to \overline{A})]$		
$\Updownarrow$	19.5	fusion
$A \to [((A \circ A) \to \overline{A}) \to \overline{A})]$		
$\Updownarrow$	24.5.a	fuzzy disjunction
$A \to [\overline{A} \vee (A \circ A)]$		
$\Updownarrow$	19.5	cotenability def'n.
$A \to [\overline{A} \vee \overline{A \to \overline{A}}]$		
$\Updownarrow$	19.3	De Morgan's laws
$A \to \overline{A \wedge (A \to \overline{A})}$		
$\Updownarrow$	14.1.i	1st contrapositive
$[A \wedge (A \to \overline{A})] \to \overline{A}$		

24.16. *Trivalence.* $\vdash A \vee \overline{A} \vee \widetilde{A}$. (Compare 23.2.c.)

Proof. To shorten the proof, note that $A \vee \overline{A}$ and $\overline{A} \vee A$ can be used interchangeably (by 14.18.g); let us abbreviate them both by u. Similarly, $A \wedge \overline{A}$ and $\overline{A} \wedge A$ can be used interchangeably; let us abbreviate them both by n. Note that $n = \overline{u}$. Then

(1)	$(u \to n) \to (A \to \overline{A})$	14.8.g(iii)
(2)	$(u \to n) \to (\overline{A} \to A)$	14.8.g(iii)
(3)	$(u \to n) \to \widetilde{A}$	(1), (2), 14.4.a
(4)	$(u \to n) \vee u$	24.14.e
(5)	$\widetilde{A} \vee u$	(3), (4), 14.7.d.δ

24.17. *Inference by trichotomy.*

$$\{A \to X,\ \overline{A} \to X,\ \widetilde{A} \to X\} \vdash X.$$

Proof. Trivalence and 14.19.b.

24.18. *Lemma.* $\vdash (A \circ A) \vee (A \to \overline{A})$.

Proof. Throughout the following derivation, let S be an abbreviation for the formula $(A \circ A) \vee (A \to \overline{A})$.

(1)	$(A \circ A) \to S$	left $\vee$-introduction 14.1.d
(2)	$(A \to \overline{A}) \to S$	right $\vee$-introduction 14.1.e
(3)	$A \to (A \to S)$	(1), fusion 19.5.δ
(4)	$\widetilde{A} \to (A \to \overline{A})$	left $\wedge$-elim. 14.1.b (and def. of $\widetilde{A}$)
(5)	$\overline{A} \to (A \to \overline{A})$	positive paradox 16.3.a
(6)	$\widetilde{A} \to S$	(4), (2), transitive 13.5.b
(7)	$\overline{A} \to S$	(5), (2), transitive 13.5.b
(8)	$\widetilde{A} \to (A \to S)$	(6), meddling 16.3.e
(9)	$\overline{A} \to (A \to S)$	(7), meddling 16.3.e
(10)	$A \to S$	(8), (9), (3), trichotomy 24.17
(11)	S	(6), (7), (10), trichotomy 24.17

24.19. *Fuzzy noncontradiction.* We have $\vdash X \to \overline{X}$ when X is any of the formulas $P \wedge \overline{P}$, $P \wedge \widetilde{P}$, $\overline{P} \wedge \widetilde{P}$.

Proof for $X = P \wedge \overline{P}$. By De Morgan's laws, $\overline{P \wedge \overline{P}}$ is interchangeable with $\overline{P} \vee P$; now apply 14.8.g(i).

Proof for $X = P \wedge \widetilde{P}$. Apply 14.8.g(ii).$\delta$ to 24.15. This yields

$$\left[P \wedge (P \to \overline{P}) \wedge (\overline{P} \to P) \right] \to \left[\overline{P} \vee \overline{P \to \overline{P}} \vee \overline{\overline{P} \to P} \right]$$

which, after unwinding the notation, is just $X \to \overline{X}$.

Proof for $X = \overline{P} \wedge \widetilde{P}$. Replace P with $\overline{P}$ in the result of the previous paragraph; we obtain $\vdash (\overline{P} \wedge \widetilde{P}) \to \overline{\overline{P} \wedge \neg \widetilde{P}}$. Now just use 24.11.a.

24.20. *Fuzzy explosion principles.*
  a. (Theorem versions.) By 24.11.d and 24.19,

$$\vdash (P \wedge \overline{P}) \rightsquigarrow Z, \qquad \vdash (P \wedge \widetilde{P}) \rightsquigarrow Z, \qquad \vdash (\overline{P} \wedge \widetilde{P}) \rightsquigarrow Z.$$

  b. (Inference rule versions.) Then by 24.12,

$$\{P, \overline{P}\} \vdash Z, \qquad \{P, \widetilde{P}\} \vdash Z, \qquad \{\overline{P}, \widetilde{P}\} \vdash Z.$$

24.21. *Tedious verifications.* The results below will be needed for 29.18 and 29.19, but some readers may prefer to skip this section in their first reading. For motivation, the reader might look at 8.17. *Exercises.* Fill in the details in the proofs sketched below.

  a. $A \vdash (A \wedge B) \leftrightarrow B$ and $\overline{B} \vdash (A \vee B) \leftrightarrow A$.
  Hints. The $(A \wedge B) \to B$ and $A \to (A \vee B)$ parts are easy. It remains to show $A \vdash B \to (A \wedge B)$ and $\overline{B} \vdash (A \vee B) \to A$. The former follows by strong adjunction 16.3.c; the latter follows by strong adjunction together with contrapositive and De Morgan laws.

  b. $A \vdash (A \to B) \leftrightarrow B$.
  Hints. The result $A \vdash (A \to B) \to B$ follows from $\vdash A \to [(A \to B) \to B]$, which is just assertion (13.8.a). The other part, $A \vdash B \to (A \to B), \vdash B \to (A \to B)$, is immediate from 24.4.d.

c. $\overline{B} \vdash (A \to B) \leftrightarrow \overline{A}$.

Hints. Replace A and B in the previous result with $\overline{B}$ and $\overline{A}$, respectively; then use contrapositive to replace $\overline{B} \to \overline{A}$ with $A \to B$.

d. $B \vee \overline{A} \vdash A \to B$. This is immediate from 24.4.d.

e. $\{\widetilde{A}, \widetilde{B}\} \vdash A \to B$.

Hints. From $\widetilde{A}$ and $\widetilde{B}$ we obtain $(B \to A) \leftrightarrow (\overline{B} \to \overline{A})$. By the contrapositive law we have $(\overline{B} \to \overline{A}) \leftrightarrow (A \to B)$. Combine using transitivity; thus $(B \to A) \leftrightarrow (A \to B)$. On the other hand, by chain ordering (24.9) we have $\vdash (A \to B) \vee (B \to A)$.

f. $\{\widetilde{A}, \widetilde{B}\} \vdash \widetilde{A \wedge B}$, $\{\widetilde{A}, \widetilde{B}\} \vdash \widetilde{A \vee B}$, $\{\widetilde{A}, \overline{B}\} \vdash \widetilde{A \vee B}$, $\{\widetilde{A}, \overline{B}\} \vdash \widetilde{A \to B}$.

Hints for 24.21.f. These four results have very similar proofs, which we outline together. In each case we want to show a result of the form $\{\widetilde{A}, X\} \vdash \widetilde{Y}$ (where X is either $\widetilde{B}$ or $\overline{B}$, and Y is either $A \wedge B$ or $A \vee B$ or $A \to B$). We shall use 24.17. Thus, it suffices to show both

$$\{\widetilde{A}, X, Y\} \vdash \widetilde{Y} \qquad \text{and} \qquad \{\widetilde{A}, X, \overline{Y}\} \vdash \widetilde{Y}$$

— eight proofs in all, but not difficult ones. For the proofs involving $\overline{Y}$, we first rewrite $\overline{Y}$ in one of these ways.

- $\overline{A \vee B}$ is interchangeable with $\overline{A} \wedge \overline{B}$, by 14.18.c;
- $\overline{A \wedge B}$ is interchangeable with $\overline{A} \vee \overline{B}$, by 19.3; and
- $\overline{A \to B}$, though not actually equivalent to anything more tractable, can be replaced in one direction: It is stronger than $B \vee \overline{A}$, by 24.4.d. The latter is interchangeable with $\overline{B} \wedge A$.

Then the eight results that we need to prove are each of the form

$$\{\widetilde{A}, X, S \vee T\} \vdash \widetilde{Y} \qquad \text{or} \qquad \{\widetilde{A}, X, S \wedge T\} \vdash \widetilde{Y}.$$

In the cases of $S \wedge T$, at least one of S, T and one of $\widetilde{A}, X$ can be combined to form an explosive pair, as in 24.20.b. In the cases of $S \vee T$, each of S, T forms an explosive pair with one of $\widetilde{A}, X$, and we can use proof by cases 14.4.b to deal with the $\vee$. (The details are left as an exercise!)

Deduction principle for Wajsberg logic

24.22. *Technical lemma*

$$\vdash (A \to \overline{A}) \to \Big\{ [A \rightsquigarrow (B \to C)] \to [(A \rightsquigarrow B) \to (A \rightsquigarrow C)] \Big\}.$$

Deduction principle for Wajsberg logic

Proof. For any formula X, we have $\overline{A} \to (A \to X)$ by 24.4.b, hence $(A \to \overline{A}) \to [A \to (A \to X)]$ by $\to$-prefixing. For any formulas Y and Z, substituting $X = Y \to (Z \to C)$ yields

$$(A \to \overline{A}) \to (A \to (A \to (Y \to (Z \to C)))).$$

By generalized permutation 13.23, then,

$$(A \to \overline{A}) \to (Y \to (Z \to (A \to (A \to C)))).$$

Now substitute $Y = A \rightsquigarrow (B \to C)$ and $Z = A \rightsquigarrow B$; this yields 24.22.

24.23. *Second technical lemma*

$$\vdash (A \circ A) \to \Big\{ [A \rightsquigarrow (B \to C)] \to [(A \rightsquigarrow B) \to (A \rightsquigarrow C)] \Big\}.$$

Proof. Start from $\rightsquigarrow$-prefixing, as presented in 24.11.e. Apply $\to$-prefixing twice, to obtain

$$\Big\{ A \to [A \to (B \to C)] \Big\} \to \Big\langle A \to \Big\{ A \to [(A \rightsquigarrow B) \to (A \rightsquigarrow C)] \Big\} \Big\rangle.$$

In the notation of 13.21, this can be restated as

$$\Big(A \to [A \to (B \to C)], \; A, \; A \Big) \to [(A \rightsquigarrow B) \to (A \rightsquigarrow C)].$$

By generalized permutation 13.23.δ,

$$\Big(A, \; A, \; A \to [A \to (B \to C)] \Big) \to [(A \rightsquigarrow B) \to (A \rightsquigarrow C)].$$

Now unwind the notation, and also apply the fusion property of cotenability 19.5. This yields 24.23.

24.24. *Distribution of $\rightsquigarrow$ over $\to$*

$$\vdash [A \rightsquigarrow (B \to C)] \to [(A \rightsquigarrow B) \to (A \rightsquigarrow C)].$$

Proof. Immediate from the preceding two lemmas, 24.18, and proof by cases 14.4.b.δ.

24.25. *Assumption.* For the deduction principle below, we assume that $\vdash$ is the syntactic consequence in a logical system satisfying these conditions:

a. Our list of axioms or theorems includes *at least* the axioms of Wajsberg logic, as listed in 24.1.

b. Our only *assumed* inference rule is detachment (δ).

In order to satisfy these conditions, we include strong adjunction as an axiom, rather than adjunction as an assumed inference rule; see 16.3.c. We emphasize that the two conditions above do not prohibit the assumption of *additional* axioms, nor the use of other *derived* inference rules.

24.26. *Wajsberg deduction principle.* With the assumptions noted above, suppose that F and G are some formulas such that $F \vdash G$. Then $F \leadsto G$ is a theorem; that is, $\vdash F \leadsto G$.

More generally, suppose F and G are some formulas and $\mathcal{H}$ is some set of formulas, such that $\mathcal{H} \cup \{F\} \vdash G$. Then $\mathcal{H} \vdash F \leadsto G$.

Remark. Note that this is a converse to 24.12. Thus we are actually showing that $\mathcal{H} \cup \{F\} \vdash G$ if and only if $\mathcal{H} \vdash F \leadsto G$.

Demonstration. The proof is similar to the proof given in 22.5. We shall just sketch some of the proof, and leave the remaining details as an exercise for the reader.

It suffices to prove the following implication for each positive integer n:

$P(n)$ $\begin{cases} \text{If } F \text{ and } G_n \text{ are any formulas and } \mathcal{H} \text{ is} \\ \text{any set of formulas, and there exists a} \\ \text{lemmaless derivation of } \mathcal{H} \cup \{F\} \vdash G_n \\ \text{in fewer than } n \text{ steps, then } \mathcal{H} \vdash F \leadsto G_n. \end{cases}$

As in 22.5, we obtain $P(1)$ for free. Now assume $P(n)$ for some $n \geq 1$; we wish to prove $P(n+1)$. Assume that some particular $F, G_{n+1}, \mathcal{H}$ are given, and some lemmaless derivation of $\mathcal{H} \cup \{F\} \vdash G_{n+1}$ is given that takes n or fewer steps. The last step of that derivation is the formula G_{n+1}, which is justified in one of four ways:

Deduction principle for Wajsberg logic

(a) G_{n+1} is an axiom (i.e., an assumed rule with no hypotheses) in our logical system,

(b) G_{n+1} is one of the hypotheses in $\mathcal{H}$,

(c) G_{n+1} is the hypothesis F, or

(d) G_{n+1} follows by Detachment from two of the earlier steps in the given lemmaless derivation.

In each of those four cases, we must describe a derivation of $\mathcal{H} \vdash F \rightsquigarrow G_{n+1}$. The methods are a bit different from those in 22.5, for a couple of reasons:

- We have changed the goal: We are now trying to prove $\mathcal{H} \vdash F \rightsquigarrow G_{n+1}$ instead of $\mathcal{H} \vdash (F \to G_{n+1})$.

- The tools available to us have changed: self-distribution is not a theorem of Wajsberg logic, but a variant of it (24.24) is.

We leave it as an exercise for the reader to fill in the details in the four cases.

Chapter 25

Classical logic

AXIOMS

25.1. Exercise. Show that the following definitions of *classical logic* are equivalent — i.e., if we assume any one of these sets of inference rules and formulas, then the other sets become provable.

a. Constructive logic (Chapter 22), plus any of the nonconstructive axioms in 22.17.

b. Relevant logic (Chapter 23), plus any of the nonrelevant axioms in 23.17–23.18.

c. Wajsberg logic (Chapter 24 including both 24.5, 24.14), plus either contraction or self-distribution (15.4).

d. This shorter list of assumptions (which is less easily compared with other logics):

inference rule $\quad \{A, A \to B\} \vdash B$;

axioms $\quad \begin{cases} \vdash A \to (B \to A), \\ \vdash [A \to (B \to C)] \to [(A \to B) \to (A \to C)], \\ \vdash (\overline{A} \to \overline{B}) \to (B \to A); \end{cases}$

definitions $\quad \begin{cases} A \lor B \text{ is an abbreviation for } \overline{A} \to B, \\ A \land B \text{ is an abbreviation for } \overline{A \to \overline{B}}. \end{cases}$

25.2. As we have noted above, classical logic includes all of constructive, relevant, and Wajsberg logic. Any theorems of any of those logics are also theorems of classical logic.

However, classical logic does not inherit all *properties* of those smaller logics. For instance, it enjoys neither the Disjunction Property of constructive logic (27.11) nor the Belnap Property of relevant logic (23.9.b).

SOUNDNESS RESULTS

25.3. It is easy to verify that any of the axiom systems listed above satisfies the two-valued interpretation 8.2, hence also the powerset interpretations and hexagon interpretation (see 11.11 and 11.12).

Completeness will be investigated in 29.7–29.15.

INDEPENDENCE OF AXIOMS

25.4. As we have indicated above, classical logic can be axiomatized in a number of different, equivalent ways. The following list will give us some interesting independence results. We take this axiomatization for constructive logic:

Positive paradox	$\vdash A \to (B \to A)$,
Self-distribution	$\vdash [A \to (B \to C)] \to [(A \to B) \to (A \to C)]$,
∧-introduction	$\vdash A \to [B \to (A \wedge B)]$,
∧-elimination	$\vdash (A \wedge B) \to A, \quad \vdash (A \wedge B) \to B$,
∨-introduction	$\vdash A \to (A \vee B), \quad \vdash B \to (A \vee B)$,
∨-elimination	$\vdash [(B \to A) \to \{(C \to A) \to [(B \vee C) \to A]\}$,
Contrapositive	$\vdash (A \to \overline{B}) \to (B \to \overline{A})$,
Explosion	$\vdash \overline{A} \to (A \to B)$.

Then we add one nonconstructive axiom:

Peirce's law $\quad \vdash [(A \to B) \to A] \to A$.

Along with these axioms, we take detachment as our only assumed inference rule.

The list given above has the advantage that it displays clearly the relation between classical logic (all eleven axioms) and constructive logic (omit just the last axiom). Also, this particular axiomatization deals *separately* with each of the connectives ∨, ∧, ¬. That is conceptually helpful for the beginner, and it also makes easier the proof given in the next few pages; see especially 25.8. That proof, and our choice of axioms, are based mostly on Robinson [1968].

In the next few pages we will prove that the eleven axioms listed above are *independent*. That means none are superfluous: it is not possible to prove one of them as a consequence of the other ten.

Our method is based on the soundness principles in 21.1. We will present eleven different semantic interpretations of the formal language. For each interpretation, detachment will be tautology-preserving,[1] and ten of the axioms will be tautologous, but one axiom will not be tautologous. This tells us that the one axiom is not provable from those ten axioms. Admittedly, the verifications for two of the interpretations will be very tedious.

Most of the interpretations developed in Chapters 8 and 9 were intuitive and natural in some sense, but the interpretations developed in the next few pages are unnatural and contrived. That is to be expected, since each of these interpretations makes false an axiom that we ordinarily consider true. Most of these contrived interpretations are "disposable," in the sense that we will use them only once.

25.5. Through 22.17 we have already established that Peirce's law is not a theorem of constructive logic; thus it is not provable from the other axioms.

25.6. We have already done much of the work to show that the self-distribution axiom,

$$[Q \to (R \to S)] \to [(Q \to R) \to (Q \to S)]$$

---

[1]In fact, for all but one of the eleven interpretations, detachment will be truth-preserving.

Independence of axioms 433

is independent of our other axioms. Indeed, with some effort we showed that the "strange" interpretation in 7.14 is tautology-preserving. Self-distribution is not tautologous for that interpretation; this follows from exercise 7.15.d. It is very simple (though a bit tedious) to verify that our other ten axioms are tautologies in that interpretation.

One way to reduce that tedious chore slightly is notice that the restrictions of $\ovee, \owedge, \ominus, \oarrow$ to the smaller set of semantic values $\{0, 1\}$ is just the two-valued interpretation. We've already verified that all our axioms are tautologies in the interpretation. Thus, it only remains to consider the cases where at least one of the variables in the formula is assigned the value $\frac{1}{2}$.

25.7. Next we shall demonstrate the independence of the "positive paradox" axiom, $A \to (B \to A)$. We use[2] five semantic values: $\Sigma = \{t, d, e, n, f\}$. The true semantic values are $\Sigma_+ = \{t, d, e, n\}$ — i.e., there is only one false value, f. The logical operators are interpreted by these rules:

$$y \ovee z = \begin{cases} f & \text{if } y \text{ and } z \text{ are both equal to } f \\ t & \text{in all other cases} \end{cases}$$

$$y \owedge z = \begin{cases} f & \text{if at least one of } y, z \text{ equals } f \\ t & \text{in all other cases} \end{cases}$$

$$\ominus y = \begin{cases} n & \text{if } y = f \\ f & \text{if } y \neq f \end{cases}$$

---

[2]Robinson [1968] used "0," "1," "2," "3," "4" as semantic values in this example, but he used those numbers only as labels; he did not use any arithmetical properties of the numbers. I have replaced those numbers with t, d, e, n, f, in an attempt to supply at least a little intuitive, classical motivation for this admittedly contrived interpretation. My t and f stand for "true" and "false," respectively, because those semantic values behave most like the classical values. I've used n for "not false," which is not quite the same thing as "true." I've used d for "don't know," because in many cases we "don't know" whether $A \to B$ is true or false. The remainder of the reasoning is fairly simple, except for e, which is "exceptional." Implication can now be described in a fashion that mimics classical logic very loosely:

Anything implies a true thing; a falsehood is the only thing that implies a falsehood; it is false that $e \to d$; and we *don't know* about any other implications.

$$y \ominus z = \begin{cases} t & \text{if } z = t & \text{or} & y = z = f \\ f & \text{if } [z = f,\ y \neq f] & \text{or} & [z = d,\ y = e] \\ d & \text{if } [z = d,\ y \neq e] & \text{or} & z \in \{e, n\} \end{cases}$$

The rule for implication is complicated, so we restate it in table form:

$y \ominus z$	$z = t$	$z = d$	$z = e$	$z = n$	$z = f$
$y = t$	t	d	d	d	f
$y = d$	t	d	d	d	f
$y = e$	t	f	d	d	f
$y = n$	t	d	d	d	f
$y = f$	t	d	d	d	t

It is now easy to verify that this interpretation is truth-preserving. The formula scheme $A \to (B \to A)$ is not a tautology for this interpretation; we can see that by taking an assignment in which $[\![A]\!] = [\![B]\!] = e$.

Finally, in a simple but very tedious fashion, it is possible to verify that all the axioms of classical syntactics except positive paradox are tautologies for this semantic interpretation. Robinson [1969] recommends carrying out this task by concentrating on the one false semantic value, f, which can only arise in a few ways; show that none of those axioms other than positive paradox can take that value.

25.8. The remaining eight axioms (those concerned with $\vee, \wedge, \neg$) can easily be shown independent by making very minor alterations in the two-valued interpretation.

In 8.2.b we described the two-valued interpretation using 14 arithmetic facts. If we change the right side of one of those 14 equations (from 0 to 1 or from 1 to 0) and leave the other 13 equations unchanged, we obtain a new (nonclassical) two-valued semantic interpretation of the formal language $\mathcal{L}$. There are 14 new interpretations obtained in this fashion (i.e., by changing just one of the 14 standard equations). Of these 14 new interpretations, it

Independence of axioms

turns out that eight are useful for demonstrating independence results. Those eight are indicated in the table below.

altered equation	axiom to be falsified	falsified by $[\![Q]\!]$	$[\![R]\!]$
$\ominus\, 0 = 0$	$(Q \to \overline{R}) \to (R \to \overline{Q})$	0	1
$\ominus\, 1 = 1$	$\overline{Q} \to (Q \to R)$	1	0
$0 \wedge 1 = 1$	$(Q \wedge R) \to Q$	0	1
$1 \wedge 0 = 1$	$(Q \wedge R) \to R$	1	0
$1 \wedge 1 = 0$	$Q \to [R \to (Q \wedge R)]$	1	1
$1 \vee 0 = 0$	$Q \to (Q \vee R)$	1	0
$0 \vee 1 = 0$	$R \to (Q \vee R)$	0	1
$0 \vee 0 = 1$	$(Q \to S) \to \{(R \to S) \to [(Q \vee R) \to S]\}$	0	0, $[\![S]\!] = 0$

All eight new interpretations are truth-preserving, since the definition of $\ominus$ is unchanged. For each interpretation, we must verify that one axiom is falsified (as indicated in the chart), and that all the other ten axioms remain tautologies. Those verifications are tedious but straightforward. They are great in number, but most of them can actually be skipped. Indeed, each interpretation changes the meaning of only one of the symbols $\vee$, $\wedge$, $\neg$, and leaves the other two unchanged, as well as leaving $\to$ unchanged. We only need to verify the tautologousness of the axioms that include a changed symbol. The unaffected axioms are already known to be tautologous in the classical two-valued interpretation 8.2, and so they remain tautologous in the altered interpretation.

For instance, the last line of the table represents the new interpretation obtained by replacing the customary equation $0 \vee 0 = 0$ with the new equation $0 \vee 0 = 1$, and leaving unchanged the other 13 arithmetic facts in 8.2.b. This new interpretation agrees with the customary two-valued interpretation in its computation of $\wedge$, $\to$, $\neg$, so any axioms involving only those symbols will still yield the same valuations and do not need to be checked. We need only to check the axioms involving $\vee$.

25.9. *Discussion.* We have now proved that the axioms of classical logic are independent of one another, but our proof used a wide variety of interpretations of the language $\mathcal{L}$. Some of these interpretations looked very different from the familiar two-valued interpretation. Indeed, some used more than two semantic values. But our conclusion is not concerned with those two semantic values, nor is it concerned with any of the interpretations used in the last few pages. Rather, it is concerned with which formulas can be used to syntactically prove which other formulas.

Chapter 26

Abelian logic

26.1. *Abelian logic* consists of basic logic plus two more axiom schemes:

- relativity, which was studied in Chapter 20, and
- centering, introduced in 26.3, below.

Unlike all the other main axiom systems considered in this book, Abelian logic is *not* a subset of classical logic (see 8.9 and 8.37).

26.2. *Motivating remarks (optional).* In Chapter 20 we added relativity to basic logic. Some consequences were that

- the identity formulas $A \to A$, $B \to B$, etc., are equivalent to one another, and all play the role of the Ackermann truth $\top$ (discussed in 20.6.c); and
- the negated identities $\overline{A \to A}$, $\overline{B \to B}$, etc. are equivalent to one another, and play the role of the Ackermann falsehood $\bot$.

Comparative logic (introduced in 8.28) is a particularly simple interpretation that fits all our axioms, if we give both $\top$ and $\bot$ the value 0. Thus, motivated by semantics, we add one more axiom below.

26.3. *Additional assumption.* To basic logic and relativity we add one more axiom scheme, which we shall call *centering*:

$$\vdash (A \to A) \leftrightarrow \overline{A \to A}.$$

Notation. Throughout the next couple of pages we shall use $\bot$ as an abbreviation for any formula of the form $A \to A$ or $\overline{A \to A}$. Such formulas are not all *equal* — i.e., they are different formulas, consisting of different strings of symbols. However, within Abelian logic such formulas are all *syntactically equivalent* (as defined in 3.42.g), and so they can be used interchangeably in substitutions (in the sense of 14.14).

26.4. *Some corollaries*
   a. $\vdash \overline{A \to A}$. (From 26.3 and 13.2.b.)
   b. $\vdash A \circ \overline{A}$. (Restatement of 26.4.a.)
   c. $\vdash B \leftrightarrow (B \circ \bot)$.

26.5. *Excluded middle.* $\vdash P \vee \overline{P}$.

Proof. As an instance of 26.4.b we have $\vdash (P \wedge \bot) \circ \overline{(P \wedge \bot)}$. By De Morgan's law, that can be restated as $\vdash (P \wedge \bot) \circ (\overline{P} \vee \overline{\bot})$. Since the Ackermann falsehood is its own negation in this logic, we have $\vdash (P \wedge \bot) \circ (\overline{P} \vee \bot)$. Now use the fact that cotenability is distributive over disjunction — i.e., use 19.6, substituting $Q = P \wedge \bot$, $R = \overline{P}$, $S = \bot$. We obtain

(1) $$\vdash \left[(P \wedge \bot) \circ \overline{P}\right] \vee \left[(P \wedge \bot) \circ \bot\right].$$

We will work on the two halves of that separately, and then combine them.

By $\wedge$-elimination, we have $\vdash (P \wedge \bot) \to \bot$. Since $\overline{P} \to \overline{P}$ is one of the formulas equivalent to $\bot$, we obtain $\vdash (P \wedge \bot) \to (\overline{P} \to \overline{P})$. By the fusion property of cotenability (19.5), that last formula can be restated as

(2) $$\vdash \left[(P \wedge \bot) \circ \overline{P}\right] \to \overline{P}.$$

From 26.4.c we have $\vdash \left[(P \wedge \bot) \circ \bot\right] \to (P \wedge \bot)$, and from $\wedge$-elimination we have $\vdash (P \wedge \bot) \to P$. Combining those by transitivity,

(3) $$\vdash \left[(P \wedge \bot) \circ \bot\right] \to P.$$

Finally, use (1), (2), (3), and 14.8.b to complete the proof.

Abelian logic

26.6. $\vdash \overline{A \to B} \leftrightarrow (B \to A)$.

Proof. Each of these formulas is equivalent to the next; the justifications for equivalences are left as an exercise.

$$\overline{A \to B}, \quad (A \to B) \to \bot, \quad (\bot \to B) \to A,$$
$$(\overline{B} \to \overline{\bot}) \to A, \quad \overline{\overline{B}} \to A, \quad B \to A.$$

26.7. *Chain ordering.* $\vdash (A \to B) \lor (B \to A)$.

Proof. Immediate from 26.5 and 26.6.

26.8. *Soundness and completeness.* It is easy to verify that the assumptions of Abelian logic are satisfied by the comparative interpretation 8.28. Thus

$$\left\{ \begin{array}{c} \text{theorems of} \\ \text{Abelian logic} \end{array} \right\} \subseteq \left\{ \begin{array}{c} \text{tautologies of the} \\ \text{comparative interpretation} \end{array} \right\}.$$

Meyer and Slaney [1989] have shown that this pairing is complete — i.e., the inclusion above is actually equality. However, this completeness result is harder to prove, and will not be proved here.

Part F

Advanced results

Chapter 27

Harrop's principle for constructive logic

Most of this chapter is devoted to proving some advanced properties of constructive logic. For motivation, the reader may want to glance ahead to Harrop's principle in 27.10, or to its simple corollaries in 27.11 and 27.13. For background, the reader may want to review Chapter 22.

Meyer's valuation

27.1. *Motivating remarks.* In 27.2 we will define a two-valued valuation $[\![\ ]\!] : \{\text{formulas}\} \to \{0,1\}$. We shall call it *Meyer's valuation*, because it is based on some ideas in several papers by R. K. Meyer. It is *not* a functional valuation — for instance, the value of $[\![A \to B]\!]$ does not depend only on the values of $[\![A]\!]$ and $[\![B]\!]$; we shall show that by example in 27.9. The valuation will be defined recursively by the rank of the formula — that is,

- we define $[\![\pi_1]\!], [\![\pi_2]\!], [\![\pi_3]\!], \ldots$ first;
- then we use those definitions in ingredients of our definition of $[\![A]\!]$ for all formulas A of rank 2;
- then use those definitions when defining $[\![A]\!]$ for all formulas A of rank 3;

and so on. There is no corresponding function $\ominus$ defined on $\{0, 1\}$. (Nevertheless, we will use the phrase "the definition of $\ominus$" as an abbreviation for "the definition of $[\![A \to B]\!]$.")

In fact, Meyer's valuation is not even strictly semantic. It is a mixture of semantic and syntactic, which happens to work quite well for proving some advanced ideas later in this chapter.

27.2. *Assumptions.* Throughout this chapter (except where explicitly noted at the very end of the chapter),

$$\vdash \text{ will mean constructive logic}$$

— i.e., the logic developed in sections 22.1–22.13. Also, we assume that $\mathcal{H}$ is some particular collection of formulas (possibly nonempty). This collection should be viewed as fixed.

Definition. First, for each propositional symbol $\pi_1, \pi_2, \pi_3, \ldots$, we define
$$[\![\pi_j]\!] = \begin{cases} 1 & \text{if } \mathcal{H} \vdash \pi_j, \\ 0 & \text{otherwise.} \end{cases}$$

Then, for longer formulas we define

$$[\![A \wedge B]\!] = \min\{[\![A]\!], [\![B]\!]\},$$
$$[\![A \vee B]\!] = \max\{[\![A]\!], [\![B]\!]\},$$
$$[\![\overline{A}]\!] = \begin{cases} 1 - [\![A]\!] & \text{if } \mathcal{H} \vdash \overline{A}, \\ 0 & \text{otherwise,} \end{cases}$$

$$[\![A \to B]\!] = \begin{cases} \max\{1 - [\![A]\!], [\![B]\!]\} & \text{if } \mathcal{H} \vdash A \to B, \\ 0 & \text{otherwise.} \end{cases}$$

We shall call this function *Meyer's valuation.*

Remark. Readers who have confused "valuation" with "functional valuation" may need to review the definitions in 7.4 and 7.5. Meyer's valuation is not defined functionally and need not turn out to be functional; see 27.9 for an example.

27.3. *Adequacy lemma.* If C is any formula and $[\![C]\!] = 1$, then $\mathcal{H} \vdash C$.

Proof. The proof will be by induction on the rank of the formula

C. For formulas C of rank 1, the result is immediate from the definition of $[\![\pi_j]\!]$.

Now suppose that C is of some rank greater than 1, and the lemma is already known for all shorter formulas. Assume $[\![C]\!] = 1$; we must show that $\mathcal{H} \vdash C$. We prove this in cases, according to how C was made from shorter formulas.

- Suppose $C = \overline{A}$ or $C = A \to B$ for some formulas A and B. Then $[\![C]\!] = 1$ implies $\mathcal{H} \vdash C$, by the definitions of $[\![\overline{A}]\!]$ and $[\![A \to B]\!]$ in 27.2.

- Suppose $C = A \wedge B$. By our assumption, $[\![A \wedge B]\!] = 1$. Then $[\![A]\!] = [\![B]\!] = 1$ by the definition of $[\![A \wedge B]\!]$ in 27.2. Since the formulas A and B are shorter than C, the induction hypothesis is applicable to them; hence both $\mathcal{H} \vdash A$ and $\mathcal{H} \vdash B$. By adjunction we obtain $\mathcal{H} \vdash A \wedge B$.

- Suppose $C = A \vee B$. Then our assumption yields $[\![A \vee B]\!] = 1$. By the definition of $[\![A \vee B]\!]$ in 27.2, at least one of $[\![A]\!] = 1$ or $[\![B]\!] = 1$ holds.[1] Those formulas are shorter than C, so the induction hypothesis is applicable to them; hence at least one of $\mathcal{H} \vdash A$ or $\mathcal{H} \vdash B$ holds. By $\vee$-introduction, it follows that $\mathcal{H} \vdash A \vee B$.

That completes the proof of adequacy.

27.4. *Semantic detachment.* If $[\![A \to B]\!] = 1$ and $[\![A]\!] = 1$, then $[\![B]\!] = 1$.

27.5. *Observation.* If $[\![A \to B]\!] = 0$ and $\mathcal{H} \vdash A \to B$, then

$$[\![A]\!] = 1, \qquad [\![B]\!] = 0, \qquad \mathcal{H} \vdash A, \qquad \mathcal{H} \vdash B.$$

Proof. First note that $1 - [\![A]\!] = [\![B]\!] = 0$. By adequacy, we know $\mathcal{H} \vdash A$. Combining that with the assumed $\mathcal{H} \vdash A \to B$ via detachment yields $\mathcal{H} \vdash B$.

---

[1] We don't necessarily know *which* of those two conditions holds, so the metamathematical reasoning that we're doing here is nonconstructive. The metasystem used throughout this book is classical, even though the formal system that we're studying might not be classical. See 2.17.

27.6. *Soundness of the constructive axioms.* Suppose that C is an instance of one of the axiom schemes of constructive logic (as listed in 25.4), and $\mathcal{H} \vdash C$. Then $[\![C]\!] = 1$.

Remarks and proof. Note that an expression such as $P \to (Q \to P)$ is an axiom *scheme*, not just an axiom. It includes infinitely many formulas — for instance, the formula $\pi_1 \to (\pi_3 \to \pi_1)$, as well as the formula $(\pi_2 \vee \overline{\pi_5}) \to \left[(\pi_1 \wedge \pi_2) \to (\pi_2 \vee \overline{\pi_5})\right]$. Those two formulas have different ranks. Our definition in 27.2 and the proof in 27.3 used increasing ranks of formulas, but the proof below is based on an entirely different set of ideas.

Most of our analysis will be done axiom by axiom, but first we make a few general observations that apply to all of the axioms.

Note that each of the axioms to be considered is of the form $A \to B$. That is, the last symbol used in the construction of the axiom is the implication symbol, not any of $\vee, \wedge, \neg$. Since it *is* an axiom, we have $\vdash A \to B$. We wish to prove $[\![A \to B]\!] = 1$, so we shall begin by assuming $[\![A \to B]\!] = 0$; it suffices to use that to generate a contradiction. Thus 27.5 is applicable, and some of its conclusions are the first steps of each proof below.

We shall use the three-column format for our proofs, but that is merely a convenient way to organize our arguments; the proofs should not be construed as syntactic derivation. All of the results of this chapter are metalogical.

We now proceed axiom by axiom.

(a) *Proof for* $[Q \to (R \to S)] \to [(Q \to R) \to (Q \to S)]$

(1)	$[\![Q \to (R \to S)]\!] = 1$	27.5
(2)	$[\![(Q \to R) \to (Q \to S)]\!] = 0$	27.5
(3)	$\mathcal{H} \vdash (Q \to R) \to (Q \to S)$	27.5
(4)	$[\![Q \to R]\!] = 1$	(2), (3), 27.5
(5)	$[\![Q \to S]\!] = 0$	(2), (3), 27.5
(6)	$\mathcal{H} \vdash Q \to S$	(2), (3), 27.5
(7)	$[\![Q]\!] = 1$	(5), (6), 27.5
(8)	$[\![S]\!] = 0$	(5), (6), 27.5
(9)	$[\![R]\!] = 1$	(7), (4), 27.4
(10)	$[\![R \to S]\!] = 1$	(7), (1), 27.4

| (11) | $[\![S]\!] = 1$ | (9), (10), 27.4 |
| (12) | contradiction | (8), (11) |

(b) *Proof for* $Q \to (R \to Q)$

(1)	$[\![Q]\!] = 1$	27.5
(2)	$[\![R \to Q]\!] = 0$	27.5
(3)	$\mathcal{H} \vdash R \to Q$	27.5
(4)	$[\![Q]\!] = 0$	(2), (3), 27.5
(5)	contradiction	(1), (4)

(c) *Proof for* $(Q \wedge R) \to Q$

(1)	$[\![Q \wedge R]\!] = 1$	27.5
(2)	$[\![Q]\!] = 0$	27.5
(3)	$\min\{[\![Q]\!], [\![R]\!]\} = 1$	(1), definition of $\wedge$ in 27.2
(4)	$[\![Q]\!] = 1$	(3)
(5)	contradiction	(2), (4)

(d) *Proof for* $(Q \wedge R) \to R$
Similar to (c); the details are left as an exercise.

(e) *Proof for* $Q \to [R \to (Q \wedge R)]$
Another exercise. *Hint.* My proof is seven steps long.

(f) *Proof for* $Q \to (Q \vee R)$
Similar to (c); my proof is five steps long.

(g) *Proof for* $R \to (Q \vee R)$
Similar to (f).

(h) *Proof for* $(Q \to S) \to \{(R \to S) \to [(Q \vee R) \to S]\}$
Another exercise; my proof is 12 steps long.

(i) *Proof for* $(Q \to \overline{R}) \to (R \to \overline{Q})$

(1)	$[\![Q \to \overline{R}]\!] = 1$	27.5
(2)	$[\![R \to \overline{Q}]\!] = 0$	27.5
(3)	$\mathcal{H} \vdash R \to \overline{Q}$	27.5
(4)	$[\![R]\!] = 1$	(2), (3), 27.5

(5)	$[\overline{Q}] = 0$	(2), (3), 27.5
(6)	$\mathcal{H} \vdash \overline{Q}$	(2), (3), 27.5
(7)	$[\overline{R}] = 0$	(4), definition of $\ominus$ in 27.2
(8)	$[Q] = 0$	(1), (7), 27.4
(9)	$[\overline{Q}] = 1$	(6), (8), definition of $\ominus$ in 27.2
(10)	contradiction	(5), (9)

(j) *Proof for* $\overline{Q} \to (Q \to R)$

(1)	$[\overline{Q}] = 1$	27.5
(2)	$[Q \to R] = 0$	27.5
(3)	$\mathcal{H} \vdash Q \to R$	27.5
(4)	$[Q] = 1$	(2), (3), 27.5
(5)	$[Q] = 0$	(1), definition of $\ominus$ in 27.2
(6)	contradiction	(4), (5)

This completes the proof of the soundness of the axioms.

HARROP'S PRINCIPLE

27.7. Definition. By a *Harrop set* we will mean a set $\mathcal{H}$ of formulas satisfying all of these properties:

a. If $P \wedge Q$ is a member of $\mathcal{H}$, then both P and Q are members of $\mathcal{H}$.

b. If $P \vee Q$ is a member of $\mathcal{H}$, then at least one of P or Q is a member of $\mathcal{H}$.

c. If $P \to Q$ is a member of $\mathcal{H}$, then Q is a member of $\mathcal{H}$.

Remarks. It is easy to see that the empty set, the set of all formulas of rank 1 (i.e., the π_j's), and the set of *all* formulas are Harrop sets. Also, any subset of

$$\{\neg P : P \text{ is a formula}\}$$

is a Harrop set, since the three rules listed above for Harrop sets merely make restrictions on formulas whose outermost operator is

∧, ∨, or →. However, not every such set is *consistent* (defined in 17.5); we shall be particularly concerned with consistent Harrop sets.

Condition 27.7.c, concerning →, may seem unnaturally restrictive. But that is not a problem, because our main interest in Harrop sets is for the investigation of ∨. In fact, condition 27.7.b is the really restrictive one, and we usually will satisfy it by simply not including in $\mathcal{H}$ any formulas of the form $P \vee Q$.

27.8. *Harrop completeness.* Let $\vdash$ denote the syntactic consequence relation (i.e., "proves") for constructive logic. Let $\mathcal{H}$ be any consistent Harrop set. Define the Meyer valuation $[\![\ ]\!]$ as in 27.2. Then for every formula C, we have $[\![C]\!] = 1$ if and only if $\mathcal{H} \vdash C$.

Proof. We already know $[\![C]\!] = 1$ implies $\mathcal{H} \vdash C$, by the Adequacy Lemma 27.3. It suffices to show that $\mathcal{H} \vdash C$ implies $[\![C]\!] = 1$. We first demonstrate this slightly weaker assertion:

(∗) $\qquad\qquad C \in \mathcal{H}$ implies $[\![C]\!] = 1.$

We prove that by induction on the rank of the formula C.

If C is a formula of rank 1, then $C = \pi_j$ for some j. If also $C \in \mathcal{H}$, then $\mathcal{H} \vdash C$, and so $[\![C]\!] = 1$ by our definition of $[\![\pi_j]\!]$ in 27.2.

Now suppose that C has rank greater than 1, and (∗) is known for all formulas shorter than C. The formula C must be of the form $P \wedge Q$ or $P \vee Q$ or $P \to Q$ or $\overline{P}$. The formulas P and Q are shorter than C, and so the induction hypothesis is applicable to them.

- Suppose $C = P \wedge Q$. By the definition of a Harrop set, both P and Q are members of $\mathcal{H}$. By the induction hypothesis, $[\![P]\!] = [\![Q]\!] = 1$. Then $[\![P \wedge Q]\!] = 1$ by our definition of $[\![P \wedge Q]\!]$ in 27.2.
- Suppose $C = P \vee Q$. By the definition of a Harrop set, at least one of P or Q is a member of $\mathcal{H}$. By the induction hypothesis, that member has Meyer valuation equal to 1. Then $[\![P \vee Q]\!] = 1$ by our definition of $[\![P \vee Q]\!]$ in 27.2.

- Suppose $C = P \to Q$. Then Q is also a member of $\mathcal{H}$, by the definition of a Harrop set. Then $[\![Q]\!] = 1$ by our induction hypothesis. By assumption, $C \in \mathcal{H}$; hence $\mathcal{H} \vdash C$. That is, $\mathcal{H} \vdash P \to Q$. Now $[\![P \to Q]\!] = 1$ by our definition of $[\![P \to Q]\!]$ in 27.2.

- Finally, suppose $C = \overline{P}$. Since $\overline{P} \in \mathcal{H}$, we have $\mathcal{H} \vdash \overline{P}$. Since $\mathcal{H}$ is consistent, we do not have $\mathcal{H} \vdash P$. By the Adequacy Lemma, then, $[\![P]\!] = 0$. Then $[\![\overline{P}]\!] = 1$ follows from the definition of $[\![\overline{P}]\!]$ in 27.2.

This completes the proof of (∗).

Finally, we consider the general case where we merely assume $\mathcal{H} \vdash C$. Then C has a lemmaless proof — i.e., a derivation in which all the justifications are axioms of constructive logic, members of $\mathcal{H}$, or Detachment. All the axioms and members of $\mathcal{H}$ have valuation 1 by the preceding argument and 27.6, and Detachment preserves that valuation by 27.4. Thus $[\![C]\!] = 1$.

27.9. *Example.* In general, Meyer's valuation need not be a functional valuation, even when $\mathcal{H}$ is a Harrop set. For instance, if we take $\mathcal{H} = \emptyset$, then $[\![A]\!]$ is 1 or 0 according as A is a theorem or a nontheorem of constructive logic. Since $\pi_1 \to (\pi_2 \to \pi_1)$ is a theorem and $\pi_1, \pi_2, \pi_2 \to (\pi_2 \to \pi_1)$ are not theorems, we get

$$[\![\pi_1 \to (\pi_2 \to \pi_1)]\!] = 1, \qquad [\![\pi_1]\!] = [\![\pi_2]\!] = [\![\pi_2 \to (\pi_2 \to \pi_1)]\!] = 0.$$

Consequently, if Meyer's valuation were a functional valuation, there would be some function $\ominus$ making this equation true:

$$\begin{aligned} 1 = [\![\pi_1 \to (\pi_2 \to \pi_1)]\!] &= [\![\pi_1]\!] \ominus [\![\pi_2 \to \pi_1]\!] \\ &= [\![\pi_2]\!] \ominus [\![\pi_2 \to \pi_1]\!] = [\![\pi_2 \to (\pi_2 \to \pi_1)]\!] = 0. \end{aligned}$$

27.10. *Harrop's principle.* Let $\vdash$ denote provability in constructive logic. Suppose $\mathcal{H}$ is a Harrop set (defined as in 27.7), and A and B are some formulas, such that $\mathcal{H} \vdash A \vee B$. Then at least one of $\mathcal{H} \vdash A$ or $\mathcal{H} \vdash B$.

Proof. If the set $\mathcal{H}$ is inconsistent, then $\mathcal{H}$ proves *every* formula, and so in particular $\mathcal{H} \vdash A$ and $\mathcal{H} \vdash B$. Now assume $\mathcal{H}$ is consistent; then the results of 27.8 are applicable. We have $\mathcal{H} \vdash A \vee B$; hence $[\![A \vee B]\!] = 1$; hence $\max\{\,[\![A]\!], [\![B]\!]\,\} = 1$. Therefore at least one of $[\![A]\!], [\![B]\!]$ is 1; hence at least one of $\mathcal{H} \vdash A$ or $\mathcal{H} \vdash B$.

THE DISJUNCTION PROPERTY

27.11. *The Disjunction property.* If $A \vee B$ is a theorem of constructive logic, then so is at least one of A or B.

Proof. The empty set is a Harrop set. Apply Harrop's principle.

27.12. *Remarks.*
a. Obviously, if either $\vdash A$ or $\vdash B$, then also $\vdash A \vee B$. Hence the "if" can be replaced with "if and only if." Thus, the disjunction property can be restated as

$\vdash A \vee B$ if and only if at least one of $\vdash A$ or $\vdash B$.

b. The Disjunction Property is *not* true for classical, relevant, or fuzzy syntactics. Indeed, in classical and relevant logics, $A \vee \overline{A}$ is a theorem scheme while A and $\overline{A}$ are not. In classical and fuzzy logics, $(A \to B) \vee (B \to A)$ is a theorem scheme while $A \to B$ and $B \to A$ are not.

ADMISSIBILITY

27.13. *Mints's principle.* We state this in several versions; all are proved below. Let $\vdash$ denote provability in constructive logic, and let A, B, C be any formulas. Then:
a. If $\overline{C} \vdash A \vee B$, then at least one of $\overline{C} \vdash A$ or $\overline{C} \vdash B$.
b. If $\overline{C} \to (A \vee B)$ is a theorem, then at least one of $\overline{C} \to A$ or $\overline{C} \to B$ is a theorem.
c. *Mints's admissibility rule*
$\vdash \overline{C} \to (A \vee B) \quad \Rightarrow \quad \vdash (\overline{C} \to A) \vee (\overline{C} \to B)$.

Proof. Part a follows from the fact that the singleton $\{\overline{C}\}$ is a Harrop set. From Part a we obtain Part b using the deduction principle 22.5. From that we obtain Part c using the Disjunction Property 27.11.

27.14. *Remark.* The negation signs on the C's might appear superfluous, since the negation signs are present on *all* the C's in Mints's property. But the negation signs are crucial; if we omit them Mints's property fails. For constructive logic, in general, $\vdash C \to (A \vee B)$ does *not* imply $\vdash (C \to A) \vee (C \to B)$. For example, try $A = \pi_1$, $B = \pi_2$, and $C = \pi_1 \vee \pi_2$.

27.15. *Mints's rule is not derivable.* That is,

$$\overline{C} \to (A \vee B) \quad \vdash \quad (\overline{C} \to A) \vee (\overline{C} \to B)$$

is not derivable in constructive logic. (*Exercise*: Prove this. *Hints*: Use the deduction principle, 10.6, and 22.14.) Thus,

> in constructive logic, not every
> admissible rule is derivable.

27.16. *Exercise.* In constructive logic:
a. If $\vdash \overline{F}$ or $\vdash G$, then $\vdash F \to G$.
b. Show that the converse fails — i.e., show that $\vdash F \to G$ does not imply at least one of $\vdash \overline{F}$ or $\vdash G$.

RESULTS IN OTHER LOGICS

27.17. *Relevant admissibility.* Relevant logic satisfies the admissibility rule of *disjunctive syllogism*:

$$\vdash \overline{A} \wedge (A \vee B) \quad \Rightarrow \quad \vdash B.$$

(The proof is long and will not be given here; see Anderson and Belnap [1975] and Dunn [1986].) However, the corresponding inference rule is not derivable in relevant logic; see 8.39.c(iii), 23.8. Thus, much as in constructive logic, in relevant logic, not every admissible rule is derivable.

Further remarks. Disjunctive syllogism can also be stated as: If $\vdash B \vee \overline{A}$, then $\vdash A \Rightarrow \vdash B$ (where the "$\Rightarrow$" is classical). One converse to that would be
$$\text{if} \quad \vdash A \Rightarrow \vdash B, \quad \text{then} \quad \vdash B \vee \overline{A}.$$
That result is true for classical logic. *Exercise*: Show that that result fails in relevant logic. *Hint*: Let B be positive paradox, and let $A = \overline{B}$. Then what?

27.18. *Preview.* In classical logic, every admissibility rule *is* derivable; we shall show that in 29.15.

Chapter 28

Multiple worlds for implications

MULTIPLE WORLDS

28.1. We now describe a general class of semantic interpretations, which has many more applications than the two given in this chapter. These interpretations may be called *multiple viewpoint interpretations*. They are also known as *multiple worlds*, *possible worlds*, *frames*, or *setups*.

The terminology and notation in this subject are not standardized throughout the literature, but most treatments use ideas similar to the ones presented below. A multiple viewpoint interpretation has the ingredients listed below, more or less. (Any imprecision in the following general description will be replaced by exact specifications in our detailed examples, which begin in 28.7.)

28.2. In each interpretation, some nonempty set V is given; we will refer to V as the set of *viewpoints*. It may be helpful to think of these different viewpoints as being held by different *observers*, or perhaps by the same observer at different times or in different possible futures or in different conceivable *worlds*. (Some mathematicians might also refer to the members of V as *universes*, but I would recommend against that terminology; I find it more convenient to use the word "universe" as in 3.19. Indeed, we

Multiple worlds 455

might even think of V as the universe because it contains all the worlds.)

We emphasize that the worlds merely have to be "conceivable" — this is a much weaker restriction than "possible," since our description of a "conceivable world" could include statements that contradict one another.

In this exposition, members of V will be denoted by $x, y, z, \ldots$. However, use of the expression "$x, y, z, \ldots$" should not be construed to mean that the members of V can be arranged into a list. Such an arrangement might be possible in some multiple viewpoint interpretations, but not in others. In our simplest examples, V is a finite set, but the theory will also apply to cases where V is an infinite set, perhaps even an uncountable set.

28.3. The set V will be equipped with some additional mathematical structure, such as an ordering or a relation or an operation. In many applications in the literature, V is equipped with some sort of three-place relation — i.e., there is some R such that $R(x, y, z)$ is either true or false for each choice of $x, y, z \in V$.

However, in this book, we will develop only two examples in any detail; both of them are atypically elementary. In both of them, V is a collection of *sets*, and so the statement "$x \subseteq y$" is either true or false for each choice of $x, y \in V$. Note that this is a two-place relation on V.

28.4. Some set Σ of "local semantic values" is given. Each viewpoint $x \in V$ is equipped with a Σ-valued valuation $[\![\ ]\!]_x$ — i.e., a mapping

$$[\![\ ]\!]_x : \{\text{formulas}\} \to \Sigma.$$

In most applications of this theory, Σ is a fairly small set — perhaps only two or three or four members.

In the two applications that we shall cover in detail later, Σ is just the set $\{0, 1\}$. However, it is best not to think of these numbers 0 and 1 as representing "false" and "true." A more accurate explanation would be

$[\![S]\!]_x = 1$ means that the observer with viewpoint x can deduce from his or her information that formula S is true; and

$[\![S]\!]_x = 0$ means that the observer with viewpoint x *cannot* deduce from his or her information that formula S is true.

Because only the values 0 and 1 are possible, the "logic of the moment" might seem to resemble the classical two-valued interpretation of 8.2, but it is really quite different. For instance, $[\![S]\!]_x$ and $[\![\neg S]\!]_x$ *could both be equal to zero*, if the observer at x is not able to deduce either of the conclusions S or $\neg S$. Keep in mind that 0 means "not yet known to be true"; this is *not* the same thing as "false."

Different logical systems may have different meanings for the phrase "can deduce from his or her information." We shall explore that phrase in more detail later; see especially 28.6.

28.5. In functional interpretations, we would valuate more complicated formulas using the valuations of simpler formulas, e.g.,

$$[\![R \to S]\!]_x \;=\; [\![R]\!]_x \ominus [\![S]\!]_x.$$

But multiple viewpoint interpretations are not functional interpretations, so that type of computation is *not applicable* here. There is no such function $\ominus$, and the value of $[\![R \to S]\!]_x$ is not determined by just the values of $[\![R]\!]_x$ and $[\![S]\!]_x$. Rather, $[\![R \to S]\!]_x$ may depend on *all* the values

$$[\![R]\!]_x, [\![R]\!]_y, [\![R]\!]_z, \ldots \quad \text{and} \quad [\![S]\!]_x, [\![S]\!]_y, [\![S]\!]_z, \ldots \; .$$

Likewise for $\vee, \wedge, \neg$. The computational procedure is as follows:

- First, some values in Σ (in our own examples, 0's and 1's) are assigned as values of the propositional variable symbols $\pi_1, \pi_2, \pi_3, \ldots$. That is, some values are assigned for $[\![\pi_j]\!]_x$, $[\![\pi_j]\!]_y$, $[\![\pi_j]\!]_z$, $\ldots$, for all positive integers j.

Multiple worlds 457

- Those values are then used to compute $[\![S]\!]_x, [\![S]\!]_y, [\![S]\!]_z, \ldots$ for all formulas S of rank 2, according to some rule. (The rules are different for different interpretations.) Any one of those values — for instance, $[\![S]\!]_y$ — may depend on not only the values of $[\![\pi_j]\!]_y$, but also the values of $[\![\pi_j]\!]_x, [\![\pi_j]\!]_z, \ldots$.
- Then *those* values are used to compute $[\![S]\!]_x, [\![S]\!]_y, [\![S]\!]_z, \ldots$ for all formulas S of rank 3.

And so on. This means, for instance, that the values of $[\![R]\!]_y$ and $[\![S]\!]_{x \cup y}$ are computed *before* the value of $[\![R \to S]\!]_x$ is computed, despite the fact that $x \cup y$ is larger and presumably more complicated than x, and the fact that y might seem to be entirely unrelated to x.

Remark. Actually, if we combine all the different observers' Σ-valued valuations $[\![S]\!]_x, [\![S]\!]_y, [\![S]\!]_z, \ldots$, the result could be viewed as one functional valuation $\langle\!\langle\ \rangle\!\rangle$ taking values in $\Sigma^\mathcal{V} =$ {functions from $\mathcal{V}$ into Σ}. For each formula S, the value $\langle\!\langle S \rangle\!\rangle$ is the function $\psi_S : \mathcal{V} \to \Sigma$ defined by $\psi_S(x) = [\![S]\!]_x$. The valuation $\langle\!\langle\ \rangle\!\rangle$ is functional, since (for instance) the value of $\psi_{S \vee T}$ is determined by the values of ψ_S and ψ_T. The drawback to this approach is that the set $\Sigma^\mathcal{V}$ is far more complicated than Σ, and may be much too abstract to provide us with any intuition.

In the very simplest cases, Σ is just $\{0, 1\}$. Then the resulting set $\Sigma^\mathcal{V}$ is just $\{0, 1\}^\mathcal{V}$, which can be identified in a natural way with $\mathcal{P}(\mathcal{V})$. In that case we consider $\langle\!\langle\ \rangle\!\rangle$ to take for its values the subsets of $\mathcal{V}$. More specifically, $\langle\!\langle S \rangle\!\rangle = \{x \in \mathcal{V} : [\![S]\!]_x = 1\}$. However, we shall not pursue this line of thought further here.

28.6. The relationship between different members of $\mathcal{V}$, indicated by that two- or three-place relationship mentioned in 28.2, may be reflected in some fashion in the valuations held by the observers.

For instance, in constructive logic, an increase in our information can only increase our power of deduction, so we have this *isotonicity* rule:

$$\text{if} \quad x \subseteq y, \quad \text{then} \quad [\![S]\!]_x \leq [\![S]\!]_y$$

for any formula S. However, that doesn't work for relevant logic. Indeed, $[\![S]\!]_x = 1$ means that statement S can be deduced *from* information x; but if we add more information, we may be adding irrelevant information, and so we may no longer be able to obtain that statement *from* our information via relevant reasoning.

IMPLICATION MODELS

28.7. We now begin development of our two detailed examples of multiple viewpoint semantics. One is for relevant implication (axiomatized as in 23.1) and the other is for constructive implication (axiomatized as in 22.2 — i.e., it also includes positive paradox). We will not consider $\vee, \wedge, \neg$ in these examples. The two examples are nearly identical, so we will cover both simultaneously, pointing out the few differences when they arise. These semantics were developed independently by Alasdair Urquhart [1972] and Richard Routley [1972].

28.8. *Definition.* By a *relevant implication model* we shall mean a collection of sets and a collection of functions, with these properties:

a. $\mathcal{V}$ is some collection of sets, with $\varnothing \in \mathcal{V}$, and with the property that
$$\text{if} \quad x, y \in \mathcal{V} \quad \text{then} \quad x \cup y \in \mathcal{V}.$$
(We emphasize that x and y are *members*, not *subsets*, of $\mathcal{V}$. It is true that x and y are subsets of *some* set, but we won't be concerned with that set right now, and so for simplicity we won't even give that set a name.)

b. For each $x \in \mathcal{V}$, some valuation $[\![\ ]\!]_x : \{\text{formulas}\} \to \{0, 1\}$ is given.

c. The valuations of any formula $P \to Q$ are related to the valuations of the shorter formulas P and Q, by this *computational rule*:

$$[\![P \to Q]\!]_x = \begin{cases} 0 & \text{if there exists at least one } y \in \mathcal{V} \\ & \text{with the property that } [\![P]\!]_y = 1 \\ & \text{and } [\![Q]\!]_{x \cup y} = 0, \\ \\ 1 & \text{if for every } y \in \mathcal{V} \text{ we have at least} \\ & \text{one of } [\![P]\!]_y = 0 \text{ or } [\![Q]\!]_{x \cup y} = 1. \end{cases}$$

Implication models

Finally, the relevant implication model is also a *constructive implication model* if it also satisfies this *isotonicity rule*:

d. $[\![P]\!]_y \leq [\![P]\!]_{x \cup y}$ for all formulas P and viewpoints $x, y \in \mathcal{V}$.

(Thus every constructive implication model is also a relevant implication model, but not conversely.)

In either type of model, we will be especially concerned with $[\![\ ]\!]_\varnothing$, the valuation associated with the empty set of information. We will call this the *principal valuation* of the model. It represents the smallest possible amount of knowledge — perhaps a completely ignorant observer, or at any rate an observer who only knows things that are "common knowledge."

28.9. *Remarks on the computational rule 28.8.c.* The computational rule is admittedly rather complicated. To understand it better, first recall that the *two-valued* interpretation would say

> $P \to Q$ is not true only in the case where P is true and Q is not true.

Now, our computational rule for multiple observers is a little like that, if we restate it this way:

> The only way that observer x will not be convinced of the truthfulness of $P \to Q$ is in the case where there is at least one observer y who has enough information to be certain that P is true, and yet the information of x and y together does not make evident the truthfulness of Q.

Instead of a very long sentence, some readers may prefer to see the computational rule for implication stated as a moderately long formula:

$$[\![P \to Q]\!]_x = \min_{y \in \mathcal{V}} \max\{1 - [\![P]\!]_y,\ [\![Q]\!]_{x \cup y}\}.$$

Two special cases that are simple enough to be noteworthy are the cases where $y = x$ or $y = \varnothing$. By considering those two cases, we see that a sufficient condition for $[\![P \to Q]\!]_x = 0$ is that $[\![Q]\!]_x = 0$ and at least one of $[\![P]\!]_x$ or $[\![P]\!]_\varnothing$ is 1.

We emphasize that this is a *sufficient* condition, but perhaps not a *necessary* condition. (If you're not familiar with that terminology, see 5.21.)

SOUNDNESS

28.10. *Soundness for implication models.* We shall demonstrate $\vdash S \Rightarrow \vDash S$ for both relevant and constructive implications. By that we mean:

- If S is a theorem of relevant implication (axiomatized in 23.1), then $[\![S]\!]_\varnothing = 1$ in every relevant implication model.
- If S is a theorem of constructive implication (axiomatized as in 22.2), then $[\![S]\!]_\varnothing = 1$ in every constructive implication model.

Demonstration. It follows easily from the definitions of the model, that its principal valuation is truth-preserving — i.e.,

$$\text{if } [\![P]\!]_\varnothing = 1 \text{ and } [\![P \to Q]\!]_\varnothing = 1, \text{ then } [\![Q]\!]_\varnothing = 1.$$

We shall apply 21.1. Thus, it suffices to show that the axioms (in 23.1 or 22.2, respectively) are true in the principal valuation. The verifications are fairly straightforward but very tedious. We shall carry out the verification only for the self-distribution axiom, which is typical. We omit the verification for the other axioms, which can be handled similarly and are left as an exercise.

Let A, B, C be any formulas. We will repeatedly use the computational rule 28.8.c.

Temporarily fix any $x, y, z \in \mathcal{V}$. We claim, first of all, that

$$\text{if } [\![B \to C]\!]_x = 1 \text{ and } [\![B]\!]_{y \cup z} = 1, \text{ then } [\![C]\!]_{x \cup y \cup z} = 1.$$

Indeed, the assumption $[\![B \to C]\!]_x = 1$ tells us (by the computational rule) that for every $w \in \mathcal{V}$, we have at least one of

Canonical models

$[\![B]\!]_w = 0$ or $[\![C]\!]_{x \cup w} = 1$. In particular, this must hold when $w = y \cup z$; our claim follows immediately.

Now continue to hold x and y fixed, but let z vary. From the preceding claim, by dividing all possibilities into cases, we find that at least one[1] of these three conditions must hold:

- some $z \in \mathcal{V}$ exists such that $[\![A]\!]_z = 1$ and $[\![B \to C]\!]_{x \cup z} = 0$, or
- some $z \in \mathcal{V}$ exists such that $[\![A]\!]_z = 1$ and $[\![B]\!]_{y \cup z} = 0$, or
- for each $z \in \mathcal{V}$, at least one of $[\![A]\!]_z = 0$ or $[\![C]\!]_{x \cup y \cup z} = 1$ must hold.

Use the computational rule to restate each of those three conditions. Thus at least one of these three conditions must hold:

$$[\![A \to (B \to C)]\!]_x = 0, \quad [\![A \to B]\!]_y = 0, \quad [\![A \to C]\!]_{x \cup y} = 1.$$

Next, hold x fixed and let y vary. (Use the computational rule for $[\![P \to Q]\!]_x$ in the case where $P = A \to B$ and $Q = A \to C$.) We find that for every $x \in \mathcal{V}$, at least one of these two conditions holds:

$$[\![A \to (B \to C)]\!]_x = 0 \quad \text{or} \quad [\![(A \to B) \to (A \to C)]\!]_x = 1.$$

Finally, use the computational rule one more time; that proves

$$[\![\, [A \to (B \to C)] \to [(A \to B) \to (A \to C)]\,]\!]_\varnothing = 1,$$

as required.

CANONICAL MODELS

28.11. Definition. There are many models that fit the specifications given in 28.8. We shall now describe two in particular, which will play a special role in subsequent proofs; we will call them the *canonical model for relevant implication* and the *canonical model*

---

[1] We don't necessarily know *which* of the three conditions holds, though a constructivist might ask that question. Our metalogic in this book is classical. See 2.17.

for constructive implication. They are nearly identical, so we will introduce them together and point out the one place where they differ.

Let $\mathcal{V}$ be the collection of *all finite sets of formulas* in our formal language. For instance, some typical members of $\mathcal{V}$ are the sets

$$\left\{ \pi_1 \to \pi_3, \quad \pi_1 \to (\pi_2 \to \pi_1) \right\},$$

$$\left\{ \pi_2 \to \pi_7, \quad \pi_1 \right\},$$

$$\left\{ \pi_4 \to \pi_4, \quad (\pi_1 \to \pi_2) \to (\pi_2 \to \pi_1), \quad \pi_1 \to \pi_7 \right\},$$

and so on. Those might be $x, y, z, \ldots$, for instance. We emphasize that we are considering sets of *particular formulas*, not formula *schemes*.

Let $\vdash$ denote the syntactic consequence for the inference system we are using — i.e., either relevant implication (axiomatized in 23.1) or constructive implication (axiomatized as in 22.2). For each $x \in \mathcal{V}$ we shall define a function

$$[\![\ ]\!]_x \quad : \quad \{\text{formulas of } \mathcal{L}\} \quad \to \quad \{0, 1\}$$

by this rule: Arrange the set of formulas x into a repetitionless and ordered *list* of formulas, which we will also denote by x. (The particular order of the list does not matter, as we shall see below.) Then define

$$[\![P]\!]_x = \begin{cases} 1 & \text{if } \vdash x \to P, \\ 0 & \text{if } \nvdash x \to P, \end{cases}$$

where the expression $x \to P$ uses the notation for iterated implications introduced in 13.21. (The reader might do well to review that notation now.)

28.12. *Justification.* We shall show that the so-called "canonical model for implication" defined in 28.11 deserves its name — i.e., it actually *is* an implication model, as defined in 28.8.

Canonical models

Verifications. First, we must remind the reader why the ordering of the list x does not matter. We saw in 13.23 that if x and x' are two different orderings of the same list of formulas,[2] then

$$\vdash x \to S \quad \text{if and only if} \quad \vdash x' \to S.$$

What about two lists x and x' that are identical except that x' has some repetitions removed? Here only half of the "if and only if" works; we have

$$\vdash x \to S \quad \text{implies} \quad \vdash x' \to S$$

by 15.6, but in general we cannot conclude the converse. Fortunately, the implication will always run in the right direction in the proof given below. The reader should check that for himself or herself each time that we need to remove repetitions (e.g., due to overlap between x and y when we form $x \cup y$).

Next, we must verify that the so-called canonical model for implication satisfies the computational rule 28.8.c. There are two cases to consider, according to whether $[\![P \to Q]\!]_x$ is 1 or 0.

First, suppose that $[\![P \to Q]\!]_x = 1$. In that case, we must verify that $[\![P]\!]_y = 1$ and $[\![Q]\!]_{x \cup y} = 0$ cannot both hold. Suppose they do both hold; we'll get a contradiction. We have

$$[\![P \to Q]\!]_x = 1, \quad [\![P]\!]_y = 1, \quad [\![Q]\!]_{x \cup y} = 0.$$

That translates to

$$\vdash x \to (P \to Q), \quad \vdash y \to P, \quad \text{but not} \quad \vdash (x \cup y) \to Q.$$

But that contradicts Prefixed Detachment 13.24. (Note that our direction of reasoning here only removes repetitions from $x \cup y$; it doesn't introduce any.)

Now for the other case; suppose that $[\![P \to Q]\!]_x = 0$. We must show that there is some $y \in \mathcal{V}$ for which $[\![P]\!]_y = 1$ and $[\![Q]\!]_{x \cup y} = 0$. Define the sequence

$$M^0(P), \quad M^1(P), \quad M^2(P), \quad M^3(P), \quad \ldots$$

---

[2] Or, in the terminology of 3.1.c, x and x' are two different sequential representations of the same multiset of formulas.

as in 13.10. The formulas in that sequence are all different, since the rank of $M^{k+1}(P)$ is greater than the rank of $M^k(P)$. Thus that sequence contains infinitely many different formulas. Since x is a finite set, some $M^k(P)$ does not belong to x. Choose any one such $M^k(P)$ to be the sole member of the singleton $y = \{M^k(P)\}$. From 13.10 we obtain $\vdash M^k(P) \to P$, which says $[\![P]\!]_y = 1$. For the other half, suppose (for contradiction) that $[\![Q]\!]_{x \cup y} = 1$. Then $\vdash (x \cup \{P_k\}) \to Q$, which can be restated as $\vdash M^k(P) \to (x \to Q)$. On the other hand, 13.10 also tells us $\vdash P \to M^k(P)$. By transitivity, combine those last two assertions to obtain $\vdash P \to (x \to Q)$. By generalized permutation 13.23, $\vdash x \to (P \to Q)$, which says $[\![P \to Q]\!]_x = 1$, contradicting our assumption. Thus we must have $[\![Q]\!]_{x \cup y} = 0$.

Finally, we must verify that the canonical model for *constructive* implication satisfies the isotonicity rule 28.8.d. That simply says that if $\vdash y \to P$, then $\vdash (x \cup y) \to P$. To prove that, suppose $x = \{A_1, A_2, \ldots, A_n\}$. By induction on k, we show that $\vdash S_k$, where S_k is the formula

$$A_k \to (A_{k-1} \to (A_{k-2} \to (\cdots \to (A_2 \to (A_1 \to (y \to P))) \cdots))).$$

Indeed, $\vdash S_0$ is given. Suppose that $\vdash S_k$ has been proved. Combine that with $\vdash S_k \to (A_{k+1} \to S_k)$ (an instance of an axiom) and detachment, to prove $\vdash A_{k+1} \to S_k$ — but that's just $\vdash S_{k+1}$.

Completeness

28.13. The following is actually two completeness results — one for relevant implication, one for constructive implication; they should be read separately.

Completeness of implication models. Let P be any formula of pure implication (i.e., not involving $\vee, \wedge, \neg$). Then, for either relevant implication or constructive implication, the following three conditions are equivalent:

(i) $\vdash P$.

(ii) $[\![P]\!]_\varnothing = 1$ in every implication model.

(iii) $[\![P]\!]_\varnothing = 1$ in the canonical model.

Demonstration. The bi-implication (iii) ⇔ (i) is actually built into the definition of the canonical model:

(iii) holds ⇔ $[\![P]\!]_\varnothing = 1$ ⇔ $\vdash \varnothing \to P$ ⇔ (i) holds.

The implication (i) ⇒ (ii) is soundness, which was demonstrated in 28.10. The implication (ii) ⇒ (iii) follows from the fact that the "canonical implication model" *is* indeed an implication model; that was demonstrated in 28.12.

28.14. *Application: irrelevance of mingle.* The formula scheme $S \to (S \to S)$ is not a theorem scheme of relevant implication. In particular, $\pi_1 \to (\pi_1 \to \pi_1)$ is not a theorem (i.e., provable) in relevant implication.

Demonstration. We already demonstrated this fact in 23.8, but following is another demonstration that may yield additional insight. We shall devise a relevant implication model for which $[\![\pi_1 \to (\pi_1 \to \pi_1)]\!]_\varnothing = 0$.

Let x and y be any two distinct sets, neither of which is a subset of the other. (For instance, we could take $x = \{1\}$ and $y = \{2\}$. However, the particular choice of x and y does not matter.)

We take $\mathcal{V} = \{\varnothing, x, y, x \cup y\}$; thus there will be four viewpoints in the model. Let an assignment be given in which

$$[\![\pi_1]\!]_x = [\![\pi_1]\!]_y = 1 \quad \text{and} \quad [\![\pi_1]\!]_{x \cup y} = 0.$$

(The value of $[\![\pi_1]\!]_\varnothing$ does not matter.) Then compute

$$[\![\pi_1 \to \pi_1]\!]_x = 0$$

since $[\![\pi_1]\!]_y = 1$ and $[\![\pi_1]\!]_{x \cup y} = 0$; and

$$[\![\pi_1 \to (\pi_1 \to \pi_1)]\!]_\varnothing = 0$$

since $[\![\pi_1]\!]_x = 1$ and $[\![\pi_1 \to \pi_1]\!]_x = 0$.

Chapter 29

Completeness via maximality

In this chapter we prove completeness results for classical, Wajsberg, and constructive logics. In each subchapter, we assume ⊢ is the inference symbol for some given syntactic logic. The three proofs make different uses of maximal unproving sets, the topic of the first subchapter.

Maximal unproving sets

29.1. *Definitions.* Let Z be a formula, and let $\mathcal{C}$ be a collection of formulas. (We emphasize that these are individual *formulas* such as $\pi_1 \to \pi_3$, not formula *schemes*.) We shall say that

$\mathcal{C}$ is *Z-unproving* if $\mathcal{C} \nvdash Z$;

$\mathcal{C}$ is *maximal Z-unproving* if $\mathcal{C}$ is Z-unproving and $\mathcal{C}$ is not a proper subset of another Z-unproving set;

$\mathcal{C}$ is *maximal unproving* if $\mathcal{C}$ is maximal Z-unproving for at least one formula Z.

Remarks. Compare with 17.5. Note that a collection of formulas is non-trivializing if and only if it is Z-unproving for at least one Z.

Maximal unproving sets 467

The term *maximal* is used in many contexts in mathematics. In most of those contexts — as in this one — a mathematical object is *maximal* for some property that it has, if the object cannot be increased and still retain that property.

29.2. *Maximal extension principle.* Suppose Z is a formula, and $\mathcal{C}_0$ is a Z-unproving collection of formulas.

Then $\mathcal{C}_0$ can be extended to a maximal Z-unproving collection. That is, there exists a maximal Z-unproving collection $\mathcal{M} \supseteq \mathcal{C}_0$.

Remarks (optional). We do not assert that we have a method for *finding* a maximal extension; we shall merely prove that at least one *exists*. Thus, the proof below is not necessarily constructive, even though we may apply the results *to* constructive logic. The distinction between formal system and "outer" system, discussed in 2.17, is important here.

Also, the extension $\mathcal{M}$ whose existence we shall prove is not really a unique or "natural" extension in any sense; the set $\mathcal{C}_0$ may have many other maximal extensions that are just as satisfactory for any purposes we might have in mind.

Our proof will use the fact, established in 6.13, that there are only countably many formulas. That rests on this book's convention, adopted in 6.2 and discussed in 6.4, that our language has only countably many symbols.

Some readers may be familiar with *Zorn's lemma*, a famous principle that is equivalent to the Axiom of Choice. Those readers will see that our "Maximal Extension Principle" is an immediate corollary of Zorn's lemma. In fact, Zorn's lemma can be used to prove a more general version of the "Maximal Extension Principle," applicable to languages that are not necessarily countable; that is of interest to some logicians. However, I prefer to avoid any development of Zorn's lemma in an introductory course. It is too abstract, and too different from any other topics considered in this book, to be readily accessible to the intended audience for this book.

At any rate, we emphasize that the recursive argument used below, or the Zorn's lemma argument that could replace it, is merely part of the metamathematical system ($\Rightarrow$) that we use in order to carry out our studies. It is not part of the formal logical system ($\vdash$) that we are presently studying.

Demonstration of the extension principle. As we noted in 6.13, our formal system has only countably many formulas, which we can represent in a list: $F_1, F_2, F_3, \ldots$.

(There are many orders in which we could form this list. The particular order that we choose will not affect the proof below, though different orders may result in different maximal extensions $\mathcal{M}$. We shall simply pick any one ordering of the formulas $F_1, F_2, F_3, \ldots$, and use that ordering throughout the rest of this proof.)

We are given a set $\mathcal{C}_0$ of formulas. Define sets $\mathcal{C}_1, \mathcal{C}_2, \mathcal{C}_3, \ldots$ recursively — i.e., each set is defined using the previous set — as follows. Assume $\mathcal{C}_{k-1}$ has already been defined (this is clear when $k = 1$). For $k = 1, 2, 3, \ldots$, define

$$\mathcal{C}_k = \begin{cases} \mathcal{C}_{k-1} & \text{if } \mathcal{C}_{k-1} \cup \{F_k\} \vdash Z, \\ \mathcal{C}_{k-1} \cup \{F_k\} & \text{if } \mathcal{C}_{k-1} \cup \{F_k\} \nvdash Z. \end{cases}$$

It follows immediately, by induction on k, that $\mathcal{C}_k \nvdash Z$ for each k. Indeed, we can describe the "construction" this way: At the kth step, we "add" the formula F_k if[1] that addition does not enable the collection to prove Z. I have placed the word "add" in quotes because if F_k is one of the formulas that was in the original collection $\mathcal{C}_0$, then we are not really changing the collection when we "add" F_k.

Let $\mathcal{M} = \bigcup_{j=0}^{\infty} \mathcal{C}_j$. We shall show that $\mathcal{M}$ has the required properties.

To show that $\mathcal{M}$ is Z-unproving, suppose on the contrary that $\mathcal{M} \vdash Z$. Then there is some derivation of Z from $\mathcal{M}$. Now, any derivation has only finitely many steps, and involves only

---

[1] We're using classical logic for our metamathematical reasoning system, so either $\mathcal{C}_{k-1} \cup \{F_k\}$ proves Z or it doesn't. However, even though we agree that one of those two alternatives must hold, we might not know *which* holds. That question is answerable in an algorithmic fashion for some inference systems, but not for others. In this respect, at least, the proof of the maximal principle is not constructive.

Moreover, to carry out the entire "construction" would take infinitely many steps. In what sense can such a procedure be carried out? We can imagine doing the first step in one second, the second step in a half second, the third step in a quarter second, and so on, but that is just imaginary. And where would we find a sheet of paper large enough for us to write the results? The proof is explicit enough to convince us that $\mathcal{M}$ exists (unless we are constructivists), but the proof does not actually *find* an example of a maximal collection $\mathcal{M}$. Perhaps this proof can be taken as evidence of the existence of God, since no one else can actually carry out the "construction." (But this speculation is intended more as humor than as serious theology.)

Maximal unproving sets 469

finitely many formulas. Hence the same derivation that shows $\mathcal{M} \vdash Z$, also shows $\{M_1, M_2, \ldots, M_j\} \vdash Z$ for some finite set $\{M_1, M_2, \ldots, M_j\} \subseteq \mathcal{M}$. Since $\mathcal{M}$ is the union of the $\mathcal{C}$'s, we have

$$M_1 \in \mathcal{C}_{p_1}, \qquad M_2 \in \mathcal{C}_{p_2}, \qquad \ldots, \qquad M_j \in \mathcal{C}_{p_j}$$

for some integers $p_1, p_2, \ldots, p_j$. Let k be the largest of those integers. Since $\mathcal{C}_0, \mathcal{C}_1, \mathcal{C}_2, \mathcal{C}_3, \ldots$ is an increasing sequence, we have $\{M_1, M_2, \ldots, M_j\} \subseteq \mathcal{C}_k$. But then our derivation also shows $\mathcal{C}_k \vdash Z$, a contradiction. Thus $\mathcal{M}$ is Z-unproving, after all.

To show that $\mathcal{M}$ is maximal, suppose that $\mathcal{N} \supsetneq \mathcal{M}$; we will show that $\mathcal{N}$ is not Z-unproving — i.e., we will show that $\mathcal{N} \vdash Z$. Since $\mathcal{N}$ is strictly larger than $\mathcal{M}$, there is at least one formula H belonging to $\mathcal{N} \setminus \mathcal{M}$; pick any such formula H. Since $F_1, F_2, F_3, \ldots$ is a list of *all* formulas, we have $H = F_k$ for some integer $k \geq 1$. Since $F_k \notin \mathcal{M}$ and $\mathcal{C}_k \subseteq \mathcal{M}$, we have $F_k \notin \mathcal{C}_k$. Now look at how $\mathcal{C}_k$ was recursively defined; since $F_k \notin \mathcal{C}_k$, we must have $\mathcal{C}_{k-1} \cup \{F_k\} \vdash Z$. Now, $\mathcal{C}_{k-1} \subseteq \mathcal{M} \subseteq \mathcal{N}$ and $F_k = H \in \mathcal{N}$, so $\mathcal{C}_{k-1} \cup \{F_k\} \subseteq \mathcal{N}$. Therefore $\mathcal{N} \vdash Z$.

29.3. *Definition.* We shall say that a set $\mathcal{R}$ of formulas is *syntactically closed* if it contains all its consequences — i.e.,

$$\text{whenever } \mathcal{R} \vdash W, \text{ then also } \mathcal{R} \ni W.$$

(Note that the converse, $\mathcal{R} \ni W \Rightarrow \mathcal{R} \vdash W$, holds for *any* set $\mathcal{R}$ of formulas.) Here are some examples:

a. The set of *all* formulas is syntactically closed.

b. The set of all *theorems* is syntactically closed. In fact, it is the smallest syntactically closed set; it is a subset of any syntactically closed set.

c. Any maximal unproving set is syntactically closed. (The proof is left as an exercise; use several sentences.)

29.4. Suppose that $\vdash$ represents an inference system that includes adjunction and $\wedge$-elimination. Let $\mathcal{R}$ be a syntactically closed set. Then for any formulas A and B,

$$A \wedge B \in \mathcal{R} \text{ if and only if both } A \text{ and } B \text{ are members of } \mathcal{R}.$$

(The proof is left as an exercise.)

29.5. *Definition.* A set $\mathcal{R}$ of formulas is *prime* if it has the property that whenever $A \vee B$ is a member of $\mathcal{R}$, then at least one of the formulas A or B is a member of $\mathcal{R}$. Here are some examples:

- The set of *all* formulas is prime.
- The empty set is prime.
- If our inference system is constructive logic, then the set of all theorems is prime. That is immediate from the Disjunction Property 27.11.
- If our inference system is classical logic, then the set of all theorems is *not* prime. For instance, $\pi_1 \vee \overline{\pi_1}$ is a theorem, but neither π_1 nor $\overline{\pi_1}$ is a theorem.

29.6. If our inference system is classical, constructive, or Wajsberg logic, then any maximal unproving set is prime.

Proof. Suppose that $\mathcal{M}$ is maximal Z-unproving and

$$A \vee B \in \mathcal{M}, \qquad A \notin \mathcal{M}, \qquad B \notin \mathcal{M};$$

we shall arrive at a contradiction. Since the set $\mathcal{M}$ is maximal Z-unproving, neither $\mathcal{M} \cup \{A\}$ nor $\mathcal{M} \cup \{B\}$ is Z-unproving. That is, we have both $\mathcal{M} \cup \{A\} \vdash Z$ and $\mathcal{M} \cup \{B\} \vdash Z$.

In the cases of classical and constructive logic, we may apply the classical deduction principle 22.5. Thus we obtain $\mathcal{M} \vdash A \to Z$ and $\mathcal{M} \vdash B \to Z$. That is, both $A \to Z$ and $B \to Z$ are members of $\mathcal{M}$. We also have $A \vee B \in \mathcal{M}$. Using "proof by cases" 14.4.b.δ, we obtain $\mathcal{M} \vdash Z$, a contradiction.

In the case of Wajsberg logic, we may apply the Wajsberg deduction principle 24.26. Thus we obtain $\mathcal{M} \vdash A \rightsquigarrow Z$ and $\mathcal{M} \vdash B \rightsquigarrow Z$. That is, both $A \rightsquigarrow Z$ and $B \rightsquigarrow Z$ are members of $\mathcal{M}$. We also have $A \vee B \in \mathcal{M}$. Using the "fuzzy proof by cases" 24.13, we obtain $\mathcal{M} \vdash Z$, a contradiction.

Classical logic

29.7. *Assumption.* Throughout this subchapter, $\vdash$ will mean

Classical logic

provability in *classical* syntactics — i.e., using the assumptions in 25.1. Thus, unproving sets and maximal unproving sets are defined as in 29.1, with classical logic for the inference system.

29.8. *Maximality and logical operators.* Let $\mathcal{M}$ be a maximal unproving set. Then, for all formulas A and B,
   a. $A \wedge B \in \mathcal{M}$ if and only if both A and B are members of $\mathcal{M}$.
   b. $A \vee B \in \mathcal{M}$ if and only if at least one of A, B belongs to $\mathcal{M}$.
   c. $\overline{A} \in \mathcal{M}$ if and only if $A \notin \mathcal{M}$.
   d. $A \to B \in \mathcal{M}$ if and only if at least one of $\overline{A}, B$ is a member of $\mathcal{M}$.

Proof. Part a is immediate from 29.3.c and 29.4. Part b is immediate from 29.6.

For part c, reason as follows: $A \vee \overline{A}$ is a theorem of classical logic, by 23.2.c. Hence $A \vee \overline{A} \in \mathcal{M}$, as noted in 29.3.b. Hence by part b, at least one of $A, \overline{A}$ is a member of $\mathcal{M}$. On the other hand, if both A and $\overline{A}$ are members of $\mathcal{M}$, then $\mathcal{M}$ can prove every formula, as noted in 22.12. But that would contradict the fact that $\mathcal{M}$ is an unproving set. Thus one, and only one, of the two formulas $A, \overline{A}$ belongs to $\mathcal{M}$.

For part d, note that in classical logic we have $\vdash (A \to B) \leftrightarrow (\overline{A} \vee B)$. Since $\mathcal{M}$ is syntactically closed, we have $A \to B \in \mathcal{M}$ if and only if $\overline{A} \vee B \in \mathcal{M}$. Now apply part b.

29.9. *Construction of valuations.* Let $\mathcal{M}$ be a maximal unproving set. Let $[\![\ ]\!]$ be its characteristic function (defined as in 3.27); thus

$$[\![A]\!] = \begin{cases} 1 & \text{if } A \in \mathcal{M}, \\ 0 & \text{if } A \notin \mathcal{M}. \end{cases}$$

Then $[\![\ ]\!]$ is a valuation in the two-valued interpretation (8.2); i.e., it satisfies the rules in 8.2.c. *Proof.* Immediate from 29.8.

29.10. The term "model" has different uses in logic. In this subchapter, a *model* for a set $\mathcal{C}$ of formulas will mean a valuation $[\![\ ]\!]$ in the two-valued interpretation (8.2) that makes every member of $\mathcal{C}$ true — i.e., that valuates every member of $\mathcal{C}$ to 1.

We do not assert that $[\![\ ]\!]$ will make *only* the members of $\mathcal{C}$ true. It will make true *at least* the members of $\mathcal{C}$, but it may also make some other formulas true.

We emphasize that the collection $\mathcal{C}$ can be any collection of formulas. They do not need to be specified by formula *schemes*.

29.11. *Gödel-Malcev characterization of non-trivializing extensions.* Let $\mathcal{C}$ be any collection of formulas. Then $\mathcal{C}$ is non-trivializing (defined in 17.5) if and only if it has a model (defined in 29.10).

Demonstration of characterization. First, suppose that $\mathcal{C}$ is Z-unproving for some formula Z. Then $\mathcal{C} \subseteq \mathcal{M}$ for some maximal Z-unproving collection $\mathcal{M}$, by 29.2. Now apply 29.9. The resulting valuation is a model for $\mathcal{M}$, and hence also a model for $\mathcal{C}$.

On the other hand, suppose $\mathcal{C}$ has a model $[\![\ ]\!]$. Thus $\mathcal{C} \ni A \Rightarrow [\![A]\!] = 1$, for all formulas A. By induction on lengths of proofs, it can be shown that $\mathcal{C} \vdash A \Rightarrow [\![A]\!] = 1$. But one of the numbers $[\![\pi_1]\!], [\![\overline{\pi_1}]\!]$ is 1, and the other is 0. Hence at least one of the statements $\mathcal{C} \vdash \pi_1$, $\mathcal{C} \vdash \overline{\pi_1}$ is false. Therefore $\mathcal{C}$ is non-trivializing.

29.12. *Classical completeness.* The theorems of classical syntactics (axiomatized as in 25.1) are the same as the tautologies of the classical two-valued interpretation (8.2). That is, $\vdash A$ if and only if $\vDash A$.

More generally, classical syntactic logic and the two-valued interpretation have the same inference rules. That is, if $\mathcal{C}$ is any collection of formulas, then $\mathcal{C} \vdash A$ if and only if $\mathcal{C} \vDash A$.

Demonstration. The first equivalence ($\Longleftrightarrow$) listed below is by 22.13.c; the second is by 29.11.

$\qquad \mathcal{C} \vdash A$
$\Longleftrightarrow \quad \mathcal{C} \cup \{\overline{A}\}$ is trivializing
$\Longleftrightarrow \quad$ there is no model of $\mathcal{C} \cup \{\overline{A}\}$
$\Longleftrightarrow \quad$ every two-valued valuation makes at least one member of $\mathcal{C} \cup \{\overline{A}\}$ false

Classical logic 473

$\iff$ every two-valued valuation that makes all the members of $\mathcal{C}$ true, must make $\overline{\overline{A}}$ false

$\iff$ every two-valued valuation that makes all the members of $\mathcal{C}$ true, must make A true

$\iff$ $\mathcal{C} \vDash A$.

29.13. *Exercise.* Suppose A and B are formulas that share no variables, and $\vdash A \to B$ in constructive logic. Then at least one of $\vdash \overline{\overline{A}}$ or $\vdash \overline{\overline{B}}$ in constructive logic.

Hint. 8.15 and 29.12. *Remark.* For a stronger result see 29.30.

29.14. *Post-completeness of classical logic.* We cannot add any new axiom schemes to classical logic and still have it remain nontrivial; it is already maximal nontrivial. More precisely:

Suppose $\mathcal{S}$ is a formula scheme, such as $A_1 \to (A_2 \vee \overline{A_2})$ — that is, an expression made from $\vee, \wedge, \to, \neg$, parentheses, and metavariables, in the usual fashion, but without any of the specific propositional variable symbols $\pi_1, \pi_2, \ldots$. Then either

- $\mathcal{S}$ is a theorem scheme of classical syntactics — i.e., it already can be proved by a derivation from the axioms, or

- adding $\mathcal{S}$ to classical syntactics as an additional axiom scheme makes every formula provable.

Remarks. A logic is called *Post-complete* (for logician Emil Post) if it is maximal nontrivial. It can be shown that *Abelian logic*, presented in Chapter 26, is also Post-complete; for details see Meyer and Slaney [1989].

Demonstration of maximality of classical logic. Assume that adding $\mathcal{S}$ does not make the system trivializing, and yet $\mathcal{S}$ is not a theorem scheme; we shall obtain a contradiction.

Say the metavariables that occur in formula scheme $\mathcal{S}$ are $A_1, A_2, \ldots, A_n$. Also, let $\mathcal{C}$ denote the axioms of classical syntactics together with all the infinitely many formulas that are represented by the formula scheme $\mathcal{S}$.

On the one hand, since $\mathcal{S}$ is not a theorem scheme, at least one of the formulas that it represents is not a theorem. By 29.12, that formula is not a tautology — i.e., there is at least one two-valued valuation $[\![\ ]\!]_0$ making that formula take the value 0. That formula is obtained by substituting some particular formulas $F_1, F_2, \ldots, F_n$ for the metavariables $A_1, A_2, \ldots, A_n$ respectively.

Each of the numbers

$$c_1 = [\![F_1]\!]_0, \quad c_2 = [\![F_2]\!]_0, \quad \ldots, \quad c_n = [\![F_n]\!]_0$$

is either 0 or 1. We can now restate that information in a fashion that refers to the numbers c_i but does not refer directly to the valuation $[\![\ ]\!]_0$:

> If we substitute numbers c_j for the metavariables A_j, and substitute numerical operators $\ovee, \owedge, \ominus, \obar$ for the symbols $\vee, \wedge, \neg, \rightarrow$, then the formula scheme $\mathcal{S}$ gets replaced by an arithmetical expression that evaluates to 0.

We will return to that fact shortly.

On the other hand, since $\mathcal{C}$ is non-trivializing, by 29.11 there exists at least one two-valued valuation $[\![\ ]\!]_1$ that makes every member of $\mathcal{C}$ equal to 1. One of the numbers $[\![\pi_1]\!]_1, [\![\overline{\pi_1}]\!]_1$ is 0 and the other is 1. Let G_0 and G_1 be the two formulas π_1 and $\overline{\pi_1}$, ordered in such a way that $[\![G_0]\!]_1 = 0$ and $[\![G_1]\!]_1 = 1$. Now, one of the formulas H represented by the formula scheme $\mathcal{S}$ can be described this way:

$$\text{replace metavariable } A_j \text{ with } \begin{cases} \text{formula } G_0 \text{ if } c_j = 0, \\ \text{formula } G_1 \text{ if } c_j = 1. \end{cases}$$

What is then the value of $[\![H]\!]_1$? We can analyze it in two ways:

On the one hand, $[\![H]\!]_1$ is computed by replacing each metavariable A_j with the number c_j, and replacing each of the symbols $\vee, \wedge, \neg, \rightarrow$ with the corresponding arithmetical operator $\ovee, \owedge, \ominus, \obar$. Analyzing in this fashion, we see that $[\![H]\!]_1 = 0$.

On the other hand, H is one of the formulas represented by the scheme $\mathcal{S}$, so H is a member of the collection $\mathcal{C}$. By our choice of $[\![\ ]\!]_1$, we have $[\![H]\!]_1 = 1$.

This contradiction completes the proof.

Classical logic 475

29.15. *Principle of classical admissibility.* In classical logic, any admissibility rule yields the corresponding inference rule. That is,

$$\text{if } \{\vdash y_1, \vdash y_2, \ldots, \vdash y_n\} \Rightarrow \vdash z$$
$$\text{then } \{y_1, y_2, \ldots, y_n\} \vdash z.$$

Remarks. We actually have an if-and-only-if result. But the other half — inference implies admissibility — follows easily from our definition of derivation.

The hard half (admissibility implies inference) fails in some nonclassical logics. See, for instance, 27.15 and 27.17.

Discussion of notation. It is our understanding that all the y_i's and z are formula schemes, involving metavariables $A_1, A_2, \ldots$ but not propositional variable symbols $\pi_1, \pi_2, \ldots$. For instance, we might have $y_1 = (A_1 \wedge A_2) \to \neg A_3$ and $z = A_2 \to (A_1 \vee A_4)$. The "if" part of the principle says that the formula schemes are related in such a way that, whenever some substitution makes all the y_i's into theorems, then the same substitution makes z into a theorem. The "then" part asserts the derivability of an inference rule.

Proof. We begin by simplifying the notation, reducing the problem to the case of $n = 1$. That can be accomplished as follows. Let x be the formula scheme $y_1 \wedge y_2 \wedge y_3 \wedge \cdots \wedge y_m$. (We can omit the parentheses, as in 14.18.g.) Then what we're trying to prove is clearly equivalent to this simpler statement:

$$\text{if } \quad \vdash x \Rightarrow \vdash z, \quad \text{then} \quad x \vdash z.$$

We now restate that in some other equivalent forms, to get a statement that will be easier to prove. By the deduction principle (22.5), that last statement is equivalent to this one:

$$\text{if } \quad \vdash x \Rightarrow \vdash z, \quad \text{then} \quad \vdash x \to z.$$

By completeness (29.12), that is equivalent to this statement:

$$\text{if } \quad \vDash x \Rightarrow \vDash z, \quad \text{then} \quad \vDash x \to z.$$

Let us restate that in slightly more detail. It says:

476   Chapter 29. Completeness via maximality

> if X and Z are formula schemes with the property that whenever some substitution makes X tautological in the two-valued interpretation, then the same substitution also makes Z tautological in that interpretation,
>
> then   no matter what formulas we substitute for the metavariables involved, the formula scheme $X \to Z$ is tautological in the two-valued interpretation.

We will prove the contrapositive of that statement. That is:

> We assume   there is at least one nontautological formula $X_0 \to Z_0$ that is an instance of the formula scheme $X \to Z$.
>
> We will show   there is at least one substitution that makes the formula scheme X into a tautological formula X_1, but makes the formula scheme Z into a nontautological formula Z_1.

By our assumption, $X_0 \to Z_0$ is nontautological. It is obtained by substituting some particular formula F_j for each A_j, where $A_1, A_2, \ldots, A_p$ are the metavariables that appear in X and Z. Our hypothesis concerns only the entire formula $X_0 \to Z_0$, but we may also consider the separate formulas X_0 and Z_0 obtained from X and Z via that substitution.

Since $X_0 \to Z_0$ is not tautological, there is at least one particular valuation $[\![\ ]\!]_0$ such that $[\![X_0 \to Z_0]\!]_0 = 0$. From the properties of the classical two-valued interpretation, it follows that $[\![X_0]\!]_0 = 1$ and $[\![Z_0]\!]_0 = 0$.

Each of the numbers

$$c_1 = [\![F_1]\!]_0, \quad c_2 = [\![F_2]\!]_0, \quad \ldots, \quad c_p = [\![F_p]\!]_0$$

is either 0 or 1. These numbers have been chosen so that they have the following property (which does not specifically refer to $[\![\ ]\!]_0$ or X_0 or Z_0):

If we substitute the numbers c_j for the metavariables A_j, and substitute numerical operators $\vee\!\!\!\bigcirc, \wedge\!\!\!\bigcirc, \ominus, \oslash$ for the symbols $\vee, \wedge, \neg, \to$, then the formula schemes X and Z are replaced by arithmetical expressions that evaluate to 1 and 0, respectively.

Now define a new substitution, as follows: Replace each metavariable A_j with the particular formula G_j, where

$$G_j = \begin{cases} \pi_1 \wedge \neg \pi_1 & \text{if } c_j = 0, \\ \pi_1 \vee \neg \pi_1 & \text{if } c_j = 1. \end{cases}$$

This substitution transforms the formula schemes X and Z into particular formulas X_1 and Y_1, which involve only the symbols $\pi_1, \vee, \wedge, \neg, \to$. Moreover, these formulas have been chosen in such a way that, in every valuation of the classical two-valued interpretation, no matter what we choose for $[\![\pi_1]\!]$, we will get $[\![X_1]\!] = 1$ and $[\![Z_1]\!] = 0$. Hence X_1 is tautologous and Z_1 is not. This completes the proof.

WAJSBERG LOGIC

29.16. *Assumption.* Throughout this subchapter, $\vdash$ will mean provability in *Wajsberg* syntactics — that is, the results of Chapter 24, including 24.14 and its consequences. Thus, unproving sets and maximal unproving sets are defined as in 29.1, with Wajsberg logic for the inference system.

29.17. *Maximality and logical operators.* Let $\mathcal{M}$ be a maximal unproving set. Then, for all formulas A and B,
   a. $A \wedge B \in \mathcal{M}$ if and only if both A and B are members of $\mathcal{M}$.
   b. $A \vee B \in \mathcal{M}$ if and only if at least one of A or B is a member of $\mathcal{M}$.
   c. For each formula A, exactly one of $A, \overline{A}, \widetilde{A}$ belongs to $\mathcal{M}$.

Proof. Part a is immediate from 29.3.c and 29.4. Part b is immediate from 29.6.

For part c, reason as follows: $A \vee \overline{A} \vee \widetilde{A}$ is a theorem, by 24.16. Hence it is a member of $\mathcal{M}$. By part b, then, at least one of $A, \overline{A}, \widetilde{A}$ is a member of $\mathcal{M}$. If two of those are members of $\mathcal{M}$, then *every* formula is consequence of $\mathcal{M}$, by 24.20.b. Since $\mathcal{M}$ is syntactically closed, $\mathcal{M}$ contains all formulas. But that contradicts our assumption that $\mathcal{M}$ is non-trivializing (as defined in 29.1).

29.18. *Tedious logical operations.* (These results are needed for 29.19, but some readers may prefer to skip this section in their first reading.)

Let $\mathcal{M}$ be a maximal unproving set. Then for all formulas A and B:
  a. If $\mathcal{M} \vdash A \leftrightarrow B$, then (by 14.14)
$$A \in \mathcal{M} \Leftrightarrow B \in \mathcal{M}, \quad \overline{A} \in \mathcal{M} \Leftrightarrow \overline{B} \in \mathcal{M}, \quad \widetilde{A} \in \mathcal{M} \Leftrightarrow \widetilde{B} \in \mathcal{M}.$$
  b. If $\widetilde{A} \in \mathcal{M}$, then $\widetilde{\neg A} \in \mathcal{M}$. (Immediate from 24.11.a.)
  c. If $\overline{A} \in \mathcal{M}$ or $B \in \mathcal{M}$, then $A \to B \in \mathcal{M}$. (By 24.21.d).
  d. If $A \in \mathcal{M}$, then
$$A \vee B \in \mathcal{M}, \quad \mathcal{M} \vdash (A \to B) \leftrightarrow B, \quad \mathcal{M} \vdash (A \wedge B) \leftrightarrow B$$
  (by $\vee$-introduction, 24.21.b, and 24.21.a).
  e. If $\overline{B} \in \mathcal{M}$, then
$$\overline{A \wedge B} \in \mathcal{M}, \quad \mathcal{M} \vdash (A \vee B) \leftrightarrow A, \quad \mathcal{M} \vdash (A \to B) \leftrightarrow \overline{A}$$
  (by De Morgan's laws, $\vee$-introduction, 24.21.a, and 24.21.c).
  f. If $\widetilde{A}, \widetilde{B} \in \mathcal{M}$, then also $\widetilde{A \wedge B}, \widetilde{A \vee B}, \widetilde{A \to B} \in \mathcal{M}$ (by 24.21.e and 24.21.f).

29.19. *Construction of valuations.* Let $\mathcal{M}$ be a maximal unproving set. Define a function $[\![\ ]\!] : \{\text{formulas}\} \to \{0, \tfrac{1}{2}, 1\}$ by

$$[\![A]\!] = \begin{cases} 1 & \text{if } A \in \mathcal{M}, \\ 1/2 & \text{if } \widetilde{A} \in \mathcal{M}, \\ 0 & \text{if } \overline{A} \in \mathcal{M}. \end{cases}$$

Then $[\![\ ]\!]$ is a valuation in the three-valued fuzzy interpretation — i.e., $[\![\ ]\!]$ has the properties listed in 8.17.

Proof. Follows easily from 29.17 and 29.18.

29.20. *Fuzzy completeness.* Let Z be any formula. Then Z is a theorem of Wajsberg logic (as developed in Chapter 24) if and only if Z is a tautology of the three-valued fuzzy semantics (as presented in 8.17).

Proof. By the soundness results of 24.2, every theorem is a tautology. Conversely, suppose Z is not a theorem. Then $\varnothing$ is a Z-unproving set. By 29.2, it can be extended to a maximal Z-unproving set, $\mathcal{M}$. Then $Z \notin \mathcal{M}$. Define $[\![\ ]\!]$ as in 29.19; then $[\![Z]\!] \neq 1$. Thus Z is not a tautology.

CONSTRUCTIVE LOGIC

29.21. *Assumption.* Throughout this subchapter, $\vdash$ will mean provability in *constructive* logic — i.e., following the development in sections 22.1–22.13. Thus, unproving sets and maximal unproving sets are defined as in 29.1, with constructive logic for the inference system.

29.22. *Definition.* A set $\mathcal{S}$ of formulas will be called *disjunction regular* if it is

- non-trivializing (see 29.1),
- syntactically closed (see 29.3), and
- prime (see 29.5).

For instance, any maximal unproving set is disjunction regular. Also, the set of all theorems is disjunction regular; moreover, it is a subset of any disjunction regular set.

We shall denote

$$\Omega = \{\text{disjunction regular sets of formulas}\}.$$

Throughout the remainder of this subchapter, most of our applications of set theory will be concerned with subsets of this set Ω. Hence $\complement S$ will mean $\Omega \setminus S$.

Notations. In the discussions below, A and B represent formulas; $\mathcal{S}$ is a set of formulas; and Ω and $\langle\!\langle A \rangle\!\rangle$ are sets of sets of formulas. We will soon develop a topology Σ on the set Ω; then Σ is a set of sets of sets of formulas.[2]

---

[2] Yes, I know those nested sets make your head hurt. Mine too. But this is the simplest proof I've found so far. We could use posets and Heyting

29.23. Suppose $\mathcal{R}$ is a disjunction regular set of formulas. Then:
a. $A \wedge B \in \mathcal{R}$ if and only if both $A \in \mathcal{R}$ and $B \in \mathcal{R}$.
b. $A \vee B \in \mathcal{R}$ if and only if at least one of A, B belongs to $\mathcal{R}$.
c. $A \to B \in \mathcal{R}$ if and only if, for each disjunction regular set $\mathcal{S} \supseteq \mathcal{R}$, we have at least one of $A \notin \mathcal{S}$ or $B \in \mathcal{S}$.
d. $\overline{A} \in \mathcal{R}$ if and only if, for each disjunction regular set $\mathcal{S} \supseteq \mathcal{R}$, we have $A \notin \mathcal{S}$.

Demonstration of a. Immediate from 29.4.

Demonstration of b. Immediate from primeness, syntactic closure, and the two axioms of $\wedge$-elimination.

Demonstration of c. On the one hand, suppose that $A \to B \in \mathcal{R}$. Then also $A \to B \in \mathcal{S}$ for every disjunction regular set $\mathcal{S} \supseteq \mathcal{R}$. Since $\mathcal{S}$ is syntactically closed, if $A \in \mathcal{S}$ then $B \in \mathcal{S}$.

On the other hand, suppose that $A \to B \notin \mathcal{R}$. Then $\mathcal{R} \nvdash A \to B$. By the deduction principle, $\mathcal{R} \cup \{A\} \nvdash B$. That is, $\mathcal{R} \cup \{A\}$ is B-unproving. By 29.2, there is some maximal B-unproving set $\mathcal{S} \supseteq \mathcal{R} \cup \{A\}$. Then $\mathcal{S} \nvdash B$ and $\mathcal{S}$ is disjunction regular.

Demonstration of d. Let B be the formula $\pi_1 \wedge \overline{\pi_1}$. Then B cannot be a member of any disjunction regular set. By 17.3, we have $\vdash \overline{A} \leftrightarrow (A \to B)$. Since $\mathcal{R}$ is syntactically closed, we have $\overline{A} \in \mathcal{R}$ if and only if $A \to B \in \mathcal{R}$. Now apply part c.

29.24. *Definitions.* Let A be any formula. By a *disjunction regular proving set for* A, we shall mean a set $\mathcal{S}$ that is disjunction regular and that satisfies $\mathcal{S} \vdash A$, or equivalently that satisfies $\mathcal{S} \ni A$. (The equivalence follows from the fact that any disjunction regular set is syntactically closed.)

Now let the collection of *all* disjunction regular proving sets for a given formula A be denoted by $\langle\!\langle A \rangle\!\rangle$. Note that $\langle\!\langle A \rangle\!\rangle$ is a subset of Ω, so it is a set of sets of formulas.

---

algebras, but that would require many definitions and examples and would reduce our set hierarchy by only one level.

Constructive logic

29.25. Observations. Let A and B be any formulas. Then:

a. $\langle\!\langle A \wedge B \rangle\!\rangle = \langle\!\langle A \rangle\!\rangle \cap \langle\!\langle B \rangle\!\rangle$.

b. $\langle\!\langle A \vee B \rangle\!\rangle = \langle\!\langle A \rangle\!\rangle \cup \langle\!\langle B \rangle\!\rangle$.

c. $\langle\!\langle \overline{A} \rangle\!\rangle$ and $\langle\!\langle A \rangle\!\rangle$ are disjoint; hence $\langle\!\langle \overline{A} \rangle\!\rangle \subseteq \complement\langle\!\langle A \rangle\!\rangle$.

d. $\langle\!\langle A \to B \rangle\!\rangle \subseteq \langle\!\langle B \rangle\!\rangle \cup \complement\langle\!\langle A \rangle\!\rangle$.

e. $\langle\!\langle A \rangle\!\rangle = \Omega$ if and only if A is a theorem.

f. $\langle\!\langle A \rangle\!\rangle = \varnothing$ if and only if the singleton $\{A\}$ is trivializing (defined in 17.5). In particular, $\langle\!\langle B \wedge \overline{B} \rangle\!\rangle = \varnothing$.

g. $\langle\!\langle A \rangle\!\rangle \subseteq \langle\!\langle B \rangle\!\rangle$ if and only if $A \to B$ is a theorem.

h. $\langle\!\langle \overline{A} \rangle\!\rangle = \langle\!\langle A \to (B \wedge \overline{B}) \rangle\!\rangle$ for any formulas A and B. (Thus $B \wedge \overline{B}$ acts as an Ackermann falsehood.)

Proof of a. Immediate from the definitions.

Proof of b. Immediate from the definitions and the fact that any disjunction regular set is prime.

Proof of c. The formulas A and $\overline{A}$ cannot both be members of a non-trivializing set.

Proof of d. Suppose not (for contradiction). Then there is some disjunction regular set $\mathcal{S} \in \langle\!\langle A \to B \rangle\!\rangle$ such that $\mathcal{S} \notin \langle\!\langle B \rangle\!\rangle \cup \complement\langle\!\langle A \rangle\!\rangle$. That is, $\mathcal{S} \notin \langle\!\langle B \rangle\!\rangle$ and $\mathcal{S} \in \langle\!\langle A \rangle\!\rangle$. Thus $\mathcal{S} \vdash A \to B$ and $\mathcal{S} \vdash A$ but $\mathcal{S} \nvdash B$ — which contradicts detachment.

Proof of e. If A is a theorem, then $\vdash A$, hence $\mathcal{S} \vdash A$ for any set of formulas $\mathcal{S}$. So for every disjunction regular set $\mathcal{S}$ we have $\mathcal{S} \vdash A$, hence $\mathcal{S} \in \langle\!\langle A \rangle\!\rangle$. Thus $\langle\!\langle A \rangle\!\rangle = \Omega$. On the other hand, if $\langle\!\langle A \rangle\!\rangle = \Omega$, then A is a member of every disjunction regular set. The set of all theorems is disjunction regular, so A is a member of that set.

Proof of f. Immediate from the definitions.

Proof of g. Immediate from the definitions.

Proof of h. Immediate from g and 17.3.

29.26. *Definition/notation.* By an "open" set we will mean a set that is equal to the union of some (or all, or none) of the $\langle\langle A \rangle\rangle$'s, for various choices of the formulas A. For instance,

- $\varnothing$ is an "open" set (since we may take the union of *none* of the $\langle\langle A \rangle\rangle$'s);
- $\langle\langle \overline{\pi_1 \to \pi_2} \rangle\rangle$ is an "open" set;
- $\langle\langle \overline{\pi_1 \to \pi_2} \rangle\rangle \cup \langle\langle \overline{\pi_3 \wedge \pi_7} \rangle\rangle \cup \langle\langle \overline{\pi_1 \vee \pi_4} \rangle\rangle$ is an "open" set;
- $\langle\langle \overline{\pi_1 \to \pi_2} \rangle\rangle \cup \langle\langle \overline{\pi_2 \to \pi_3} \rangle\rangle \cup \langle\langle \overline{\pi_3 \to \pi_4} \rangle\rangle \cup \cdots$ is an "open" set;
- Ω is an "open" set (since it is equal to $\langle\langle \pi_1 \to \pi_1 \rangle\rangle$; see 29.25.e).

The collection of all "open" sets will be denoted by Σ.

29.27. *Lemma.* The set Σ is a topology on Ω.

(Thus the "open" sets are actually the open sets of a topological space; our terminology was justified. After we prove this lemma we will be able to drop the quotation marks.)

Proof. We already noted that $\varnothing$ and Ω are members of Σ. And obviously the union of arbitrarily many members of Σ is also a member of Σ, since the union of unions is just a union. So we just need to show that the intersection of two given members of Σ is also a member of Σ. Say the two given members of Σ are

$$\bigcup_{\lambda \in L} \langle\langle A_\lambda \rangle\rangle \quad \text{and} \quad \bigcup_{\mu \in M} \langle\langle B_\mu \rangle\rangle$$

where $\{A_\lambda : \lambda \in L\}$ and $\{B_\mu : \mu \in M\}$ are two sets of formulas. Their intersection is the set

$$\left(\bigcup_{\lambda \in L} \langle\langle A_\lambda \rangle\rangle \right) \cap \left(\bigcup_{\mu \in M} \langle\langle B_\mu \rangle\rangle \right) = \bigcup_{\lambda \in L,\, \mu \in M} \left(\langle\langle A_\lambda \rangle\rangle \cap \langle\langle B_\mu \rangle\rangle \right) = \bigcup_{\lambda \in L,\, \mu \in M} \langle\langle A_\lambda \wedge B_\mu \rangle\rangle.$$

Thus it is a union of values of $\langle\langle\ \rangle\rangle$, so it is also a member of Σ.

Constructive logic

29.28. *Proposition.* The function $\langle\!\langle\ \rangle\!\rangle$ is one of the valuations $[\![\ ]\!]$ for the topological interpretation given by the topology defined in 29.22–29.27. (The *principal valuation* would be a good name for it.)

Proof. We must show that the function $\langle\!\langle\ \rangle\!\rangle$ satisfies the four equations in 10.1.c. The two equations for $\wedge$ and $\vee$ were easy — they didn't even involve the topology, and we already showed them in 29.25. It remains to show that the function $\langle\!\langle\ \rangle\!\rangle$ satisfies

$$\langle\!\langle A \to B \rangle\!\rangle = \mathrm{int}\Big(\langle\!\langle B \rangle\!\rangle \cup \mathsf{C}\langle\!\langle A \rangle\!\rangle\Big), \qquad \langle\!\langle \overline{A} \rangle\!\rangle = \mathrm{int}\Big(\mathsf{C}\langle\!\langle A \rangle\!\rangle\Big).$$

We saw in 29.25 that $\langle\!\langle A \to B \rangle\!\rangle \subseteq \langle\!\langle B \rangle\!\rangle \cup \mathsf{C}\langle\!\langle A \rangle\!\rangle$. The left side is an open set; hence it is contained in the interior of the right side. This proves $\langle\!\langle A \to B \rangle\!\rangle \subseteq \mathrm{int}(\langle\!\langle B \rangle\!\rangle \cup \mathsf{C}\langle\!\langle A \rangle\!\rangle)$.

To prove $\mathrm{int}(\langle\!\langle B \rangle\!\rangle \cup \mathsf{C}\langle\!\langle A \rangle\!\rangle) \subseteq \langle\!\langle A \to B \rangle\!\rangle$, reason as follows: Suppose (for contradiction) that that inclusion is false. Then there exists some disjunction regular set $\mathcal{R}$ such that

$$(1) \qquad \mathcal{R} \in \mathrm{int}\Big(\langle\!\langle B \rangle\!\rangle \cup \mathsf{C}\langle\!\langle A \rangle\!\rangle\Big) \setminus \langle\!\langle A \to B \rangle\!\rangle.$$

Since $\mathcal{R} \notin \langle\!\langle A \to B \rangle\!\rangle$, we have $A \to B \notin \mathcal{R}$. By 29.23, there is some disjunction regular set $\mathcal{S} \supseteq \mathcal{R}$ that satisfies $A \in \mathcal{S}$, $B \notin \mathcal{S}$, which can be restated as

$$(2) \qquad \mathcal{S} \in \langle\!\langle A \rangle\!\rangle \cap \mathsf{C}\langle\!\langle B \rangle\!\rangle.$$

On the other hand, from (1) we have $\mathcal{R} \in \mathrm{int}(\langle\!\langle B \rangle\!\rangle \cup \mathsf{C}\langle\!\langle A \rangle\!\rangle)$. Therefore $\mathcal{R}$ is a member of some open subset of $\langle\!\langle B \rangle\!\rangle \cup \mathsf{C}\langle\!\langle A \rangle\!\rangle$. In view of our definition of open sets, there is some formula P such that

$$\mathcal{R} \in \langle\!\langle P \rangle\!\rangle \subseteq \langle\!\langle B \rangle\!\rangle \cup \mathsf{C}\langle\!\langle A \rangle\!\rangle.$$

Since $\mathcal{R}$ is a disjunction regular set that proves P, any larger disjunction regular set also proves P. In particular, $\mathcal{S}$ proves P. That is, $\mathcal{S} \in \langle\!\langle P \rangle\!\rangle$. But then

$$(3) \qquad \mathcal{S} \in \langle\!\langle B \rangle\!\rangle \cup \mathsf{C}\langle\!\langle A \rangle\!\rangle,$$

which contradicts (2).

We have proved $\langle\!\langle A \to B \rangle\!\rangle = \text{int}\big(\langle\!\langle B \rangle\!\rangle \cup \mathsf{C}\langle\!\langle A \rangle\!\rangle\big)$ for all formulas A and B. Specializing this result, we also know

$$\Big\langle\!\!\Big\langle A \to (B \wedge \overline{B}) \Big\rangle\!\!\Big\rangle = \text{int}\Big(\langle\!\langle B \wedge \overline{B} \rangle\!\rangle \cup \mathsf{C}\langle\!\langle A \rangle\!\rangle\Big).$$

By 29.25.h and 29.25.f, this simplifies to $\langle\!\langle \overline{A} \rangle\!\rangle = \text{int}(\mathsf{C}\langle\!\langle A \rangle\!\rangle)$, completing our proof.

29.29. *Constructive completeness.* Let P be any formula. Then the following conditions are equivalent:

a. P is a theorem of constructive logic.

b. $\langle\!\langle P \rangle\!\rangle = \Omega$, in the notation of the last few pages.

c. P is a tautology for the topological interpretation given by the topology defined in 29.22–29.27..

d. P is a tautology for every topological interpretation.

Using some results that were marked "optional":

e. P is a tautology for every finite topology.

f. P is a tautology for every topology on a finite set.

Not proved in this book:

g. P is a tautology for the usual topology on the real line.

Proof. The implication a $\Rightarrow$ d was proved in 22.14. Obviously d $\Rightarrow$ c. The implication c $\Rightarrow$ b follows from 29.28. The implication b $\Rightarrow$ a follows from 29.25.e.

The implications d $\Rightarrow$ e $\Rightarrow$ f and d $\Rightarrow$ g are obvious. We proved e $\Rightarrow$ d in 10.11, and f $\Rightarrow$ e in 4.26.

That condition g implies the other conditions is much harder, and is not proved in this book. A proof can be found in Tarski [1938], but it is not recommended for beginners. Tarski not only considered the real line, but characterized *all* topologies whose topological interpretations characterize constructive logic; the real line happens to be the most elementary of those topologies. A slightly shorter proof (still too complicated for beginners) is given in Bezhanishvili and Gehrke [2005]. That proof concentrates on the real line, but generalizes from constructive logic to the modal logic S4.

29.30. *A corollary on variable sharing.* Suppose A and B are formulas that share no variables, and $A \to B$ is a theorem of constructive logic. Then at least one of $\overline{A}$ or B is a theorem of constructive logic.

Remarks, in lieu of proof. This result would follow immediately from 10.10 and 29.29.g, but we did not prove the latter. We did give a proof of a slightly weaker result, in 29.13.

NON-FINITELY-AXIOMATIZABLE LOGICS

29.31. Each of the logics that we have studied is determined by finitely many axiom schemes and assumed inference rules. That is merely because I have selected the most elementary logics for this introductory textbook. A logic does not need to be finitely axiomatized. In fact, several examples in the literature show that, even if the logic can be characterized by a finite functional interpretation, it might not have a corresponding finite axiomatization. We shall describe just one particularly simple example.

Let $\Sigma = \{0, 1, F, T\}$ and $\Sigma_+ = \{T\}$. Define the interpretation by

$$x \ominus y = \begin{cases} F & \text{if } x = F \text{ and } y = 0, \\ F & \text{if } x = T \text{ and } y = 1, \\ T & \text{otherwise.} \end{cases}$$

The example is pure implicational logic — i.e., we do not consider $\vee \wedge \neg$. The resulting set of tautologies is nontrivial, i.e., neither empty nor equal to the set of all formulas; for instance, $\pi_1 \to \pi_1$ is tautological but $\pi_1 \to \pi_2$ is not. Dziobiak [1991] has shown that the set of tautologies of this finite functional interpretation cannot be characterized by a finite set of axiom schemes and assumed inference rules. (The proof is long and will not be given here.) For further related results and references see Palasińska [1994].

References

R. Ackermann, *Introduction to Many-Valued Logics*, Routledge & Kegan Paul, London, 1967.

P. H. G. Aczel, Saturated intuitionistic theories, pp. 1–11 in: *Contributions to Math. Logic (Colloquium, Hannover, 1966)*, North-Holland, Amsterdam, 1968.

A. R. Anderson and N. D. Belnap, Jr., *Entailment: the Logic of Relevance and Necessity*, Princeton University Press, Princeton, 1975.

K. Appel and W. Haken, Every planar map is four colorable. Part I. Discharging, *Illinois J. Math.* **21** (1977), 429–490.

K. Appel, W. Haken and J. Koch, Every planar map is four colorable. Part II. Reducibility, *Illinois J. Math.* **21** (1977), 491–567.

Aristotle, *The Complete Works of Aristotle, Revised Oxford Translation*, ed. J. Barnes, Princeton University Press, 1984.

J. Barwise and J. Etchemendy, *The Liar: An Essay on Truth and Circularity*, Oxford University Press, New York, 1987.

J. Barwise and L. Moss, *Vicious Circles: On the Mathematics of Non-Wellfounded Phenomena*, CSLI Publications, Stanford, 1996.

N. Belnap, Entailment and relevance, *J. Symbolic Logic* **25** (1960), 144–146.

G. Bezhanishvili and M. Gehrke, Completeness of S4 with respect to the real line: Revisited, *Annals of Pure and Applied Logic* **131** (2005), 287–301.

E. Bishop, *Foundations of Constructive Analysis*, McGraw-Hill, New York, 1967 (later revised as *Constructive Analysis* by Bishop and Bridges).

J. M. Borwein, Brouwer-Heyting sequences converge, *Mathematical Intelligencer*, volume **20**, number 1 (winter 1998), 14–15.

R. Brady, ed., *Relevant logics and their rivals, volume II*, Ashgate Publishing, Hants, England, 2003.

D. Bridges and F. Richman, *Varieties of Constructive Mathematics*, London Math. Soc. Lecture Note Ser. **97**, Cambridge University Press, Cambridge, 1987.

B. Bryson, *The Mother Tongue: English and How It Got That Way*, Wm. Morrow and Company, New York, 1990.

W. A. Carnielli, M. E. Coniglio, I. M. L. D'Ottaviano, eds., *Paraconsistency: the logical way to the inconsistent (Proceedings of the world congress held in São Paulo)*, Lecture Notes in Pure and Applied Mathematics **228**, Marcel Dekker Inc., New York, 2002.

L. Carroll, *Symbolic Logic*, edited with annotations and introduction by W. W. Bartley, Clarkson N. Potter Publishers, NY, 1977.

E. Casari, Logica e comparativi (Logic and comparatives), pp. 392–418 in: *Scienza e filosofia: Saggi in onore di Ludovico Geymonat* (*Science and philosophy: Essays in honor of Ludovico Geymonat*), ed. by Corrado

Mangione, Garzanti, Milan, 1985.

E. Casari, Comparative logics and abelian ℓ-groups, pp. 161–190 in: *Logic Colloquium '88: Proceedings of the colloquium held at the University of Padova, Padova, August 22–31, 1988*, ed. by R. Ferro, C. Bonotto, S. Valentini and A. Zanardo. *Studies in Logic and the Foundations of Mathematics* **127**, North-Holland Publishing Co., Amsterdam, 1989.

C. C. Chang, Proof of an axiom of Łukasiewicz, *Transac. Amer. Math. Soc.* **87** (1958), 55–56.

J. R. Chidgey, *On entailment*, University of Manchester doctoral dissertation, 1974.

B. J. Copeland, ed., *Logic and Reality: Essays on the Legacy of Arthur Prior*, Clarendon Press, Oxford, 1996.

K. Devlin, *Sets, Functions and Logic*, 2nd edtn., Chapman and Hall, London, 1992.

M. R. Diaz, *Topics in the Logic of Relevance*, Philosophia Verlag, München, 1981.

K. Dŏsen, The first axiomatization of relevant logic, *J. Philosophical Logic* **21** (1992), 339–356.

A. G. Dragalin, *Mathematical Intuitionism : Introduction to Proof Theory*, Translations of Mathematical Monographs **67**, American Mathematical Society, Providence, 1988.

M. Dummett, *A propositional calculus with denumerable matrix*, Journal of Symbolic Logic **24** (1959), 97–106.

J. M. Dunn, Relevant logic and entailment, pp. 117–224 in: *Handbook of Philosophical Logic, Volume III: Alternatives to Classical Logic*, ed. by D. Gabbay and F. Guenthner, D. Reidel Publishing Co., Dordrecht, 1986.

W. Dziobiak, A finite matrix whose set of tautologies is not finitely axiomatizable, *Rep. Math. Logic* **25** (1991), 105–112.

R. Epstein, *The Semantic Foundations of Logic, volume 1: Propositional Logics*, Nijhoff International Philosophy Series **35**, Kluwer Academic, Dordrecht, 1990.

M. C. Fitting, *Intuitionistic Logic, Model Theory, and Forcing*, Studies in Logic and the Foundations of Mathematics, North-Holland Publishing Co., Amsterdam, 1969.

R. J. Fogelin and W. Sinnott-Armstrong, *Understanding Arguments: An Introduction to Informal Logic*, Harcourt Brace Jovanovich, San Diego, fourth edition, 1978.

B. Grünbaum, Venn diagrams and independent families of sets, *Mathematics Magazine*, Jan.–Feb. 1975, 12–23.

A. G. Hamilton, *Logic for Mathematicians*, Cambridge University Press, Cambridge, 1978.

J. v. Heijenoort. *From Frege to Gödel: A Source Book in Mathematical Logic, 1879–1931*. Harvard University Press, Cambridge, 1967.

J. Herbrand, *Recherches sur la théorie de la démonstration*. PhD thesis, Université de Paris, 1930. (Translation of Chapter 5 appears in v. Heijenoort

References

1967.)

S. Hirokawa, Y. Komori, and M. Nagayama, A lambda proof of the P-W theorem, *J. Symbolic Logic* **65** (2000), 1841–1849.

D. Jarden, A simple proof that a power of an irrational number to an irrational exponent may be rational, *Scripta Mathematica* **19** (1953), 229.

P. T. Johnstone, The point of pointless topology. *Bull. Amer. Math. Soc. (N.S.)* **8** (1983), 41–53.

I. Kaplansky, *Set Theory and Metric Spaces*, Allyn and Bacon, Boston, 1972. Later reprinted by other companies. Currently available from American Mathematical Society, Providence.

C. K. Kielkopf, *Formal Sentential Entailment*, University Press of America, Washington, D.C., 1977.

M. Kline, *Mathematics: the Loss of Certainty*, Oxford University Press, NY, 1980.

T. S. Kuhn, *The Structure of Scientific Revolutions*, Chicago University Press, 1962.

A. Margaris, *First Order Mathematical Logic*, Blaisdell Publishing Co., Waltham, Mass., 1967. Later reprinted by Dover Publications, NY.

E. P. Martin, *The P-W problem*, Ph.D. Thesis, Australian National University, Canberra, 1978.

C. A. Meredith, The dependence of an axiom of Łukasiewicz, *Transac. Amer. Math. Soc.* **87** (1958), 54.

R. K. Meyer, Metacompleteness, *Notre Dame J. of Formal Logic* **17** (1976), 501–516.

R. K. Meyer, Peirced clean through, *Bulletin of the Section of Logic* **19** (1990), 100–101.

R. K. Meyer and J. K. Slaney, Abelian logic (from A to Z), pp. 245–288 in Priest et al. [1989].

R. K. Meyer and J. K. Slaney, A, still adorable, pp. 241–260 in Carnielli et al. [2002].

J. Norman and R. Sylvan, eds., *Directions in Relevant Logic* (based on the International Conference on Relevant Logic, St. Louis, 1974), Kluwer Academic Publishers, Dordrecht, 1989.

I. E. Orlov, The calculus of compatibility of propositions (in Russian), *Matematicheskiĭ Sbornik* **35** (1928), 263–286.

K. Palasińska, Three-element nonfinitely axiomatizable matrices, *Studia Logica* **53** (1994), 361–372.

M. Pavičić and N. D. Megill, Non-orthomodular models for both standard quantum logic and standard classical logic: repercussions for quantum computers, *Helv. Phys. Acta* **72** (1999), 189–210.

M. Pei, *The Story of Language*, New American Library, New York, 1984.

G. Priest, R. Routley, J. Norman, eds., *Paraconsistent Logic: Essays on the Inconsistent*, Philosophia Verlag, München, 1989.

H. Rasiowa, *An Algebraic Approach to Non-Classical Logics*, Studies in Logic and the Foundations of Mathematics **78** North-Holland Pub. Co., Am-

sterdam, 1974.
H. Rasiowa and R. Sikorski, *The Mathematics of Metamathematics*, Monografie Matematyczne **41**, Polska Akademia Nauk, Warsaw, 1963.
M. Reghiş and E. Roventa, *Classical and Fuzzy Concepts in Mathematical Logic and Applications*, CRC Press, Boca Raton, 1998.
N. Rescher, *Many-valued Logic*, McGraw-Hill Book Co., New York, 1969.
G. Restall, *An Introduction to Substructural Logics*, Routledge, 2000.
J. W. Robbin, *Mathematical Logic: A First Course*, W. A. Benjamin Inc., New York 1969.
T. T. Robinson, Independence of two nice sets of axioms for the propositional calculus, *J. Symbolic Logic* **33** (1968), 265–270.
A. Rose and J. B. Rosser, Fragments of many-valued statement calculi, *Transac. Amer. Math. Soc.* **87** (1958), 1–53.
R. Routley, A semantical analysis of implicational system I and of the first degree of entailment, *Mathematische Annalen* **196** (1972), 58–84.
R. Routley, with V. Plumwood, R. K. Meyer, and R. T. Brady, *Relevant Logics and Their Rivals, Part I: The Basic Philosophical and Semantical Theory*, Ridgeview Publishing Co., Atascadero, California, 1982.
L. I. Rozonoer, Proving contradictions in formal theories, I, *Automat. Remote Control* **44** (1983), no. 6, part 2, 781–790; translated from *Avtomat. i Telemekh.* **1983**, no. 6, 113–124.
E. Schechter, *Handbook of Analysis and its Foundations*, Academic Press, 1996.
E. Schechter, Constructivism is difficult, *Amer. Math. Monthly*, January 2001.
R. Shelly, Sing His Praise! A Case For A Capella Music as Worship Today, *20th Century Christian*, Nashville, 1987, pp. 33-34.
S.-J. Shin, *The Logical Status of Diagrams*, Cambridge University Press, 1994.
S. Smale, Mathematical problems for the next century, *Mathematical Intelligencer* **20** (2) 1998, 7–15.
A. Tarski, Sentential calculus and topology, pp. 421–454 in *Logic, Semantics, Metamathematics: Papers from 1923 to 1938 by Alfred Tarski*, translated by J. H. Woodger; second edition 1983 by Hackett Publishing Co. Translated from "Der Aussagenkalkül und die Topologie," *Fundamenta Mathematicae* **31** (1938), 103-34.
R. Taylor and A. Wiles, Ring-theoretic properties of certain Hecke algebras, *Ann. of Math.* **141** (1995), 553–572.
R. H. Thomason, On the strong semantical completeness of the intuitionistic predicate calculus. *J. Symbolic Logic* **33** (1968), 1–7.
S. K. Tollefson and K. S. Davis, *Reading and Writing about Language*, Wadsworth Publishing Company, Belmont, California, 1980.
A. S. Troelstra and D. van Dalen, *Constructivism in Mathematics: An Introduction*, North-Holland, Amsterdam, 1988.
A. Urquhart, Semantics for relevant logics, *Journal of Symbolic Logic* **37**

References

(1972), 159–169.

D. van Dalen, *Logic and Structure*, Universitext series, Springer-Verlag, Berlin, first and later editions, 1979, 1985, 1994.

S. Wagon, *The Banach-Tarski Paradox*, Encyclopedia Math. Appl. **24**, Cambridge Univ. Press, Cambridge, 1985.

A. N. Whitehead and B. Russell, *Principia Mathematica*, Cambridge University Press, three volumes, 1910–1913; second edition, 1925–1927.

B. L. Whorf, *Language, Thought, and Reality: Selected Writings of Benjamin Lee Whorf*, edited by J. B. Carroll, M.I.T. Press, Cambridge, Massachusetts, 1956.

Symbol list

Delimiters
( ) sequence, ordered pair 3.1
{ } set 3.2
[ ) interval 3.7.f
⟦ ⟧ semantic valuation 7.4
⟨⟨ ⟩⟩ regular proving sets 29.24

Some special sets
∅ empty set 3.6.e, 3.9
ℕ natural numbers 3.7.d
ℤ integers 3.7.d
ℚ rational numbers 3.7.d
ℝ real numbers 3.7.d
Ω universe 3.19
Σ topology 4.2

Set relations
∈ element or member 3.2
∋ contains 3.15
∉ nonmember 3.2
⊆ subset 3.13
⊇ supserset 3.13
⊊ proper subset 3.13
≼ partial order 3.20
≈ equivalence relation 3.42

Set operations
$X \times Y$ product of sets 3.7.g
→ function 3.22 (also see "implication" under logic)
$\mathcal{P}()$ powerset 3.34
∪ union 3.36
∩ intersection 3.36
\ relative complement 3.36
∁ complement 3.43
△ symmetric difference 3.46
| | cardinality 3.2, 3.62

Logic
∨ disjunction (or) 5.9
∧ conjunction (and) 5.12
¬ negation (not) 5.13
— negation (not) 5.13
→ implication (if–then) 5.21 (see also "function" under set operations)
↔ biimplication 5.23
∘ cotenability 5.32
& fusion 5.33
⊤ truth 5.34
⊥ falsehood 5.34
π_j proposition symbols 6.2
$A, B, C, \ldots$ metavariables 6.6
$\mathcal{D}_n$ Dugundji formula 11.13
⊢ derivable 12.5
δ detach. corollary 13.16
⇝ fuzzy implication 24.10
∼ fuzzy half-negation 24.10
⟨⟨ ⟩⟩ regular proving sets 29.24

Semantic logic
Σ semantic values 7.3
Σ_+ true semantic values 7.3
Σ_- false semantic values 7.3
⟦ ⟧ valuation 7.4
Ⓐ and, conjunction 7.5.b
Ⓥ or, disjunction 7.5.b
⊖ implication 7.5.b
⊖ negation 7.5.b
⊙ cotenability 7.8

⊨ semantic inference 7.9–7.10
⊭ nontautologous 7.9
⊥ Ackermann falsehood 5.34
ⓐ fusion 11.2.h

Miscellaneous

⇒ metalogical implies 2.17
⇔ metalogical iff 2.17
□ modal operator 2.19.d
◇ modal operator 2.19.d
∞ infinity 3.62
int interior 4.10
⊕ aut 5.9
x, y, z individual variables 5.41
∀ universal quantifier (for each) 5.41
∃ existential quantifier (there exists at least one) 5.41
$n!$ factorial 5.53
$\binom{n}{k}$ binomial coefficient 5.67
f_n Fibonacci number 5.69

Index

∀, *see* quantifier
Abelian logic, *see* comparative logic
AC, *see* Axiom of Choice
Ackermann constants, 172, 260, 298, 341, 362, 366, 377, 380, 437, 481
actual infinity, 117
adequacy, 48, 51, 444, *see also* complete
adjunction, 242, 299, 336, 338, 339, 346, 385, 388, 394, 407, 428, 469
  irrelevant, 360
  strong, 338, 359, 385, 394, 428
adjunctive explosion, *see* explosion
admissibility, 35, 44, 45, 242, 359, 451, 475
Age of Reason, 52
alethic, 38
algebra, 20, 42, 56, 83, 102, 106, 111, 174, 197, 232, 238, 302, 306
  avoidance of, 5, 57, 131, 295, 480
algorithm, 52, 53, 92, 393, 468
all, 184, *see also* quantifier
alphabet, 206

ambiguity, 7, 66, 73, 74, 80, 81, 152, 158, 191, 210, 215, 216, 220, 322
amoral, 176
and, 155, 206, 336, 394, *see also* adjunction
  -elimination, 336, 394
  -introduction, 336, 339, 394
and/or, 152
antecedent, 231
antilogism, 265
antisymmetric, 83
antitone, 88
applied logic, 38
arbitrariness, classical logic, 14
argument, 84
Aristotle, 58, 59, 111
artificial intelligence, 28, 79
artificiality, classical logic, 14
assertion, 324
assignment, 238
associative, *see* commutative and associative
assumed inference rules, 314, 385, 387, 407, 428
assumptions, 175, 313, *see also* background, hypotheses
atoms (in set theory), 69
audience, 4
aut, 152

automatic theorem-proving, 29
automobile cruise controls, 59
Axiom of Choice, 92, 118, 120, 467
axiomatic set theory, 25
axioms, 43, 313

background assumption, 165, 169
Banach-Tarski paradox, 93
bang head against wall, 31, 33
basic implication, 318, 393
basic logic, 8, 336
Belnap relevance property, 409
Beltrami's geometries, 49
Bernstein's cardinality, 120
bias, 26, 146
binary operator, 90, 95
binomial coefficient, 203
boldface, 164
bottom, 173
boundary, 126, 138
bounded quantifiers, 187
braces, 65, 68, 218
brackets, 65, 218
Brouwer, 5, 6, 148, 162, 341

C, see complement
cancellation, 61, 158, 263, 340, 378, 380, see also involution
canonical model, 461
Cantor, Georg, 42, 116, 125
cardinality, 68, 118
careless, 28
Carroll, Lewis, 39, 112

cases, 58, 64, 82, 106, 161, 166, 167, 170
　excluded third, see excluded middle
　proof by, 339, 346, 352, 422
causality, 166
centering, 437
certainty, 28–30, 52, 55
certification, 30, 32, 34, 176, 177
chain order, 9, 83, 376, 402, 410, 414, 419
　nontautological, 275, 280, 282
　paradoxical, 189
　tautological, 260, 277, 284, 421, 439
Chaitin, Gregory, 54
characteristic function, 87, 143, 471
choice function, see Axiom of Choice
choose, 92, 203
Church decontractioned, 278
Church's chain, 277
Church's constants, 173
Church's diamond, 278
churches, pianos in, 33
classical logic, 3, 57
　admissibility, 475
　arbitrariness, 14–15
　completeness, 472
　implication, 161
　informal, 146
　Post-completeness, 473
　semantics, 245
　syntactics, 430

Index 497

classical set theory, 69
clopen, 128
closed
   complement of open, 128
   syntactically, 469
   under operations, 129, 287
clothes dryers, 59
coassertion, 378
codomain, 84
Cohen, Paul, 93
commutative and associative
   ∘, 171, 297
   &, 172, 297, 370
   ∨ ∧, 217, 238, 339, 346
   ∪ ∩, 96
   △, 102
comparative logic, 59, 61, 157, 258, 262, 437
complement, 95, 101, 282
complete, *see also* adequacy
   for negation, 157
   functionally, 250
   many meanings in math, 48
   ordered field, 42
   pairing of logics, 46, 53, 454–485
completed infinity, 117
composite number, 202
computability theory, 32
computers, 18, 28–29, 34, 54, 79, 91, 123, 393
concatenation, 209
conclusion, 15, 44, 113, 114, 196, 241, 315, 319, 359, *see also* consequence
conjunction, 150, *see also* "and"

conjunctive explosion, *see* explosion
connectives, 17, 150, 207, 432
   nonredundant, 286
   redundant, 249
consequence, 44, 161, 174, 241, 315, 339, 449, 469, *see also* conclusion, disjunctive
consequent, 231
consistent, 50, 52, 54, 157, 368, 449
constructive
   AC isn't, 93
   axioms, 399
   completeness, 464–465, 484
   Disjunction Property, 451
   empty set, 76
   implication, 386–393
   implication model, 459
   indirect proofs aren't, 176
   Manifesto, 56
   negation, 158
   philosophy, 62
   sound in topologies, 397
   syntactics, 385–404
contains, 81
continuous, 127, 180, 187, 188, 192
contraction, 351, 355, 417
contradiction, 157, 354, 368, *see also* proof by
contrapositive, 166, 174, 341, 372, 399
convention, 36, 65, 93, 139, 215, 216, 315
converse, 164

corollary, 37, 330
cotenable, 170, 240, 297, 374
countable, 121, 214, 467
countably infinite, 120
counterexample, 46
cruise controls, 59
crystal logic, 273, 411
cultural relativism, 14

δ (delta), *see* continuous, detachmental corollary
Δ (uppercase delta), *see* symmetric difference
$\mathcal{D}$, *see* Dugundji formulas
De Morgan's laws, 109, 110, 160, 342, 373
decimal expansions, 41
Dedekind cut, 42
deduction principle
   classical, 16, 387
   fuzzy, 258, 428
   relevant, 260, 407
   several versions, 45
defined symbols, 219
definitions, use *if* for *iff*, 164
demodalizer, 325
deontic, 38
derivation, 40, 43, 314–317, 320
designated elements, 152
detachment, 45, 242, 298, 319, 342, 386, 422, 445
detachmental corollary, 330
diabolical, 176
diagram, *see also* graph
   $\bot$ and $\top$, 173
   boundary region, 126

generalization/specialization, 227
inclusion, 129, 131, 272, 274, 277
nested quantifiers, 188
order of proof, 352, 358, 360, 400, 415
Pascal's triangle, 204
related logics, 9, 365, 410
same cardinality, 119
traditional vs. pluralist, 3
tree, 211, 230
Venn, 73, 102–111, 113, 144, 145
dichotomy, *see* proof by
difference of two sets, 95
direct proof, 174
discovery, 30, 177
discrete topology, 130, 282
dishwashers, 59
disjoint, 99
disjunction, *see also* "or"
   principle, 396
   Property, 153, 293, 451
   regular, 479
   regular proving set, 480
disjunctive
   consequence, 265, 406
   implication, 399
   syllogism, 265, 452
distributive, 92, 105–107, 110, 247, 271, 273, 336–338, 347, 351, 355, 374, 386, 395, 427, 432
domain, 84, 184
double elliptic geometry, 49
double negative, 158, 160

doubleton, 71
doubt, 15
Dr. Seuss, 122
duality
  modal logic, 38
  propositions, 160, 191, 338
  set theory, 110
Dugundji formulas, 307
Dummett's LC, 402

ε (epsilon), *see* continuous
$\in$ (epsilon), *see* element
$\exists$, *see* exists, quantifier
each, 184, *see also* quantifier
Einstein, Albert, 30
either/or, 152
elements, 68
elimination rules, 337
ellipsis, 201
emotions, 22
empirical, 49, 55
empty set, 26, 72, 75, 76, 81, 98
English, 26, 35, 77, 80, 113, 123, 146–205, 208, 220, 336
Enlightenment, 52, 53
epistemic, 38
equals sign, 207
equipollent, 119
equivalence class, 42, 99, 142, 306
equivalence relation, 99, 163
essence of reason, 147, 149
Eubulides paradox, 79
Euclidean, 15, 49
every, *see* quantifier

examples, 16, 46, 66, 76, 93, 124, 162, 176
excluded middle, 57, 62, 64, 126, 156, 284, 355, 399, 406, 418, 438
excluded third case, *see* excluded middle
exclusive or, 152
existential quantifier, *see* quantifier
exists, 27, 51, 53, 78, 83, 92, 113, 117, 119, 176, 467, 468, *see also* quantifier
explosion, 157, 365, 367, 415
extended real line, 74
exterior, 138
extremes, 260, 275, 277, 280, 376, 412, 414

f_n, *see* Fibonacci numbers
factorial, 195
facts, isolated, 15
faith, 55
false, *see* semantic values
falsehood-suffixing, 341
Fermat's Last Theorem, 162
Fibonacci numbers, 204
fiction interpretation, 270
figure, *see* diagram, graph
finite, 120, *see also* topology
finite interpretation, 237, 238
  constructive, 290, 398
  fuzzy, 310
  relevant, 411
  Sugihara, 266, 310
first contrapositive, 336, 341
first-order theory, 38
flying pigs, 112

for all, for each, *see* quantifier
forbidden, 33
formal
   language, 34, 206–232
   set theory, 25
   system, 34
formalism, formalizable, 27
formula, 43, 210, 215
   countably many, 214
   scheme, 43, 222
formula-valued function, 228
forward proof, 174
foundations, 21
Four Color Theorem, 29
fourth contrapositive, 372
frames, 131, 454
free will, 59
function, 84
functional interpretation, 237
functionally complete, 250
fusion, 172, 297, 369, 374, 395
fusion detachment, 298
fuzzy logics, 251, 417, 477
fuzzy thinking, 59

general position, 103
general topology, 22
generalization, 225
generated topology, 139
geometry, 49
Glivenko, V., 342, 402, 403
God, 56, 468
Gödel, K., 23, 53, 93, 398, 472
Goldbach's Conjecture, 63
Golden Ratio, 205
good taste, 62
grammar, 22, 34, 39, 210
graph, *see also* diagram

function, 87, 88, 90
fuzzy nontautology, 257
   implication, 253, 264
   intervals, 74, 283
   open sets, 133
great circle, 50
greatest integer function, 88
grouping, 65, 215

ham sandwich, 75, 164
Harrop set, 448
Harrop's principle, 450
Hartogs's cardinality, 120
heap paradox, 253
Heisenberg's uncertainty, 53
Herbrand Principle, 16, 387
hexagon logic, 272, 305
Heyting, 5, 6, 162
Heyting algebras, 5, 131, 479
higher system, 35
Hilbert problems, 52
Hilbert's program, 53, 55
history, 49
homomorphism, 56, 238, 302
hypotheses, 44, 196, 241, 315,
   see also assumptions

idempotency, 339
identity, 102, 319, 326, 386
if, *see* implication
iff (if and only if), 163
illogical behavior, 23
illustration, *see* diagram, graph
image, 86, 301
imaginary number, 118
implication, 161, 458
   basic, 318, 393
   completeness, 464

Index 501

constructive, 386
iterated, 332
model, 458
symbol, 207, 315
implicative
conjunction, 413
disjunction, 417, 419
explosion, 365, 415
explosion, weak, 362
imprecise data, 59
imprecise language, 35
includes, 81
inclusive or, 152
incompleteness, *see* complete pairing of logics
inconsistent, *see* consistent
increasing, 88
independence, 51, 431
index set, 86
indicator function, *see* characteristic function
indirect proof, *see* proof
indiscrete topology, 130
individual variables, 37
induction, 195
inf topology, 139
inference rule, 44, 241, 315, *see also* assumed
inference system, 313
infimum topology, 139
infinite, 120
infix notation, 220
informal, 25, 35
informal set theory, 25, 65, 69
inhabited set, 76
initialization step, 196
inner system, 34

instance, 222
int, *see* interior
integers, 72
intended audience, 4
interior, 133
intermediate logic, 401
interpretation, 39, 49, 237
intersection, 95, 98
interval, 65, 73
introduction rules, 337
intuitionistic logic, 5, 385
inverse image, 86
involution, 91, 101, 373
irrational
behavior, 22
numbers, *see* rational
irrelevant, *see* adjunction, expansion, mingle, positive paradox, relevant
isolated facts, 15
isomorphic, 42
isotone, 88, 457, 459
italics, 164
iterated implication, 332

Jabberwocky, 39
Jarden's Proof, 63–64
joke, 75, 160, 164, 203, 468, *see also* self-referencing
justification, 33, 179, 315, 316, 320, 326, 385, 389

knowledge, causality of, 167
known, 38

language, *see* English, formal
larger, 68, 80
largest, 80

502   Index

Law, *see* associative, Brouwer's, commutative, contrapositive, De Morgan, distributive, excluded middle, noncontradiction
least upper bound, 41
Leibniz's language, 52
lemma, lemmaless, 326
length, 209
logicism, 27
love, 75, 148
lower set topology, 130
lower system, 34
Lowth's grammar, 159
Łukasiewicz logic, 251

MaGIC, 19
manifesto, 56
map, mapping, 84
material implication, 161
matrix interpretations, 237
maximal, 467
   classical logic is, 473
   extension, 467
   unproving set, 466
maximum, 89
meaning, 39, 51, 235
meet, 99
members, 68
Meredith, 263, 324, 378, 419
meta-, 35
metametavariables, 224
metavariables, 43, 209
Meyer's valuation, 443
mingle, 265, 275, 277, 357, 358, 405, 409–411, 465
minimum, 89
Mints's principle, 451

minus, 91, 95
mnemonic, 95, 184
modal logic, 38
modalizer, 325
model, 458, 471
modus ponens, *see* detachment
monotone, 88, 101, 316
Moore method, 22
most true or most false, 172
multiple consequences, 339, 346
multiple viewpoints, 454
multiple worlds, 454
multiset, 66
multivalued logics, 59
myth, 147

$\mathbb{N}$, *see* natural numbers
naive set theory, 69
nand, 250
natural numbers, 72
necessary, 38, 161
necessary and sufficient, 163
negation, 156, 206
   -elimination, *see* not-
   in basic logic, 341
   in modal logic, 38
   -introduction, *see* not-
   of quantifiers, 191
   -prefixing, 341
Newton, Isaac, 52
non-trivializing, 367, 472
nonconstructive, *see* constructive
noncontradiction, 157, 355
nonempty, 76
non-Euclidean, 49
nonlogical symbols, 38

Index 503

nonmember, 68
nonmonotonic logics, 316
nonredundant, *see* redundancy
non-self-inclusive, 77
nonsense, 14, 39, 61, 168, *see also* relevant, irrelevant
normal, 5
not, *see also* negation
   -elimination, 158, 284, 372, 399, 402
   -introduction, 158, 341
   -prefixing, 341
nothing, 75
novelist, 22
null set, *see* empty set

object system or language, 34
obligatory, 38
observers, 454
one-directional reasoning, 176
1-to-1 correspondence, 118
only if, 161
open, 128
operations
   closed under, 129, 287
   on sets, 94
   order of, *see* order
   unary, binary, 90
opposites, 156, *see also* involution
or, 152, 206, 395
   -elimination, 336, 339, 395
   -introduction, 336, 395
order
   of operations, 67, 216, 220
   preserving, 88, 101, 136, 229
   reversing, 88, 97, 101, 229
ordered pair or triple, 66

ordinals, 118
ordinary mathematics, 21
outer system, 35
overview, 20

$\mathcal{P}$, *see* powerset
π (pi), *see* propositional variable symbols
pairwise disjoint, 99
paracomplete, 157
paraconsistent, 157
paradigm, 18
paradox, 77, 79, 93, 189, 253, *see also* positive paradox
parallel postulate, 49
parentheses, 65, 211, 214, 215, 218, 346
partially ordered set, 83
partition, 99
Pascal's Triangle, 203
pedagogical advantages of pluralism, 13
Peirce's law, 399, 415, 423, 431, 432
permitted, 33, 38
permutation, 319, 334, 393
Philo of Megara, 58
physical universe, *see* real world
pianos in churches, 33
plane, as ordered pairs, 75
plane, usual topology, 132
plausible alternatives, 15
pluralism, 4, 13, 57
point set topology, 22
points, 68
poset, 83
positive logic, 267, 339

positive paradox, 261, 286, 359–365, 385, 386, 405, 412, 415–417, 431, 433
positively designated, 152
possible, 38
possible worlds, 454
Post-complete, 473
postulates, *see* axioms
potential infinity, 116
powerset, 94
   cardinality, 125
   interpretation, 270, 304
predetermined, 58
predicate logic, 37
prefix cancellation, *see* cancellation
prefix notation, 220
prefixing, 319, 333, 340, 341, 343, 363, 364, 386, 421
prescriptive grammar, 159
prime number, 63, 202, 203
prime set of formulas, 153, 470
primitive symbol, 206, 219
principal valuation, 459, 483
Principia Mathematica, 27
principle of teaching, 14, 16
principles, 37, *see also* admissibility, conjunction, conservation of mass, disjunction, duality, Deduction, explosion, finite interpretation, finite representation, finite refutation, Glivenko, Gödel, Harrop, Herbrand, image, Incompleteness, induction, maximal extension, Mints, restriction, soundness, strong induction, substitution, Tarski, uncertainty
problems, Hilbert, 52
product of sets, 74
program, Hilbert's, 53
prohibited, 33
proof
   by cases, *see* cases
   by contradiction, 24, 175, 258, 265, 352, 354, 399
   by contrapositive, 174
   by dichotomy, 258, 352, 354, 399, 406
   by example, 290, 327
   direct or indirect, 174
proper, 80, 225
propositional logic, 37, 150
propositional variable symbols, 206, 207
provable, *see* derivation
proves, 315
psychiatric problems, 23

$\mathbb{Q}$, *see* rational numbers
quantifier, 6, 17, 23, 37, 113, 183–194
quantitative logics, 59
question mark, 177
quotation marks, 164

$\mathbb{R}$, *see* real numbers
R, *see* relevant logic
R-W logic, 372
range, 84–87
rank, 212, 239

Index

rational numbers, 48, 63, 72, 122
real numbers, 72
    complete ordered field, 42
    completion of rationals, 48
    interval notation, 73
    uncountable, 123
    usual topology, 131, 484
real world, 27, 30, 49, 93
reason, essence of, 147, 149
recursive, recursion, 52, 195
reductio, 352, 399, 406
redundancy
    of an axiom, 51, 375, 432
    of connectives, 249, 250, 286
reflexive, 83, 99
regular, 5
relative complement, 95
relativity, 378
relevant, *see also* irrelevant, symbol sharing
    axioms, 405
    completeness, 464
    implication, 405
    implication model, 458
    introduction, 61, 168
restriction, 301
reversible, 177
RM, 405, 412, *see also* mingle
Rose-Rosser logic, 417
rules, 37, *see also* inference rule
Russell's paradox, 77

Σ (Sigma)
    collection of sets, 269
    semantic values, 235, 281
    topology, 127, 281

scheme, 222
Schroeder's cardinality, 120
second contrapositive, 341
self-distribution, 355
self-inclusive, 77
self-referencing, 78, 505
semantic, 39–49, 235–310, 444
    consequence, 241
    equivalence, 100
    values, 152, 235
sentential logic, 37, 150
sequences, 65
sequential representations, 67
sets, 26, 66, 78, *see also* cardinality, complement, countable, De Morgan's laws, difference, disjoint, duality, empty set, equivalence relation, foundations, infinite, intersection, members, partition, powerset, product, subset, uncountable, union, Venn diagram
setups, 454
share, *see* symbol sharing
Sheffer stroke, 250
simple set, 77
singleton, 71
slithy, 39
sloppy, 28, 35
smaller, 81
solvability, 53
some, 113
Sorites Paradox, 253
soundness, 48, 51, 383
    Abelian logic, 439

constructive logic, 397
fuzzy logics, 417
implication models, 460
relevant logics, 408
space, topological, 128
spacetime events, 147
specialization, 225
specialized
asssertion, 324
conjunctive detachment, 423
contraction, 354, 417, 423
De Morgan, 284
mingle, 320
Peirce, 399, 423
square root, 63
Star Trek, 22
stays away, 192
Stone's formula, 284, 402
strict, 68, 80, 88, 225
string, 208
strong
adjunction, *see* adjunction
completeness, 48
induction, 198
not-elimination, 378
proof by dichotomy, 399, 406
reductio, 399, 406
strongest, 173, 261
strongly increasing, 88
strongly irrelevant, 415
subformulas, 211
subset, 26, 79, 83
substitution, 225, 343–345
subtraction, 38, 91, 95
sufficient, 161
suffix cancellation, *see* cancellation
suffixing, 319, 320, 340, 341, 343, 364
Sugihara logic, 258, 263
superset, 79
switch, 399
syllogism, 111
symbol sharing, 221, 250, 261, 267, 276, 289, 409, 412, 473, 485
symmetric, 99
symmetric difference, 102
syntactic, 39–49, 313–439, 444
closure, 469
consequence, 315
equivalence, 100

Tarski, 16, 93, 387, 484
tautology, 37, 43, 46, 240
-preserving, *see* admissibility
Tertium Non Datur, *see* excluded middle
Tertullian (Quintus Septimius Florens Tertullianus), 33
then, 161
theorem, 37, 43, 46, 314
there is, *see* exists, quantifier
thermostats, 59
third contrapositive, 372
thought, 14
three-column proof, 179, 316, 446
top, 173
topological space, 128
topology, 5, 126–145
as ordinary math, 22
completeness, 484

Index 507

constructive soundness, 397
examples, 129–133
finite, 128, 131, 142, 290
generated, 139
interpretation, 281
usual on the reals, 131
tove, 39
transition step, 196
transitivity, 83, 99, 323, 324
tree diagram, 211, 230
trial-and-error, 31
triple negation, 341
trivializing, 367
trivially true, 169
true, *see* semantic values
true love, 75
truth-preserving, 241
two-valued interpretation, 57, 245–251, 304
and powersets, 269, 304
completeness, 472
equivalent to hexagon, 305

unary operator, 90
Uncertainty Principle, 53
uncountable, 121, 123, 467
undefined symbols, 206, 219
underlining, 164
underlying assumption, 165
union, 95, 98
universal quantifier, *see* quantifier
universe, universal set (Ω), 83, 101, 103, 184, 269, 281
unknowable, 53
unproving, 466

unrelated extremes, *see* extremes
upper set topology, 131
usual topology
on the plane, 132
on the reals, 131, 484

vacuously true, 82, 169
vague, 28
valuation, 65, 237–240, 443, 458, 483
value, 84, 152, 235
variable sharing, *see* symbol sharing
vel, 152
Venn, *see* diagram
viewpoints, 454

Wajsberg logic, 417, 422
weak completeness, 48
weak contraction and other weak formulas, 351–355
weakest, 173, 261
weakly increasing, 88
weakly irrelevant, 411
weakly nonconstructive, 401
whenever, 161
Whorf, Benjamin, 147
word, 208
working backwards, 31, 177
worlds, 270, 454

xor, 152

$\mathbb{Z}$ (integers), 72
Zadeh logic, 251
Zorn's lemma, 467